Leishmaniasis

This book covers the history, biology, pathogenesis, clinical manifestations, and the latest advancements in diagnosis, treatment, and control strategies against leishmaniasis. Chapter 1 presents the life cycle of *Leishmania* parasites, focusing on vector biology and transmission dynamics, while elucidating the mechanisms of infection and disease progression. The clinical manifestations of cutaneous, visceral, and mucocutaneous leishmaniasis are examined in detail, alongside the immune response of the host and the evasion strategies by the parasite. It also investigates the genetic and molecular diversity of *Leishmania*, highlighting their virulence mechanisms, and reviews diagnostic methods from traditional techniques to advanced molecular diagnostics. The chapter also covers the current therapeutic approaches, including potential drugs and combination therapies along with integrated vector management strategies for effective disease control. The book explores preventive measures, community interventions, and public health strategies, addressing challenges in leishmaniasis research and identifying knowledge gaps. It presents real-world applications and outcomes through case studies and clinical trials. This book is useful for researchers and students of infectious diseases, tropical medicine, and parasitology.

Key Features:

- Explores the pathogenesis and epidemiology of leishmaniasis, providing a global perspective on its public health impact
- Details the life cycle of *Leishmania* parasites, focusing on vector biology, transmission dynamics, and mechanisms of infection
- Examines the clinical manifestations of cutaneous, visceral, and mucocutaneous leishmaniasis, including the host immune response and parasite evasion strategies
- Reviews current therapeutic approaches, including potential drugs, combination therapies, and integrated vector management strategies for effective control
- Presents preventive measures, community interventions, public health strategies, and the challenges in leishmaniasis research

Leishmaniasis
Pathogenesis and Potential Treatment

Edited by
Veer Singh, Ashish Kumar,
and Kumar Abhishek

CRC Press
Taylor & Francis Group
Boca Raton London New York

CRC Press is an imprint of the
Taylor & Francis Group, an **informa** business

Designed cover image: **Veer Singh, Ashish Kumar, Kumar Abhishek**

First edition published 2026
by CRC Press
2385 NW Executive Center Drive, Suite 320, Boca Raton FL 33431

and by CRC Press
4 Park Square, Milton Park, Abingdon, Oxon, OX14 4RN

CRC Press is an imprint of Taylor & Francis Group, LLC

ISBN: 978-1-032-97975-5 (hbk)
ISBN: 978-1-032-97976-2 (pbk)
ISBN: 978-1-003-59645-5 (ebk)

DOI: 10.1201/9781003596455

Typeset in Sabon
by KnowledgeWorks Global Ltd.

To those who care, conserve, and protect but do not destroy the beauty and unique characteristics of

Kshit (Earth)
Jal (Water)
Pawak (Fire)
Gagan (Sky)

and

Sameera (Air)

Contents

Preface

Leishmaniasis remains one of the world's most challenging and neglected tropical diseases, affecting millions of people, particularly in areas of Asia, Africa, and Latin America. Caused by protozoan parasites of the genus *Leishmania* and transmitted by infected phlebotomine sandflies, the disease presents in diverse clinical forms such as cutaneous, mucocutaneous, and visceral, each with unique pathophysiological features, diagnostic complexities, and treatment requirements. The burden of leishmaniasis extends beyond clinical symptoms, impacting social and economic conditions in endemic regions and straining public health systems.

This book, *Leishmaniasis: Pathogenesis and Potential Treatment*, brings together a wide spectrum of current knowledge, ranging from basic biology and disease mechanisms to advanced diagnostic technologies, therapeutic strategies, and public health approaches. It is designed to serve as a comprehensive resource for researchers, healthcare professionals, academicians, and policymakers engaged in combating this multifaceted disease.

The volume begins with an overview of the historical background, global distribution, and epidemiological burden of leishmaniasis. A deeper understanding of the *Leishmania* parasite's life cycle and its transmission through sandfly vectors lays the foundation for comprehending the complex host–pathogen interactions that drive infection and disease progression. Detailed discussions on the clinical manifestations of the disease provide insights into the varied symptoms and severity observed across different geographical regions and patient populations.

A major focus of this book is the molecular and genetic basis of *Leishmania* virulence, which influences disease severity, treatment response, and transmission dynamics. The text also emphasizes recent advancements in diagnostic methodologies, including traditional parasitological techniques and emerging molecular tools that promise rapid and accurate detection. Treatment approaches are reviewed comprehensively, covering both existing therapeutic regimens and novel drug candidates under investigation. Attention is given to challenges such as drug resistance, treatment failures, and the need for combination therapies.

Recognizing that treatment alone is insufficient to curb the spread of leishmaniasis, the book also highlights strategies for integrated vector management and community-based prevention programs. Environmental factors, behavioral interventions, and public health education are considered essential components of a successful disease control strategy. An emerging area of interest—plant-based compounds for leishmaniasis therapy—is also explored, reflecting the growing demand for sustainable and accessible treatment options.

Real-world experiences are captured through case studies and clinical trials, providing practical perspectives on diagnosis, treatment efficacy, and patient outcomes. In addition, statistical modeling and time-series data analysis offer valuable tools for understanding disease trends and predicting future outbreaks, particularly in endemic regions such as Bihar, India. The application of *in silico* approaches for drug design and disease modeling represents a cutting-edge frontier in the search for more effective therapeutic solutions.

Finally, the book presents a forward-looking view of future directions, highlighting research priorities, potential solutions, and the importance of global collaboration to eliminate leishmaniasis as a public health problem. It is our hope that this volume will inspire continued scientific inquiry and support evidence-based strategies to mitigate the impact of this enduring disease.

Aim and Scope: The aim of this book is to provide a comprehensive understanding of leishmaniasis, covering its pathogenesis, clinical manifestations, diagnostic methods, treatment strategies, and preventive measures. It integrates current research, case studies, and innovative approaches, including plant-based therapies and statistical modeling, to serve as a valuable resource for researchers, clinicians, and public health professionals working on parasitic and neglected tropical diseases.

Acknowledgment

It gives us immense pleasure to acknowledge Manipal University Jaipur, Jaipur (India) and ICMR—Rajendra Memorial Research Institute of Medical Sciences, Agamkuan, Patna. Veer Singh would like to acknowledge the Department of Health Research (DHR), Ministry of Health and Family Welfare, Government of India, for support through the Young Scientist Fellowship Grant R.12014/37/2022-HR. Ashish Kumar would like to thank the Indian Council of Medical Research (ICMR), Department of Health Research, Ministry of Health and Family Welfare, Government of India, for providing workplace and necessary research support. The editors of this book are also grateful to the Director, ICMR-RMRIMS, for his guidance and support during completion of this book. It is also our great pleasure to acknowledge and express our enormous debt to all the contributors who have provided their quality material to prepare this book. We are grateful to our beloved family members, who joyfully supported and stood with us in many hours of our absence to finish this book project. We are also grateful to our friends and colleagues who offered their support and encouragement throughout the writing process. We would also like to acknowledge the publishing team at CRC Press for their professionalism, enthusiasm, and belief in this project. Their expertise in design, production, and marketing has been essential in bringing this book to life.

Veer Singh
Ashish Kumar
Kumar Abhishek

About the Editors

Veer Singh is currently working as Assistant Professor (Research) in the Department of Biosciences at Manipal University Jaipur, Jaipur. Prior to joining MUJ, he served as a Young Scientist at the ICMR–Rajendra Memorial Research Institute of Medical Sciences, Patna, India. He completed his Ph.D. in Biochemical Engineering from IIT (BHU), Varanasi. He also received funding/grant from the Department of Health Research (DHR), Government of India of ₹48.65 lakhs. His research is focused on wastewater treatment, development of eco-friendly adsorbents, material characterization, bioremediation, adsorption, environmental toxicity and human health, and nanomaterial for wastewater treatment. He has published 36 research/review articles in SCI/Scopus-indexed journals, five patents, and contributed 17 book chapters in books. He serves as an editor for several reputed journals and has edited multiple books published by reputed international publishers.

Ashish Kumar is currently working as Scientist-D and Head of Department at the Department of Biochemistry, Rajendra Memorial Research Institute of Medical Sciences, Patna, India. He has published more than 45 publications in international/national journals. He has received several awards and received several funding (sponsored projects) from many national and international agencies. Currently, four doctoral and two post-doctoral students are working under the supervision of Dr. Ashish Kumar. Dr. Kumar is working in the interdisciplinary area of life science. His work is mainly focused on visceral leishmaniasis (VL) and post-kala-azar dermal leishmaniasis (PKDL). He is also working on VL and PKDL diagnosis, treatment, and environmental impacts on disease development.

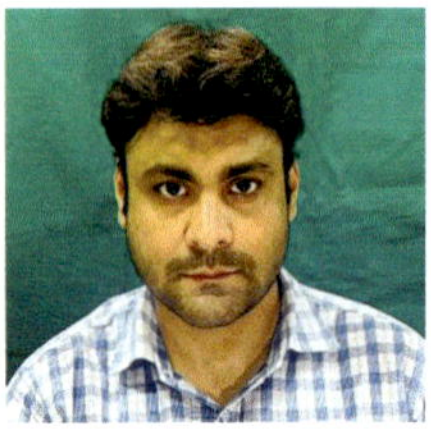

Kumar Abhishek is currently working as Scientist-C, Department of Health Research, Model Rural Health Research Unit (MRHRU), Kurhani, Muzaffarpur, under ICMR-Rajendra Memorial Research Institute of Medical Sciences, Patna, (Bihar) India. He got his PhD from the Department of Microbiology, University of Calcutta, West Bengal. He has more than 26 publications in peer-reviewed journals. He works in the field of visceral leishmaniasis (VL), post-kala-azar dermal leishmaniasis (PKDL), parasitic diseases, and on the epidemiology of communicable and non-communicable diseases.

Chapter 1

Introduction to leishmaniasis

History, epidemiology, and global health impact

Ravi Ranjan
Department of Biochemistry, ICMR-Rajendra Memorial Research Institute of Medical Sciences, Patna, India
Department of Biochemistry, University of Calcutta, Kolkata, India

Sonali Vedika and Kundan Kunal
Department of Biochemistry, ICMR-Rajendra Memorial Research Institute of Medical Sciences, Patna, India

Amit Priyadarshi
P.G. Department of Zoology, Veer Kunwar Singh University, Ara, Bihar, India

*Ashish Kumar**
Department of Biochemistry, ICMR-Rajendra Memorial Research Institute of Medical Sciences, Patna, India

1.1 INTRODUCTION

Leishmaniasis is a disease caused by a protozoan parasite—*Leishmania*. This disease occurs mostly in the tropical regions and subtropical regions of the world, as the effect of certain environmental conditions supports the growth and multiplication of the carrier vector—sandfly (1). The terminology "neglected" is often used with tropical diseases when referring to leishmaniasis, because over a million are affected annually due to leishmaniasis, which still lacks attention (2).

Leishmania sneaks around to avoid being noticed while slowly taking over the host's immune system. The effect of *Leishmania* on a person depends on several factors like the health and immune status of a person bitten by a

* Corresponding author.

DOI: 10.1201/9781003596455-1

sandfly. The effect also depends on the type of parasite and how well the person's immune system fights it. For instance, *Leishmania donovani* and *Leishmania infantum* go to the internal organs, causing a serious disease called visceral leishmaniasis (VL), while *Leishmania major* and *Leishmania tropica* stay in the skin, causing skin sores (3).

Clinical forms document various conditions, from skin sores to systemic collapse; the most common form, cutaneous leishmaniasis (CL), manifests as painless ulcers at the bite site, often resembling a volcano crater with raised edges. These lesions typically heal spontaneously within months but leave disfiguring scars (4). In regions like Afghanistan and Syria, CL is endemic, earning nicknames such as the "Baghdad Boil." A 2017 study of Syrian refugees in Lebanon found that 82% of CL patients suffered social stigmatization, with women being disproportionately affected due to cultural beauty norms (5) (Figure 1.1).

Mucocutaneous leishmaniasis (MCL) is caused primarily by *Leishmania braziliensis* in Latin America; MCL begins as a skin lesion but metastasizes to mucous membranes of the nose, mouth, and throat. Left untreated, it erodes cartilage and soft tissue, leading to severe disfigurement (6).

The lethal form, VL, attacks the spleen, liver, and bone marrow, causing fever, anaemia, and wasting. If it is not treated, mortality exceeds 95%. This form of disease is called lethal because it can cause death if the health condition of a patient becomes worse and a properly guided treatment is not received. The attack of VL is being reported by different countries. In India the most endemic regions are mainly Bihar, West Bengal, Jharkhand, and Assam. The disease has an indirect effect of malnutrition and poor housing (7).

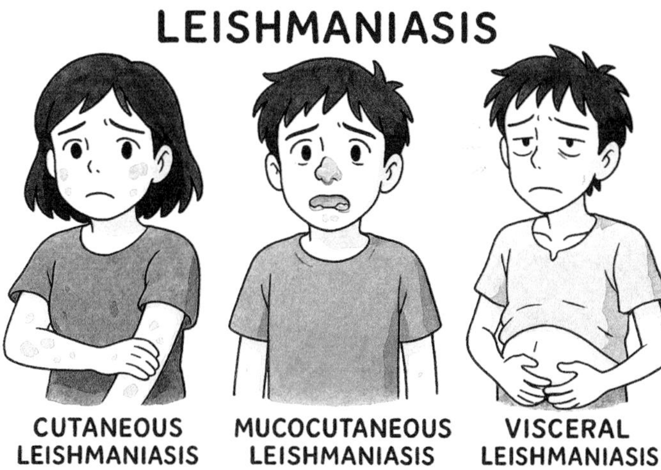

Figure 1.1 Three clinical forms of leishmaniasis—cutaneous leishmaniasis (CL), mucocutaneous leishmaniasis (MCL), and visceral leishmaniasis.

A study in Bihar found that 75% of patients are classified as poor, with most living in mud houses and engaged in agriculture. Poverty is also a major determinant of VL in Bihar (8).

Apart from these three main forms, leishmaniasis disease also has a fourth form which often develops the post treatment of VL (kala-azar), thus called post kala-azar dermal leishmaniasis (PKDL). The disease manifests as skin lesions, such as macules, nodules, papules, and depigmented patches, predominantly on the face, arms, and trunk. PKDL functions as a reservoir for the *Leishmania* parasite, facilitating disease transmission. The diagnosis is established through a combination of clinical evaluation, skin tests, and laboratory confirmation. Effective treatment entails prolonged anti-leishmanial therapy to eradicate the parasite and prevent the progression of the disease (9).

1.1.1 Life cycle of leishmaniasis

Leishmania life cycle is an impressive example of how living things adapt to survive (Figure 1.2).

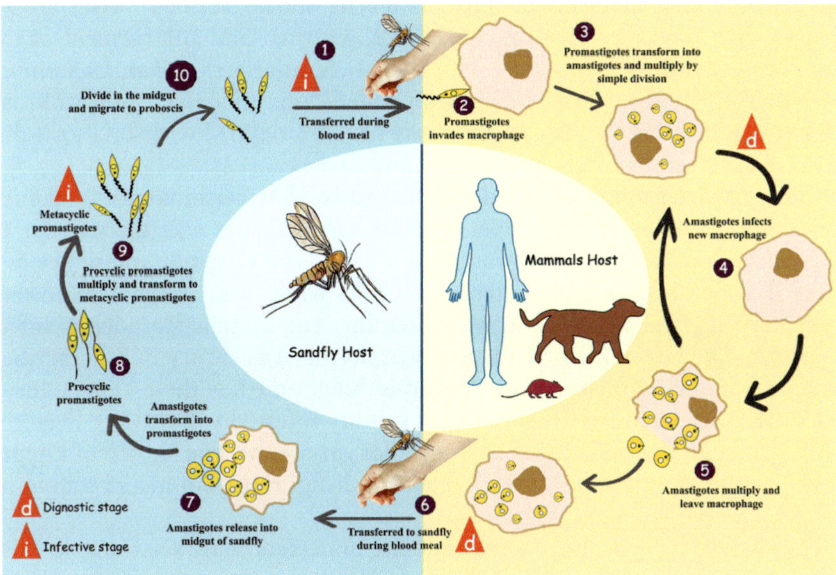

Figure 1.2 The life cycle of leishmaniasis. 1. The infected sandfly takes the blood meal on mammals, and simultaneously the *Leishmania* amastigotes are transferred to mammalian host. 2. Promastigotes invade the macrophage. 3. Promastigotes transform into amastigotes and multiply by simple diffusion. 4, 5. Amastigotes keep infecting new neighbouring macrophages. 6, 7. Upon next sandfly bite the amastigote form is transferred to the sandfly midgut. 8, 9. Procyclic promastigotes multiply and transform into metacyclic promastigotes. 10. These metacyclic promastigotes divide and migrate to the proboscis of the sandfly.

This parasite has two main forms: promastigote, which can move around because it has a whip-like tail, and amastigote, which is round and cannot move. The promastigote lives in the gut of sandflies. When a sandfly that carries the parasite bites a person, it injects these promastigotes into the skin. Once inside the body, these parasites are quickly taken up by macrophages, which are cells that help fight off infections. Instead of being killed, the promastigotes lose their tails, change into amastigotes, and start to multiply. Over time, the macrophages that are filled with parasites burst open, releasing many new parasites. These new parasites can then infect nearby cells or are picked up by another sandfly when it bites, continuing the cycle (10).

1.2 HISTORY OF LEISHMANIASIS

The story of leishmaniasis spans the annals of human civilization, interwoven into the tapestry of history through pandemics, conflicts, and scientific advancements. The earliest documented references to a disease resembling kala-azar appear in ancient Indian medical texts. The *Sushruta Samhita* (circa 600 BCE) describes a chronic febrile illness termed "visarpa" or "pandu roga," marked by prolonged fever, wasting, and enlargement of the abdomen—symptoms consistent with VL (11). Similarly, the *Charaka Samhita* (300 BCE–200 CE) details a condition called "rajayakshma" (tuberculosis), which some scholars argue that they may have overlapped with kala-azar due to shared symptoms like splenomegaly and cachexia (12).

Some of archaeological evidence suggests that leishmaniasis plagued humans long before microscopes or modern medicine was discovered. In 2014, paleo-pathologists examining 3,000-year-old Egyptian mummies discovered skeletal lesions consistent with VL, while pre-Columbian pottery from Peru depicts facial deformities reminiscent of mucocutaneous infections (13). The Greek historian Herodotus, in the 5th century BCE, described Persian warriors afflicted with "noxious boils" during their campaigns—likely CL contracted in Mesopotamia's sandfly-infested valleys (14). Even the Old Testament may hold clues: some scholars argue that the "botch of Egypt" mentioned in Deuteronomy 28:27 refers to disfiguring skin ulcers caused by *Leishmania* parasites (5) (Figure 1.3).

The disease's persistence across epochs underscores its relationship with human societies. Leishmaniasis thrived where poverty, environmental disruption, and conflict intersected. Medieval Persian physicians documented the "Baghdad Boil" (CL) in texts like the *Canon of Medicine* by Avicenna (980–1037 CE), prescribing remedies ranging from cauterization to poultices of crushed beetles (16). In the 16th century, Spanish conquistadors in South America wrote of indigenous tribes disfigured by espundia (MCL), which they attributed to divine punishment (17). By the Industrial Revolution, colonial expansion and urbanization created new hotspots: deforestation in India's Brahmaputra Valley displaced sandfly habitats, triggering kala-azar epidemics that claimed over lakhs of lives between 1824 and 1900 (18).

Figure 1.3 (A) Moche ceramic vessel depicting someone with mucocutaneous leishmaniasis and (B) ancient pottery with a mutilation of nose and upper lip, suggestive of mucosal leishmaniasis. (Source: (A) Logan Museum of Anthropology, (B) Altamirano-Enciso et al. (15)).

During the scorching summer season of the late 19th century in colonial India, a perplexing illness began to afflict British soldiers stationed in the Bengal region. At that time, British physicians in India were already clinically dealing with outbreaks of this new disease "kala-azar" based on the symptom-based treatment, but they initially misidentified this new ailment as either malaria or tuberculosis. The affected soldiers experienced severe and unrelenting fevers, significant weight loss, and noticeable swelling of the spleen and liver. The local population referred to this disease as kala-azar, which translates to "black sickness" in Hindi, a name derived from the ashen, greyish pallor that characterized the skin of those nearing death. The medical officers of the British Raj were at a loss, as the contemporary treatments available, including quinine and mercury-based tonics, were ineffective against this condition (19) (Figure 1.4).

By 1870, kala-azar had become one of the British Empire's most feared adversaries, killing an estimated 25% of European troops stationed in Assam (20).

A Russian army doctor, named Peter Borovsky, was the first to identify and recognize protozoa bodies in skin lesions (21). In 1901, Charles Donovan, an Irish doctor in Madras, identified *Leishmania* amastigotes in spleen smears, a breakthrough paralleled by William Boog Leishman's work on British soldiers in 1903. Under his microscope, Leishman observed clusters of

Figure 1.4 Peter Borovsky, William Boog Leishmaniasis, and Charles Donovan. (Image Source: www.wikipedia.com.)

tiny, oval-shaped organisms infesting the man's cells—parasites later named *L. donovani* in honour of Leishman and his contemporary, Charles Donovan, who independently identified the same pathogen in Madras (22, 23).

Colonial narratives often framed leishmaniasis as a "tropical disease" tied to the "uncivilized" conditions of native populations (24). These discoveries marked a watershed moment in tropical medicine, yet it was only the latest chapter in a story-spanning time. The vector identification for leishmaniasis was equally important for better understanding of the disease. The transmission by *Phlebotomus argentipes* sandflies was confirmed in 1922 by Indian physician Upendranath Brahmachari (25) (Figure 1.5).

During World War II, Allied troops stationed in North Africa and Sicily faced outbreaks of "Aleppo Evil" (CL), prompting urgent drug trials of antimonial drugs like sodium stibogluconate. Yet, despite these advances, the disease remained a "silent killer" in the Global South, neglected by policymakers and pharmaceutical companies alike. By 1945, antimonial drugs reduced mortality, but their toxicity sparked debates over "acceptable risks" for soldiers versus civilians (26, 27).

In 1990, the World Health Organization (WHO) classified it as a "neglected tropical disease," a label reflecting its association with poverty (28). Today, leishmaniasis persists as a disease of poverty, haunting slums, refugee camps, and conflict zones from Syria to South Sudan. Its story is a mirror-reflecting humanity's triumphs and failures—a testament to the enduring interplay between parasites and people.

1.3 EPIDEMIOLOGY

Leishmaniasis is a significant global health concern, impacting over 350 million people across 88 tropical, subtropical, and temperate countries. Roughly

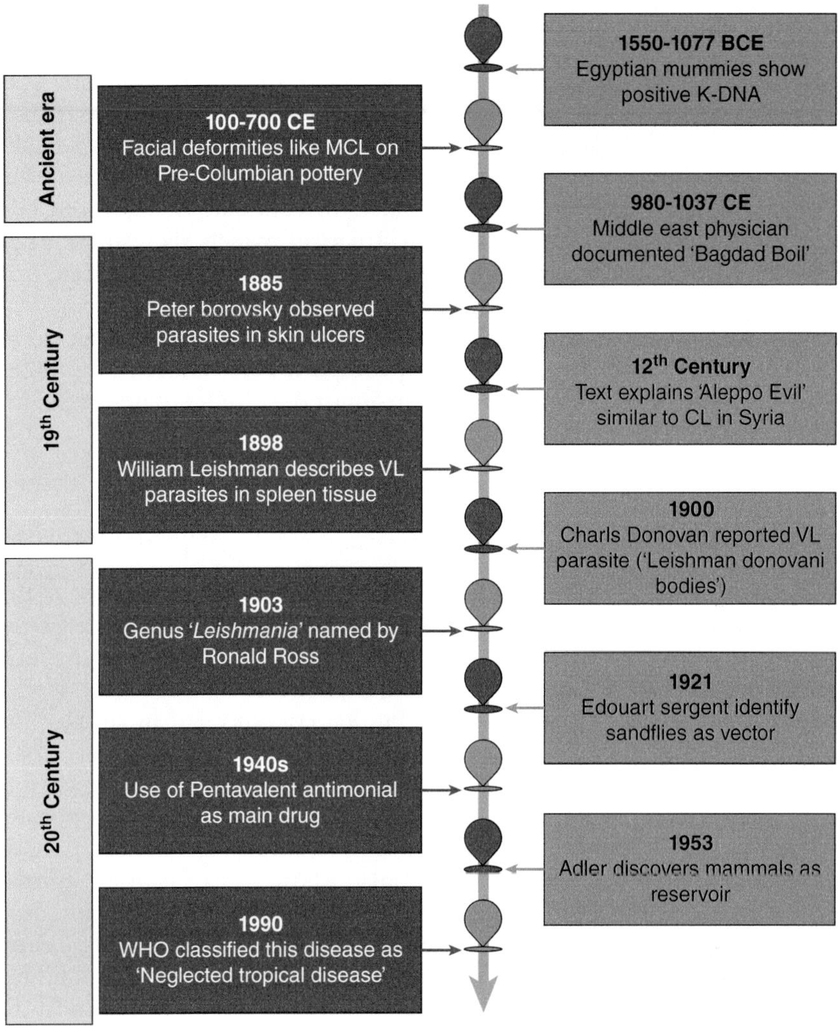

Figure 1.5 Historical timepoints of leishmaniasis.

12 million individuals currently suffer from the disease, with an estimated 0.2–0.4 million new cases of VL and 0.7–1.2 million new cases of CL, occurring annually worldwide (29). The parasite has been reported to infect the world's most impoverished populations in over 90 countries spanning Asia, Africa, the Middle East, and Central and South America (30). While likely underestimated, the current annual incidence of CL ranges from 700,000 to 1.2 million cases, with roughly 95% occurring in the Americas, the Mediterranean basin, the Middle East, and Central Asia. Estimates for annual cases of VL are currently under 100,000. This marks a significant decrease from previous estimates of 400,000, with more than 95% of reported cases to the WHO,

originating from Brazil, China, Ethiopia, India, Kenya, Nepal, Somalia, and Sudan (31).

The specific type of parasite, geographical location, host characteristics, and symptoms of the disease all vary. For instance, *L. donovani* causes VL in South Asia (Bangladesh, India, and Nepal) and East Africa (Sudan, Somalia, Ethiopia, and Kenya). This form often impacts younger individuals, while older adults may have acquired immunity. In contrast, *L. infantum* (also known as *Leishmania chagasi* in Latin America), which also causes VL, is typically found in the Mediterranean region, the Middle East, Pakistan, Iran, and Brazil (32). Anthroponotic transmission, meaning transmission from humans to humans, is the characteristic of the *L. tropica* complex (found in the Old World) and the *L. donovani* complex, particularly in the Indian subcontinent. *L. donovani* is prevalent in South Asia (India, Bangladesh, and Nepal) and East Africa (Sudan, Ethiopia, Kenya, and Somalia) (33).

1.4 GLOBAL HEALTH IMPACT

Throughout history, arthropod-borne diseases have caused immense suffering and mortality. The 17th to early 20th centuries were particularly devastating, marked by widespread outbreaks of diseases like plague, malaria, and yellow fever (34). Leishmaniasis disproportionately affects the world's poorest populations, with the highest burden concentrated in tropical and subtropical regions (35). While *L. donovani* can cause clinical disease at any age, in areas with consistent transmission, older adults often show lower incidence due to acquired immunity. In East Africa, *L. donovani* is responsible for both human-to-human (anthroponotic) and animal-to-human (zoonotic) transmission. Conversely, *L. infantum* (also known as *L. chagasi*) primarily causes VL in the Mediterranean, the Middle East, Afghanistan, Iran, Pakistan, and Brazil, although isolated cases have been reported in Central Asia, China, Mexico, and parts of Central and Latin America (36). Children and immunosuppressed adults face a higher risk of developing clinical disease from *L. infantum* compared to adults with healthy immune systems. To simplify the discussion of its many species, leishmaniasis is broadly categorized by geography into the Old World and New World. The Old World encompasses Asia, the Middle East, Africa, and Southern Europe, while the New World refers to the Western Hemisphere, specifically Mexico, Central America, South America, and the USA (37).

Reliable data on neglected tropical diseases (NTDs) like leishmaniasis is notoriously hard to come by. Leishmaniasis often appears in isolated outbreaks and remote areas, making it difficult to accurately estimate its true impact from official reports. It's widely accepted that reported case numbers for leishmaniasis significantly underestimate the actual burden, and studies measuring this underreporting are rare (38). For example, VL is fatal if left untreated, and many deaths go unrecorded. Even with access to treatment,

VL can still have a case-fatality rate of 10–20% (39). To improve the understanding of leishmaniasis, the WHO organized a series of regional meetings. During these meetings, national programme managers and expert professionals provided detailed information on the disease's epidemiology, ecology, geographical distribution, trends, drug access, and management within their respective countries (29).

The varying distribution of different clinical forms of leishmaniasis is influenced by a combination of factors. These include climatic elements like temperature, rainfall, relative humidity, and precipitation, as well as physical barriers in geography. Crucially, biotic factors, such as the distribution and abundance of vertebrate hosts, also play a significant role. Beyond these, socioeconomic conditions, environmental changes (like land degradation and deforestation due to urbanization), and even regional conflicts have all been linked to how *Leishmania* parasites and their clinical manifestations spread. Additionally, factors, like living standards and the host's immune response, are major contributors. Poor living conditions in many rural and semi-urban areas—including inadequate housing, poor waste management, and insufficient sewage and drainage—create ideal breeding and resting grounds for sandflies, the vectors that transmit the disease. Leishmaniasis appears to affect adults more often than children. This could be because adults, especially those in rural farming communities, have greater exposure to the parasites and their vectors. While this can vary, men generally show higher susceptibility to infection. The spread of leishmaniasis is further exacerbated by people moving from rural to urban areas for economic reasons. Additionally, a significant risk of infection exists in homes where animals are kept in close proximity to people. This is because certain animals, like dogs and felines, can act as reservoir hosts for the disease, particularly for VL. All these elements collectively contribute to the emergence and re-emergence of leishmaniasis outbreaks (40). The distribution and severity of leishmaniasis are highly sensitive to climate, as it directly influences the life cycles, distribution, survival, and population sizes of both the parasite's vectors (like sandflies) and its reservoir hosts. For instance, in India, sandflies, especially *P. argentipes*, are particularly active and abundant between June and September, when temperatures typically range from 27 to 31°C. However, significant fluctuations in temperature can negatively impact the development of the *Leishmania* promastigote within sandflies. This disruption can lead to the parasite spreading to areas not previously known for the disease. Furthermore, environments with dense vegetation are associated with a higher risk of leishmaniasis infection and transmission. There's also a clear link between forest fragmentation and deforestation and an increased risk of the disease (41).

Global health is crucial to national interests because it directly impacts poverty and social instability. The well-being of people worldwide is tied to development, as poor health can undermine economic progress, weaken institutions, and threaten families. The rapid spread of emerging infectious diseases further highlights this global interconnectedness (42). Despite being estimated as the ninth largest disease burden among infectious diseases, leishmaniasis often gets overlooked in discussions about tropical disease priorities. This neglect stems from its complex epidemiology and ecology, the lack of simple tools for managing cases, and scarce up-to-date incidence data. Consequently, policymakers often fail to recognize its importance.

In response to the World Health Assembly Resolution 2007/60.13, the WHO convened the Expert Committee on Leishmaniasis in March 2010. This committee subsequently released the first updated technical report on leishmaniasis in over two decades. Both the WHA Resolution and the Expert Committee report emphasized the critical need to update epidemiological data to develop effective leishmaniasis control strategies. Policymakers and funding organizations widely use disease burden estimates, often expressed in disability-adjusted life years (DALYs) lost, to set priorities. DALYs, first promoted in the 1993 World Development Report, have been a subject of intense scrutiny ever since (29). The accuracy of disease burden measurements hinges on reliable data concerning incidence, duration, severity, and mortality, as well as the assumptions used in calculations. Even though a new round of global disease burden estimation is underway, it doesn't include or support empirical data collection and field validation. Instead, conservative assumptions are used for underreporting rates, leading to the belief that the true incidence of leishmaniasis may be substantially higher. Estimates for less common forms of leishmaniasis are often unavailable due to a lack of data (43). Mortality estimates are even more uncertain than incidence figures, as studies show that a large proportion of kala-azar (VL) deaths occur outside of healthcare facilities and are often unrecognized, making accurate reporting impossible.

1.4.1 Challenges in disease control

A major challenge is the re-emergence of previously controlled or eliminated diseases in new geographic locations. Current regional and international capacities for detecting, identifying, and effectively responding to these diseases are insufficient. Surveillance and vital records in leishmaniasis-affected countries are often incomplete, highlighting significant data limitations (44). Furthermore, the emergence of drug resistance in vector-borne pathogens and insecticide resistance in their vectors poses a significant obstacle. The declining number of professionals trained in key fields like medical entomology, vector ecology, and tropical medicine further threatens our ability to address vector-borne diseases now and in the future (45).

1.4.2 Urgent need for better data and control

Improved surveillance systems are urgently needed, especially in disease hotspots targeted for more intensive control or elimination efforts. Accurate measures of progress, such as validating trends in surveillance data and

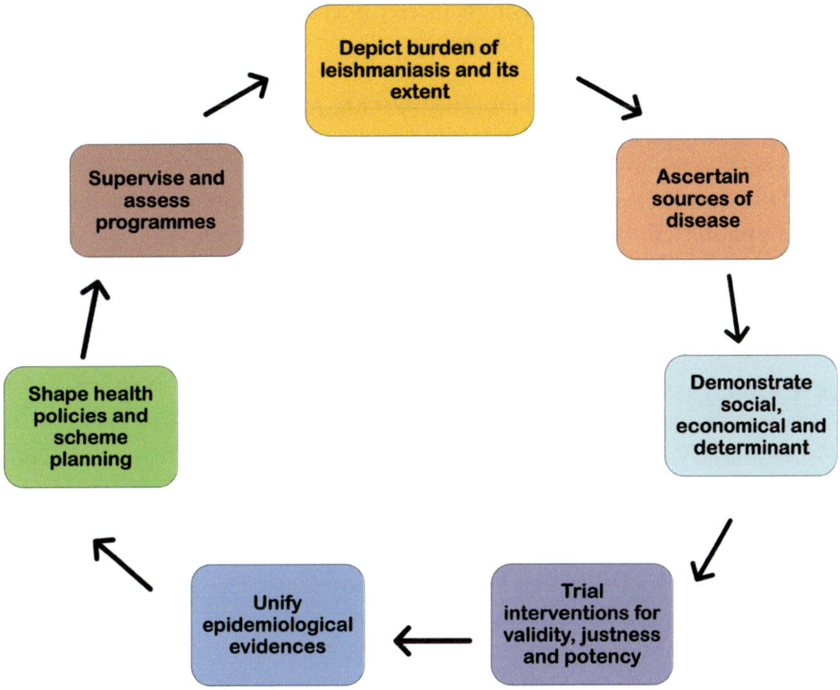

Figure 1.6 Considerations for global implications of leishmaniasis.

determining precise case-fatality rates, can only be achieved through active, community-based data collection.

The data gathered should allow for a more nuanced interpretation of published disease burden estimates. The uncertainties in current data should also spur efforts to improve the evidence base for leishmaniasis and other neglected diseases. In areas where leishmaniasis transmission is anthroponotic (human -to-human), the effective treatment of patients can help control the spread of the parasite (46). It's also recommended that individuals, living in or returning from endemic areas, undergo serological screening before receiving immunosuppressive treatment (Figure 1.6).

1.4.3 Global health initiatives and partnerships

To gain a deeper understanding of and to address the complex challenges posed by vector-borne diseases, the Forum on Microbial Threats held workshops. These gatherings explored the intricate connections between hosts, pathogens, vectors, and their ecosystems as well as discussed about the recent advancements in confronting these evolving threats. A workshop in June 2007 initiated this effort. Subsequently, in September 2014, another workshop was convened to analyse trends in the incidence and prevalence of vector-borne diseases (29) within an increasingly interconnected and ecologically disrupted world; initiatives include the following:

- WHO Roadmap for NTDs (2021–2030): Aims to eliminate VL as a public health problem by 2030 (target: <1 case/10,000 people in endemic areas) (47).
- DNDi and Gavi: Developing affordable treatments (e.g., thermostable amphotericin B) and advocating for vaccine research (48).
- Cross-Border Collaboration: Essential for regions like East Africa and South Asia due to migration and climate-induced vector spread (49).

1.5 CONCLUSION

Leishmaniasis with its complex clinical manifestations affects the world's poor population across the globe. It has a considerable medical, economic, and social burden on vulnerable people. The existing factors, like varying epidemiological and clinical forms besides socioeconomic reasons, different *Leishmania* species with a complex life cycle, resistance to drugs, cause a complete elimination of this disease tricky and insufficient, despite the implementation of planned efforts and measures. Having said that the disease impacts the public health and quality of life globally, both qualitative and quantitative emphases on disease control and management are the need of the hour. To gain comprehensive perception, historical epidemiology may play an important role in assessing lasting outcomes of disease management efforts. Their relevance is in terms of their role in shaping the pattern of disease and its spread across various regions. Addressing future global health challenges and devising strategies for them can be driven by accomplishments as well as setbacks found in prior inferences. Indeed, the dynamics of leishmaniasis can be understood on the foundation of integrated social, economic, and ecological considerations.

ACKNOWLEDGEMENT

The authors express their deepest gratitude to the ICMR-Rajendra Memorial Research Institute of Medical Sciences, Patna, for the invaluable support they provided throughout the course of this study.

REFERENCES

1. Bhunia, G. S., V. Kumar, J. Kumar, P. Das, and S. Kesari. "The use of remote sensing in the identification of the eco-environmental factors associated with the risk of human visceral leishmaniasis (kala-azar) on the Gangetic plain, in north-eastern India." *Annals of Tropical Medicine & Parasitology* 104, no. 1 (2010): 35–53.

2. Muniaraj, M. "The lost hope of elimination of kala-azar (visceral leishmaniasis) by 2010 and cyclic occurrence of its outbreak in India, blame falls on vector control practices or co-infection with human immunodeficiency virus or therapeutic modalities?" *Tropical Parasitology* 4, no. 1 (2014): 10–19.
3. Chappuis, F., S. Sundar, A. Hailu, H. Ghalib, S. Rijal, R. W. Peeling, J. Alvar, and M. Boelaert. "Visceral leishmaniasis: What are the needs for diagnosis, treatment and control?" *Nature Reviews Microbiology* 5, no. 11 (2007): 873–882.
4. Reithinger, R., J.-C. Dujardin, H. Louzir, C. Pirmez, B. Alexander, and S. Brooker. "Cutaneous leishmaniasis." *The Lancet Infectious Diseases* 7, no. 9 (2007): 581–596.
5. Oumeish, O. Y. "Cutaneous leishmaniasis: A historical perspective." *Clinics in Dermatology* 17, no. 3 (1999): 249–254.
6. Marsden, P. D., and R. R. Nonata. "Mucocutaneous leishmaniasis: A review of clinical aspects." *Revista da Sociedade Brasileira de Medicina Tropical* 9 (1975): 309–326.
7. Ranjan, R., S. Kumari, and A. Kumar. "Geographical Distribution of VL and PKDL in India." In Veer Singh and Ashish Kumar (eds.), *Visceral Leishmaniasis and Post-kala-azar Dermal Leishmaniasis*, pp. 18–26. CRC Press, 2025.
8. Singh, V., and A. Kumar, eds. *Visceral Leishmaniasis and Post-kala-azar Dermal Leishmaniasis: Pathogenesis, Treatment and Disease Control*. CRC Press, 2025.
9. Zijlstra, E. E., A. M. Musa, E. A. G. Khalil, I. M. El Hassan, and A. M. El-Hassan. "Post-kala-azar dermal leishmaniasis." *The Lancet Infectious Diseases* 3, no. 2 (2003): 87–98.
10. Lazar, L. T. Y., and K. S. Abass. "Morphology, life cycle, pathogenesis and virulence factors of genus *Leishmania*: A review." *Plant Archives* 20, no. 2 (2020): 4057–4060.
11. Kumar, A., R. Milan, and J. Kumar. "A literary review on pandu roga." *Journal of Ayurveda and Integrated Medical Sciences* 9, no. 7 (2024): 131–137.
12. Lochan, K. *Medicines of Early India: With Appendix on a Rare Ancient Text*. Vol. 49. Chaukhambha Publications, 2003.
13. Nicholson, S. "The Depiction of Pathology in Pre-Columbian Ceramic Pottery and Figurines." *PhD diss.*, California State University, Northridge, 2008.
14. Adler, S., and O. Theodor. "The distribution of sandflies and leishmaniasis in Palestine, Syria and Mesopotamia." *Annals of Tropical Medicine & Parasitology* 23, no. 2 (1929): 269–306.
15. Altamirano-Enciso, A. J., M. C. A. Marzochi, J. S. Moreira, A. O. Schubach, and K. B. F. Marzochi. "On the origin and spread of cutaneous and mucosal leishmaniasis, based on pre-and post-Colombian historical sources." *História, Ciências, Saúde-Manguinhos* 10 (2003): 853–882.
16. Lafta, R. K., and A. Hussain. "Baghdad Boil is reemerging in Iraq." *Indian Journal of Public Health Research & Development* 10, no. 8 (2019): 1475.
17. Al-Kamel, M. A. "Stigmata in cutaneous leishmaniasis: Historical and new evidence-based concepts." *Our Dermatology Online* 8, no. 1 (2017): 81.
18. Lisboa, K. M. "Unhealthiness, disease, and immigration: German views of Brazil." *História, Ciências, Saúde-Manguinhos* 20 (2013): 119–139.
19. Gupta, P. C. S. "History of kala-azar in India." *The Indian Medical Gazette* 82, no. 5 (1947): 281.
20. Dodds-Price, J. "Kala azar in Europeans in the Nowgong District of Assam." *The Indian Medical Gazette* 55, no. 3 (1920): 87.
21. Hoare, C. A. "Early discoveries regarding the parasite of oriental sore." *Transactions of the Royal Society of Tropical Medicine and Hygiene* 32, no. 1 (1938): 66–92.
22. Leishman, W. B. "On the possibility of the occurrence of trypanosomiasis in India." *British Medical Journal* 1, no. 2213 (1903): 1252.

23. Donovan, C. "Kala-azar in Madras, especially with regard to its connexion with the dog and the bug (Conorrhinus)." *The Lancet* 174, no. 4499 (1909): 1495–1496.
24. Schonebaum, A. "Vectors of contagion and tuberculosis in modern Chinese literature." *Modern Chinese Literature and Culture* 23, no. 1 (2011): 17–46.
25. Mathew, J. "Ronald Ross to UN Brahmachari: Medical research in colonial India." *Indian Journal of History of Science* 53, no. 4 (2018): 115–122.
26. Klaus, S., and S. Frankenburg. "Cutaneous leishmaniasis in the Middle East." *Clinics in Dermatology* 17, no. 2 (1999): 137–141.
27. Borborema, S. E. T., J. A. Osso, H. F. de Andrade, and N. do Nascimento. "Biodistribution of meglumine antimoniate in healthy and *Leishmania (Leishmania) infantum chagasi*-infected BALB/c mice." *Memórias do Instituto Oswaldo Cruz* 108, no. 5 (2013): 623–630.
28. Hunt, P., R. Steward, J. B. de Mesquita, and L. Oldring. *Neglected Diseases: A Human Rights Analysis*. Vol. 6. World Health Organization, 2007.
29. Alvar, J., I. D. Vélez, C. Bern, M. Herrero, P. Desjeux, J. Cano, J. Jannin, M. den Boer, and WHO Leishmaniasis Control Team. "Leishmaniasis worldwide and global estimates of its incidence." *PLoS One*, 7, no. 5 (2012): e35671. https://doi.org/10.1371/journal.pone.0035671
30. Tabbabi A., T. Vasconcelos Dos Santos, K. Bekhti, N. Haouas, P. Junxiong, M. A. Garcia-Bereguiain. Editorial: Vector-borne diseases and consequences on human health: A multidisciplinary approach. *Frontiers in Public Health* 11 (2023): 1326243. https://doi.org/10.3389/fpubh.2023.1326243
31. Mann, S., K. Frasca, S. Scherrer, A. F. Henao-Martínez, S. Newman, P. Ramanan, and J. A. Suarez. A review of leishmaniasis: Current knowledge and future directions. *Current Tropical Medicine Reports*, 8, no. 2 (2021): 121–132. https://doi.org/10.1007/s40475-021-00232-7
32. Makarani, N., K. Bharadava, A. Kaushik, A. Dave, A. K. Gangawane, R. S. Kaushal. Leishmaniasis: A multifaceted approach to diagnosis, maladies, drug repurposing and way forward. *The Microbe* 6 (2025), 100239. https://doi.org/10.1016/j.microb.2025.100239
33. Georgiadou, S. P., K. P. Makaritsis, and G. N. Dalekos. Leishmaniasis revisited: Current aspects on epidemiology, diagnosis and treatment. *Journal of Translational Internal Medicine*, 3, no. 2 (2015): 43–50. https://doi.org/10.1515/jtim-2015-0002
34. Gubler, D. J. Resurgent vector-borne diseases as a global health problem. *Emerging Infectious Diseases*, 4, no. 3 (1998): 442–450. https://doi.org/10.3201/eid0403.980326
35. Okwor, I., and J. Uzonna. Social and economic burden of human leishmaniasis. *The American Journal of Tropical Medicine and Hygiene* 94, no. 3 (2016): 489–493. https://doi.org/10.4269/ajtmh.15-0408
36. Hong, A., R. A. Zampieri, J. J. Shaw, L. M. Floeter-Winter, and M. F. Laranjeira-Silva. One health approach to leishmaniases: Understanding the disease dynamics through diagnostic tools. *Pathogens (Basel, Switzerland)* 9, no. 10 (2020): 809. https://doi.org/10.3390/pathogens9100809
37. Akuffo, H., C. Costa, J. van Griensven, S. Burza, J. Moreno, and M. Herrero. New insights into leishmaniasis in the immunosuppressed. *PLoS Neglected Tropical Diseases* 12, no. 5 (2018): e0006375. https://doi.org/10.1371/journal.pntd.0006375
38. Relman, D. A., and E. R. Choffnes, eds. *The Causes and Impacts of Neglected Tropical and Zoonotic Diseases: Opportunities for Integrated Intervention Strategies*. National Academies Press, 2011.
39. Yeshaw, Y., A. T. Tsegaye, and S. G. Nigatu. Incidence of mortality and its predictors among adult visceral leishmaniasis patients at the University of Gondar Hospital: A retrospective cohort study. *Infection and Drug Resistance* 13 (2020): 881–891. https://doi.org/10.2147/IDR.S245991

40. El Omari, H., A. Chahlaoui, F. Z. Talbi, K. E. Mouhdi, and A. El Ouali Lalami. Impact of climatic factors on the seasonal fluctuation of leishmaniasis vectors in central Morocco (Meknes Prefecture). *The Canadian Journal of Infectious Diseases & Medical Microbiology = Journal canadien des maladies infectieuses et de la microbiologie medicale* 2020 (2020): 6098149. https://doi.org/10.1155/2020/6098149
41. Walsh, J. F., D. H. Molyneux, and M. H. Birley. "Deforestation: Effects on vector-borne disease." *Parasitology* 106, no. S1 (1993): S55–S75. https://doi.org/10.1017/S0031182000086121
42. Baker, R. E., A. S. Mahmud, I. F. Miller, et al. Infectious disease in an era of global change. *Nature Reviews Microbiology* 20 (2022): 193–205. https://doi.org/10.1038/s41579-021-00639-z
43. Torres-Guerrero, E., M. R. Quintanilla-Cedillo, J. Ruiz-Esmenjaud, and R. Arenas. Leishmaniasis: A review. *F1000Research* 6 (2017): 750. https://doi.org/10.12688/f1000research.11120.1
44. Sutherst R. W. Global change and human vulnerability to vector-borne diseases. *Clinical Microbiology Reviews* 17, no. 1 (2004): 136–173. https://doi.org/10.1128/CMR.17.1.136-173.2004
45. Murray, H. W., J. D. Berman, C. R. Davies, and N. G. Saravia. Advances in leishmaniasis. *Lancet (London, England)* 366, no. 9496 (2005): 1561–1577. https://doi.org/10.1016/S0140-6736(05)67629-5
46. Stockdale, L., and R. Newton. A review of preventative methods against human leishmaniasis infection. *PLoS Neglected Tropical Diseases* 7, no. 6 (2013): e2278. https://doi.org/10.1371/journal.pntd.0002278
47. Casulli, A. New global targets for NTDs in the WHO roadmap 2021–2030. *PLoS Neglected Tropical Diseases* 15, no. 5 (2021): e0009373. https://doi.org/10.1371/journal.pntd.0009373
48. Pagliusi, S., Y. Che, and S. Dong. The art of partnerships for vaccines. *Vaccine*, 37, no. 40 (2019): 5909–5919. https://doi.org/10.1016/j.vaccine.2019.07.088
49. Negev, M., S. Paz, A. Clermont, N. Groag Pri-Or, U. Shalom, T. Yeger, and M. S. Green. "Impacts of climate change on vector borne diseases in the Mediterranean Basin—Implications for preparedness and adaptation policy. *International Journal of Environmental Research and Public Health* 12, no. 6 (2015): 6745–6770. https://doi.org/10.3390/ijerph120606745

Chapter 2

Leishmania life cycle

Vector biology and parasite transmission dynamics

*Himanshu Nilesh Panchal**

Department of Pharmaceutical Chemistry, Parul Institute of Pharmacy, Parul University, Vadodara, Gujarat, India

Sapna Desai

Department of Pharmacology at Pioneer Pharmacy College, Vadodara, Gujarat, India

Satchitkrishna Chandrashekhar Radharapu

St. Lawrence College, Kingston, Canada

Divya J. Tandel

Department of Pharmacology, Parul Institute of Pharmacy, Parul University, Vadodara, Gujarat, India

2.1 INTRODUCTION

Leishmaniasis is one of a large range of neglected tropical illnesses linked to poverty that significantly affects global health (1). As a protozoan illness that causes misery to humans, leishmaniasis is perhaps the second most important disease after malaria (2). In both the Old and New Worlds, leishmania is transmitted by insects, obligatory intracellular protozoa (family *Trypanosomatidae*) that cause diseases of the skin, mucosa, and viscera (3). Except for *Leishmania donovani* and *Leishmania tropica*, leishmaniasis is mostly zoonotic; however, there is some evidence that both species have animal reservoirs in Asia and Africa. These parasites reproduce intracellularly in humans and typically manifest as cutaneous or visceral diseases (4).

Visceral leishmaniasis (VL), commonly referred to as kala-azar, cutaneous leishmaniasis (CL), and mucocutaneous leishmaniasis (ML) are the three primary types of the illness. In more than 95% of cases, kala-azar, another name for VL, is lethal if treatment is not received. Frequent episodes of fever, loss

* Corresponding author: hnnh12501@gmail.com

 DOI: 10.1201/9781003596455-2

of weight, liver and spleen enlargement, and anemia are its hallmarks. Brazil, East Africa, and India account for the majority of instances. Only 25–45% of the estimated 50,000–90,000 new cases of VL that occur worldwide each year are reported to the WHO. It may cause an epidemic or cause death. The most prevalent kind, CL, results in skin lesions, primarily ulcers, on exposed body areas. These can cause severe impairment or stigma, as well as permanent scarring. The Americas, the Mediterranean region, the Middle East, and central Asia account for around 95% of CL cases. Between 600,000 and 1 million new cases are thought to arise globally each year, although only around 200,000 are reported to the WHO. The mucous membranes of the mouth, throat, and nose can be partially or completely destroyed by ML. Bolivia, Brazil, Ethiopia, and Peru account for more than 90% of cases of ML (5).

Leishmania species vary in the severity of their clinical manifestations, which can range from self-curing skin lesions to potentially fatal visceral illness. Parasite traits, vector biology, and host variables all interact to determine the result, with immunological responses among the host factors playing a major role. The most severe, systemic type of leishmaniasis, known as VL, is typically deadly if left untreated. It is caused by *Leishmania infantum* in the basin of the Mediterranean Sea, the Middle East, central Asia, South America, and Central America, and by *L. donovani* throughout Asia and Africa. Following treatment for VL, otherwise healthy individuals may develop a skin manifestation known as post-kala-azar dermal leishmaniasis (PKDL). Although CL often just results in an ulcer that cures on its own in three to 18 months, it can also cause stigmatization, deformity, and scarring as disabling effects. Up to 10% of cases of CL develop more severe symptoms, depending on the parasite type. ML, diffuse CL, disseminated CL, and leishmaniasis recidivans are the names given to these severe symptoms (6).

The fact that *Leishmania* is a broad genus indicates that it has a complicated life cycle with two primary stages: promastigote and amastigote, which vary depending on the host—either a vertebrate host (human or animal) or an invertebrate vector (sandfly). The female sandfly (*Phlebotomus* and *Lutzomyia*) mediates transmission of this parasite, which lives in cells of the mammalian monocytic phagocytic system and can be either anthroponotic or zoonotic, depending on the location (7–9). Approximately 350 million individuals are at risk for the illness, which is prevalent in over 98 countries (9). *Leishmania* parasites are known to be spread by more than 90 sandfly species (5).

2.2 THE *LEISHMANIA* PARASITE

2.2.1 Morphological forms of the parasite: promastigote and amastigote

Leishmania parasites exhibit two main morphological forms that alternate between the host and sandfly. In the midgut of the female phlebotomine sandfly, promastigotes are grown, which are elongated and flagellated, so that they can move in the digestive tract of the sandfly. During a nocturnal blood meal, this form is injected into the mammalian host to cause the infection. Therefore, it is considered under the infective stage. After entering the host body, promastigotes are engulfed by the mononuclear phagocytes and convert into amastigotes that are non-flagellated, round, or oval intracellular forms. Amastigotes also multiply in the phagocytes, and then they cause the clinical symptoms of leishmaniasis in the host. They can spread through different routes, such as hematogenous and lymphatic, leading to mucocutaneous and visceral diseases depending upon the species (9).

2.2.2 Life cycle stages and their characteristics

It is a digenetic lifecycle which involves a vertebrate host and an invertebrate vector (female sandfly). The cycle starts when the sandfly injects the promastigotes into the host's blood, and then these promastigotes are engulfed by the phagocytes that convert these promastigotes into amastigotes through the reticuloendothelial system. Sandfly gets infected by ingesting the amastigotes through the blood meal from the host, and then the amastigotes are converted into the promastigotes in the midgut. They multiply and travel to the proboscis and become metacyclic promastigotes that are again ready to infect the next vertebrates. The vector is typically nocturnal, silent flying, and most active, which aligns with the time of transmission. Both humans and animals serve as the reservoirs, while in anthroponotic transmission, humans serve as the primary reservoir; some species that do this type of transmission are *L. donovani* and *L. tropica*. In zoonotic cycles, dogs, rodents, monkeys, and other wild animals serve as the reservoirs, while dogs are the most important reservoirs for *L. infantum* transmission (9).

2.2.3 Genetic diversity and its impact on disease severity and transmission

All the leishmania species exhibit genetic and biological diversity, which influences disease presentation, transmission, and response to treatment. Over 20 main species infect humans, which are all divided into two subgenera: *Leishmania* and *Viannia*. These species are geographically distributed based on their different pathogenic profiles, such as *L. donovani*, which causes VL in India and East Africa. On the other hand, *Leishmania mexicana* and *Leishmania braziliensis* cause cutaneous and mucocutaneous leishmaniasis in the Americas. In disease severity, a notable factor is viral coinfection, such as *Leishmania* RNA virus in species like *Leishmania (Viannia) guyanensis* and *Leishmania (V.) braziliensis*. Immune responses are increased due to this viral coinfection, which leads to severe mucosal damage and metastatic spread. Other strains, such as *Leishmania major*, do not show this effect. To determine disease severity, the host immune response interacts with the parasite genetics. Mucosal complications are increased by a strong TH1 response that may control parasites. During a high parasite load and the visceral disease, a TH2 response is generated. Diagnosis, treatment decisions, and vaccine development are impacted by diversity. Therefore, species identifications become important for clinical management (9).

2.3 THE SANDFLY VECTOR

2.3.1 Taxonomy of sandfly vector

Inaccurately suggesting to non-experts that they may be at danger of catching a vector-borne disease, while on vacation at the shoreline, the name "sandfly" can be misleading. In reality, the English name is known to the beetle's pale tint. Confusion is exacerbated by the fact that flies of the genera Culicoides (Diptera: Ceratopogonidae) and blackflies (Diptera: Simuliidae) share a common name in several parts of the world. The vectors of leishmaniases and other diseases of public health significance, which are correctly called "phlebotomine sandflies," must thus be differentiated (1). There are currently thought to be more than 800 species spread over the globe. They belong to the family Psychodidae, subfamily Phlebotominae, and suborder Nematocera of the order Diptera. The non-vector, non-biting moth flies (subfamily Psychodinae), which are frequently observed near shower drains, belong to the same family as phlebotomine sandflies, Psychodidae (10, 11).

Phlebotomine sandflies are still contentious, difficult, and distant from being definitively classified. Lewis et al. (11, 12) suggested three genera for New World species (*Lutzomyia Franca*, *Brumptomyia Franca* and *Parrot*, and *Warileya Hertig*) and two genera for Old World species (*Phlebotomus Rondani* and *Sergentomyia Franca*) based on the groundbreaking categorization of Theodor (1948, 1958). Some Chinese sandfly species with primitive features belong to a separate taxon known as *Chinius* Leng, 1987 (13). Modern Old World taxonomists generally agree with these three genera: *Phlebotomus*, *Sergentomyia*, and *Chinius*. Although a few more species have been or will soon be described, their medicinal significance is still unclear (14). There are 96 species, 11 subgenera, and 17 subspecies identified in the genus *Phlebotomus*. Lewis (15) said. The majority of entomologists currently classify Neotropical sandflies according to Lewis et al. (16), which were later evaluated by Young and Duncan (17). These authors identify the three genera *Lutzomyia*, *Brumptomyia*, and *Warileya*, which comprise 11 species groups and 15 subgenera. Although there have been more recent amendments put forth, none of them have gained widespread acceptance. Galati (18), the most recent and thorough, identified 464 species of Neotropical phlebotomine sandflies, which were arranged into 23 genera, 20 subgenera, three species groupings, and 28 series (14).

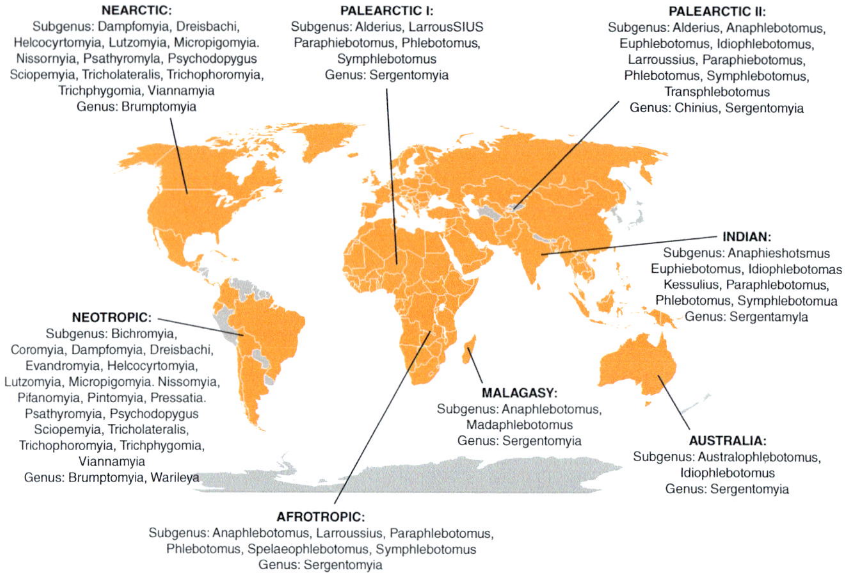

Figure 2.1 Distribution map of sand flies by genera and subspecies (23).

2.3.2 Distribution

Sand flies lack a water stage in their life cycle, unlike mosquitoes and other Diptera (19). However, along with temperature, humidity is a significant component that negatively affects and influences the growth of sand flies (20–22). This explains why the sandfly's range is restricted to regions with temperatures above 15.6°C for at least three months of the year (19), which still includes the largest area of the world, from latitude 50°N to latitude 40°S (New Zealand and the Pacific islands are not affected) (Figure 2.1).

2.3.3 Biology

The four phases of development that phlebotomine sandflies go through are egg, larva (four instars), pupa, and adult. Although they require moderately damp and warm settings, the immature stages do not require standing water to finish their development, in contrast to mosquitoes. Adult females lay their eggs in an appropriate environment that is high in organic matter, such as soil and animal waste, which gives the newly born larvae moisture, food, and shelter. Initially white or light grey in color, eggs (0.3–0.5 mm in length) frequently become dark brown or black after a few hours following laying. The temperature has a significant impact on egg hatching, and larval growth is often sluggish. Nine sandfly species from six genera or subgenera were recently studied in 15 laboratory colonies to assess their embryonic and larval development times over a one-year period. First-instar larvae appear

Figure 2.2 Larva of the sandfly *Phlebotomus perniciosus* (25).

12–19 days after the female has finished oviposition and had a blood meal, followed by pupae 25–59 days later and adults 35–69 days later (24).

Caterpillar-shaped larvae have tiny antennae that resemble leaves and head capsules. Although they are infrequently collected in nature, their lengthy caudal setae can aid in identifying them as sandfly larvae (25) (Figure 2.2) (12). However, these are not often used in taxonomy. The larvae barely leave the location of oviposition. Similar to tiny chrysalises, pupae have a solid substrate connected to one end of the fourth-stage larval exuvia.

Adults are tiny, seldom growing more than 3.5 mm (26). When at repose, they keep their wings in a distinctive "V" form on the backs and are coated in thick hairs (Figure 2.3) (12). Their hues vary from nearly white to nearly black. The legs are delicate and quite lengthy. It consumes honeydew made by homopterous aphids (Hemiptera: Aphidoidea) or sweet secretions from plants. For the development of egg batches to be completed, females need at least one blood meal. Without a bloodmeal, only a small number of phlebotomine sandfly species can lay viable eggs. They assault the host silently, in contrast to mosquitoes. Despite they can bite throughout the day if disturbed, adults are most active in the late hours of the day, at night, and early in the morning.

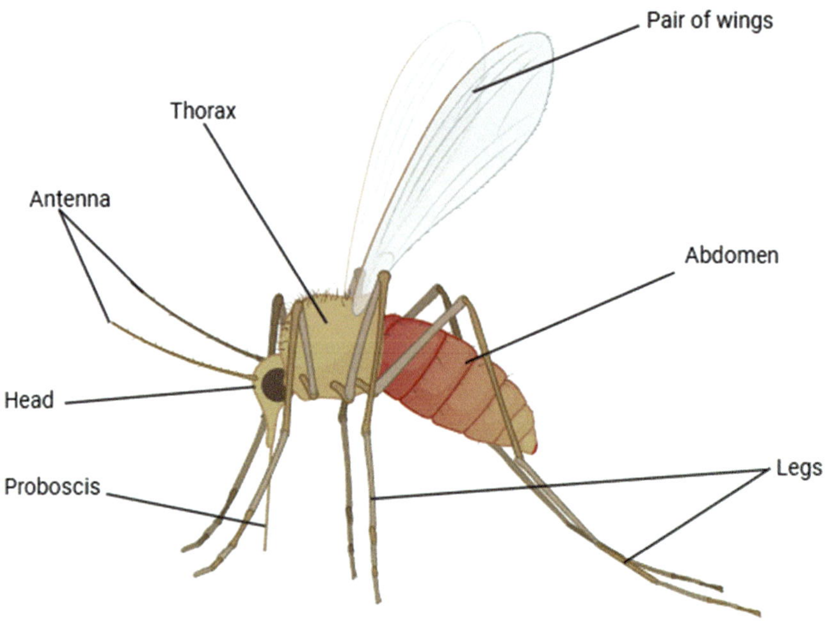

Figure 2.3 Female sandfly (*Phlebotomus papatasi*) (27).

Phlebotomines have a flying speed of less than 1 m/s, which is significantly slower than that of mosquitoes (27). Their incapacity to fly at wind speeds higher than this is the main factor limiting their dispersion range. Adult activities are largely limited to the area surrounding larval breeding sites because of their normally relatively small flying range (about 300 m). Forest species spread over shorter distances than peridomestic ones, according to evidence from mark-release-recapture experiments. *Phlebotomus ariasi*, for instance, may spread out over a distance of 2 km (28, 29). In contrast, species found in Neotropical forests rarely seem to spread out across more than 1 km (14). Mating takes place at or close to the host (Figure 2.4) (12). The males create sex pheromones and gather in leks on or close to the host. Males' wing vibrations can play a significant role in luring females to mate (30). Resting places are often dark, damp, and chilly micro-habitats that are close to larval breeding grounds (10). Rainfall and temperature have the most effect on adult sandflies' seasonal activity.

2.4 THE LIFE CYCLE *LEISHMANIA*

Within the sandfly vector, *Leishmania* parasites develop in a highly complex manner (Figure 2.5), necessitating certain differentiation processes to produce a successful infection.

Figure 2.4 Phlebotomus perniciosus mating (30).

2.4.1 Infection of sandfly: ingestion of infected mammalian host cells

Mainly the transformation of amastigotes to promastigotes takes place in the sandfly of the genus *Phlebotomus*, which acts as the intermediate vector and host. The main carrier of the disease is the female sandfly. The infected host consists of *Leshmania* parasite in their macrophages. Female sandfly is on the search for the blood meal from the host. In the host, they remain in the form of amastigotes in the host macrophage cells. Amastigotes are the same size and are spherical to avoid detection. Their size varies between 2.5 and 5.0 µm (31). Amastigotes are the intracellular forms of parasites present in the phagolysosomes of macrophages. Female flies suck the amastigotes by inserting their saw-like structure from the peripheral circulation of the skin. They are not able to access the macrophages present in the organs like the spleen and liver. During piercing, they produce a wound on the skin, which releases amastigotes into the blood circulation that is sucked by a fly and taken to its abdomen. As the amastigotes move to the sandfly abdomen, various conditions change, such as an increase in pH and a decrease in temperature, which trigger the growth of the parasite in the vector. At the gut, pH and temperature amastigotes are transformed into promastigotes with flagella beating at the anterior end, which provides motility to the promastigotes (32).

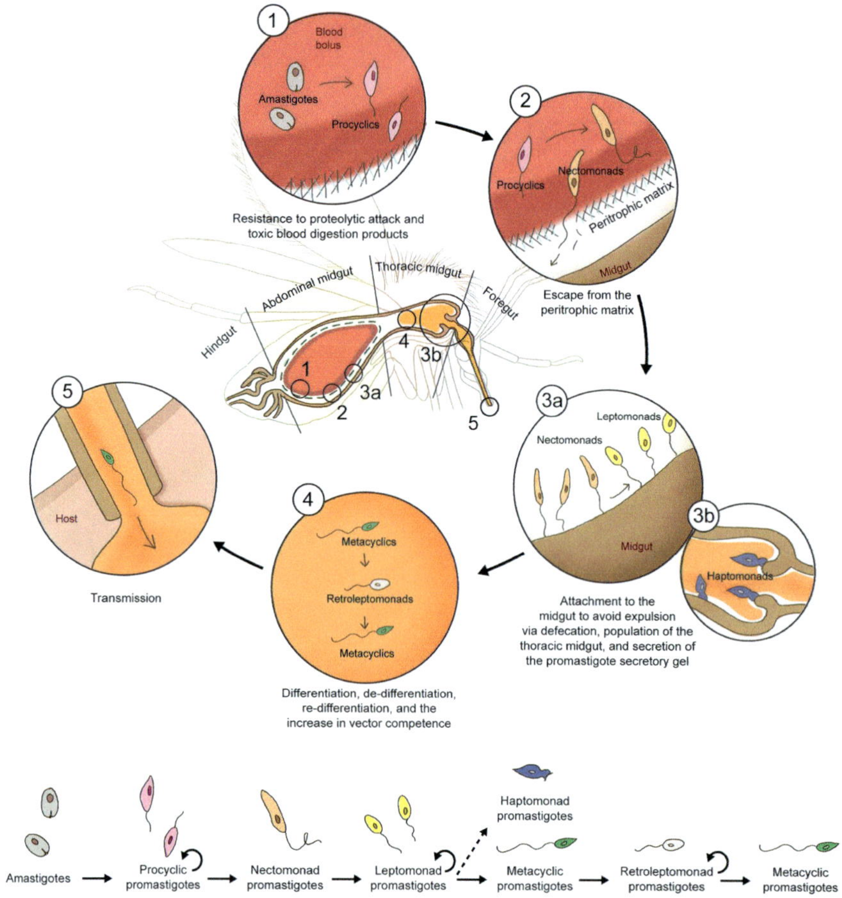

Figure 2.5 Leishmania development in the midgut of sand flies (34).

2.4.2 Development in the sandfly

2.4.2.1 Multiplication of promastigotes in the midgut

Mostly *Leshmania* species are suprapylarian (development in the midgut), while the subgenus *Viannia* grows in the hindgut instead of the midgut. After the bloodmeal, the first differentiation starts due to changes in conditions such as an increase in pH and a decrease in temperature. In this first differentiation, amastigotes are transformed into weakly motile forms known as procyclic promastigotes. Before multiplication starts to avoid excretion with blood remnants, procyclic promastigotes have to bind with the midgut epithelium. By escaping from the endoperitrophic region and inserting their anterior end of flagella between the epithelial microvilli, they attach to the midgut (33). Attachment to the midgut occurs between the lipophosphoglycan of the parasite and the galectin of the sandfly epithelium (32). The first step of differentiation is important because parasites have to resist the digestive protease attack on them. It was seen that these promastigotes are more resistant to proteolytic activity during the translational state because dynamic changes in the glycocalyx are seen during the translational period. Procyclic promastigotes decrease their replication and begin differentiating after 48 hours of bloodmeal ingestion, giving rise to long, highly motile forms called nectomonad promastigotes. These types have the ability to enter the midgut lumen from the peritrophic matrix. After moving into the anterior midgut, these nectomonads undergo differentiation into leptomonad promastigotes. This leptomonad is responsible for the population of the anterior midgut of the sandfly, and they also secrete the promastigote secretory gel. After this stage, the leptomonad promastigotes undergo another differentiation process known as metacyclogenesis, which forms metacyclic promastigotes from the leptomonad promastigotes. These metacyclic promastigotes are the parasites that become infective to the vertebrate hosts. It is seen that metacyclic forms come into existence due to purine deprivation during the differentiation process. Morphologically, metacyclic forms have a smaller body with a long flagellum, which provides fast motility to the parasites. Additionally, leptomonad promastigotes attach to the stomodeal valve of sandflies and give birth to the haptomonad promastigotes. After the first meal, flies have a second meal, which is a non-infected one; metacyclics go under de-differentiation and form retroleptomonad in the presence of newly available nutrients. Retroleptomonad, are similar to the leptomonads and opposite to the metacyclics, which is reflected in nature. As soon as the stress increases in the midgut of the sandfly, retroleptomonads go under redifferentiation and form metacyclic promastigotes. Retroleptomonads' replication has better vector infections both qualitatively and quantitatively. It shows that the development of the parasites in the sandfly is a gradual process. Parasite amplification is dependent on the multiple blood meals taken by the sandfly. Even though the parasite number is less in the initial stage, the second meal will increase the amplification of the leptomonads to the metacyclics. Additionally, de-differentiation of metacyclics to retroleptomonads will boost the development of better infections and vector competence (33).

2.4.2.2 Migration of promastigotes to the foregut and proboscis

Migration of the metacyclic promastigotes from the midgut to the foregut is done by the regurgitation process in the stomach of the sandfly. Leptomonad promastigotes secrete a proteophosphoglycan filament, which acts as a gel-like plug known as PSG Plug. Due to this plug, the sandfly feeding process gets impaired, and to facilitate blood intake, the sandfly itself regurgitates. It is also seen that due to the PSG Plug, the stomodeal valve gets damaged, due to which sandflies are not able to take full blood meals, so they often take small blood meals from various hosts, which increases infection (33).

2.4.2.3 Transmission to mammalian host

2.4.2.3.1 Inoculation of metacyclic promastigotes into the skin during a blood meal

The egestion of the metacyclic promastigotes into the host is done by probing. Ten-fold more metacyclic promastigotes are induced in the host in comparison to the parasite present in the foregut of the sandfly. The sandfly ingests the probe into the host skin, which releases the inoculums into the host blood, and the inoculums consist of the metacyclic promastigotes, saliva, and PSG. It is egested because sandfly clears their midgut and foregut, so that the imbibed blood from the host can go into the midgut; simultaneously, parasites are also deposited into the host body. Saliva plays a very important role in causing the CL and also in blood feeding. Co-inoculation of the saliva and the parasites shows alteration in the immune response of the host to protect the parasite because saliva contains potent vasodilatory and antihemostatic properties (32).

2.4.2.3.2 Transformation of promastigotes into amastigotes within host cells

To enter host cells, parasites use mannose-fucose, fibronectin, and toll-like receptors, and also use CR1 and CR3 complement receptors (35). As soon as metacyclic-promastigotes enter the host, they encounter neutrophils that engulf the promastigotes. After this they enter the macrophages with the help of the CR-3 receptor. Through focal exocystosis of the host-cell membrane, starting from the endosome, lysosome, and endoplasmic reticulum, forms the promastigote-containing phagosome. All the supply from the various membranes makes up the nascent parasitophorous vacuole, and at the initial time, parasites live in this vacuole. To evade the microbicidal consequences of the host cell, promastigotes inhibit the synthesis of phagolysosomes, and this is done by the accumulation of periphagosomal F-Actin. It is done to protect the phagosome from the degradation of acidic pH and lysosomal hydrolases. Mainly, acidification is mediated by the vacuolar proton ATPase. It also has glycoconjugate lipophosphoglycan on promastigotes, and it is absent in the amastigotes. It also saves promastigotes from the acidic environment. There are various factors that contribute to converting the promastigotes into amastigotes, one of them is low pH in the phagosome, which is between 4.7 and 5.2. Another factor is temperature; sandfly temperature is 25°C, and host temperature is 37°C. All these factors activate certain gene expression changes in the parasites, which bring structural and metabolic changes in the parasite. Amastigotes are the small rounded structure with no flagellum, and they have learned to survive in the acidic environment of phagosome through various ways such as resisting lysososmal enzymes and hydrolytic degradation, and to avoid detection from the host cell, they interfere into the hosts immune response and another way is by interfering with production of reactive oxygen species (ROS), preventing oxidative damage. To support amastigote replication, the parasitophorous vacuole undergoes various modifications by acquiring various markers. As soon as the amastigotes fully differentiate, they start multiplying inside the parasitophorous vacuole. After this, through host-cell lysis and direct transfer, they spread to the neighboring cells. The cycle continues in the sandfly and gets converted into promastigotes (36).

2.5 THE VECTOR LIFE CYCLE

The four main phases of the sandfly life cycle are eggs, larvae, pupae, and adults (Figure 2.6). A female sandfly typically lays around 50 eggs in sheltered areas that are selected depending on moisture and the availability of organic matter around it (37, 38). Although this period may be extended in cooler weather—eggs may become dormant under unfavorable circumstances—the eggs usually hatch four to 20 days after oviposition (38, 39). There are four

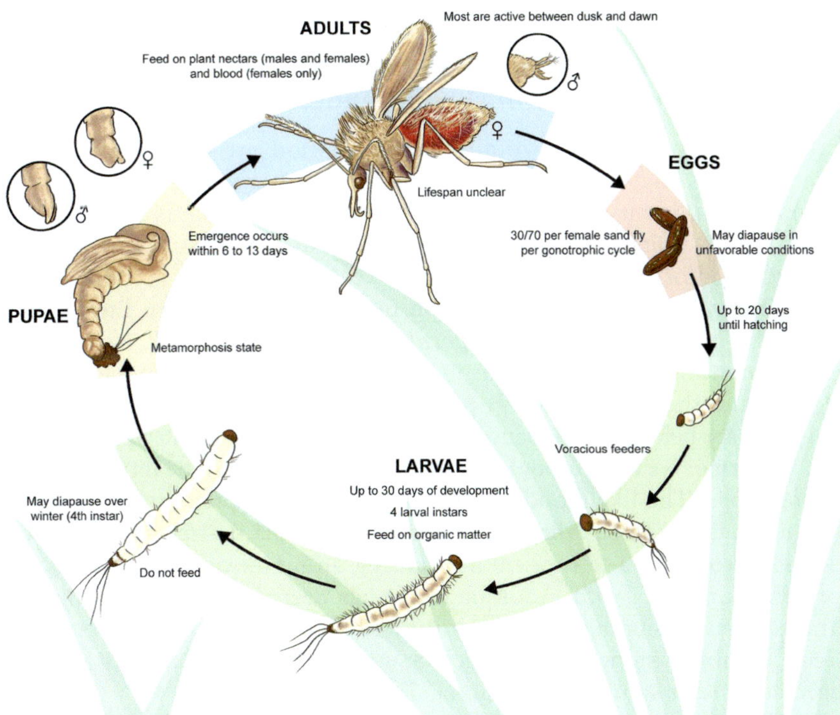

Figure 2.6 Diagrammatic illustration of the life cycle of sand flies (10).

larval instars, and depending on the type of sandfly, temperature, and food availability, larval development typically takes 20–30 days (40). However, in sandfly species that diapause to cope with winter, this time may be extended to many months (only fully developed larvae diapause—instar four) (37, 38). Adults emerge during the dark hours, frequently soon before dawn, after a pupation period of 6–13 days (38). Typically, men appear before females. Very little information is known about the adult life expectancy of sand flies in the wild, especially when it comes to females. In the wild, females may survive longer because they go through many gonotrophic cycles—some up to three—whereas males may only live for a week or so (41). Although certain species have been reported to have mated several times prior to laying fertile eggs, oviposition typically takes place five to eight days following blood-feeding (41). Although certain sandfly species are known to be endophagic and endophilic, known as domestic species, the majority of sandfly species are exophagic. A more thorough explanation of the biology, behavior, and morphology of sand flies—which are used to identify different species—can be found elsewhere (37, 38, 41, 42–45). However, there is still a lot to learn about these topics.

2.6 FACTORS AFFECTING TRANSMISSION CYCLES

2.6.1 Risk assessment of sandfly-transmitted parasite spreading in previously non-endemic European areas

If the sandfly population size and activity period are inadequate, vector-borne diseases may not establish themselves in a particular environment (the minimum density needed for *Leishmania* transmission is 50 vectors/m^2 sticky trap). Despite the existence of Phlebotomus *perniciosus*, *Phlebotomus perfilewi*, and/or *Phlebotomus neglectus* vectors, this may account for the lack or rarity of *L. infantum* infection in central and Eastern Europe (45). The farther north spreading of *P. perniciosus*, *P. ariasi*, and *P. neglectus*, as well as canine leishmaniasis in the north of Italy, Spain, and southern France (46–49), and the discovery of *P. perniciosus* at higher altitudes in southern Spain (50) are believed to be caused by conditions that facilitate vector spread, such as warmer weather brought on by climate change and the introduction of infected dogs. The likelihood that leishmaniasis may spread to sandfly marginal areas is increased when sick dogs enter leishmania-free areas, and dog adoption programs and tourism are causing an influx of diseased dogs into northern Europe (51–53). Additionally, it has been reported that congenital parasite transmission has caused canine leishmaniasis in three generations of Czech Republic canines (54). In the US, congenital transmission is acknowledged as a major source of infection (55).

Despite the widespread distribution of its vector, *Phlebotomus papatasi*, the low prevalence of CL in southern Europe is due to *L. major*, which lacks a suitable rodent reservoir host (56). However, a hybrid of *L. infantum* and *L. major* was found in Portuguese AIDS patients, and it was shown to grow and disseminate in *P. papatasi* (57, 58). The naturally occurring vector of *Leishmania (Sauroleishmania) tarentolae* in reptiles, the sandfly *Sergentomyia minuta* (59), was later discovered to have *L. major* DNA. These results highlight the significance of continuous surveillance by highlighting the danger of *L. major* importation from North Africa as well as the possibility of hybrids appearing and spreading throughout the native *P. papatasi habitat* (60).

Since many cases are found among migrants from endemic areas into Europe, it is possible that the *L. tropica* cycle will occur in southern European countries where the vector Phlebotomus *sergenti* is common (61). Up until the 1950s, *L. tropica* was mostly endemic on the island of Greece; since the early 2000s, cases have been reported there as well as in the Ionian Islands (62). The vector in Crete is thought to be Phlebotomus *similis*, which is common throughout Greece and Asia Minor (63, 64). Additionally, rodent species from the Middle East and North Africa that are not seen in Europe are part of the zoonotic cycle of *L. tropica*. Only in Cyprus, Europe, has anthropogenic VL caused by *L. donovani* been documented. Due to its endemicity in Turkey, eastern Africa, and the Indian continent, this species has the most influence on public health worldwide (65). *L. donovani* and *L. infantum* are closely related, and *Phlebotomus tobbi* and *P. perniciosus* have successfully created a hybrid of *L. infantum* and *L. donovani* (66). One of the top priorities is to stop *L. donovani* from spreading into Europe (62).

2.6.2 Factors influencing the transmission of sandfly-borne pathogens in endemic areas

The transmission frequency in endemic places fluctuates regionally and worldwide and is highly seasonal, depending on environmental conditions that impact vector and infected host abundance. The incubation period of the pathogen—at least two or three months for VL and only a few days for phleboviruses—determines the occurrence of clinical cases. The Naples and Sicilian viruses cause sandfly fever, which reflects the life cycle of *P. papatasi*, starting in April and peaking in September (45). During the warm months, the Toscana virus produces neurological illnesses. However, in Europe, the fall and winter months are often when the number of clinical cases of VL in dogs and humans peaks. The Fuenlabrada outbreak's sandfly infection rates with *L. infantum* peaked around the conclusion of the transmission period, suggesting that people are more susceptible to infection in the late summer (67).

The number of infected vertebrate hosts determines the likelihood of *Leishmania* infection, and many domestic and sylvatic host species—aside from dogs—carry chronic subclinical infections that may be passed on to the vector (68, 69). Although viremia is temporary and no vertebrate reservoir host has been found as of yet, phleboviruses may infect a variety of host species besides humans (61). Despite the fact that viruses are congenitally transmitted within the vector, infection rates in the following generations progressively decline, indicating that the vector cannot support the viruses in humans indefinitely on its own (45). Because of this, this is an epidemiological conundrum.

In the late 1980s and early 1990s, immunosuppression in HIV-positive people was the primary cause of the *Leishmania*–HIV virus coinfection pandemic in Mediterranean nations. In contrast, needle sharing among HIV positive patients for intravenous purposes was found to be a unique *L. infantum* transmission pathway (70), with no indication of more contact with infectious vectors. The first significant epidemic of human leishmaniasis in Spain since the middle of the 20th century was the *Leishmania*–HIV virus coinfection pandemic. Before this, leishmaniasis was a common illness in Spain, especially in children who had not yet established immunity, and it was associated with the Spanish Civil War, poor sanitation, and famine (71).

ACKNOWLEDGMENT

The authors are thankful to Parul University, Vadodara, Gujarat, India, for necessary support for this work.

REFERENCES

1. Mukhopadhyay D, Dalton JE, Kaye PM, Chatterjee M. Post kala-azar dermal leishmaniasis: an unresolved mystery. Trends in parasitology. 2014 Feb 1;30(2): 65–74.
2. Claborn DM. The biology and control of leishmaniasis vectors. Journal of global infectious diseases. 2010 May 1;2(2):127–34.
3. Mann S, Frasca K, Scherrer S, Henao-Martínez AF, Newman S, Ramanan P, Suarez JA. A review of leishmaniasis: current knowledge and future directions. Current tropical medicine reports. 2021 Jun;8:121–32.
4. Cecílio P, Cordeiro-da-Silva A, Oliveira F. Sand flies: basic information on the vectors of leishmaniasis and their interactions with *Leishmania* parasites. Communications biology. 2022 Apr 4;5(1):305.
5. WHO. Leishmaniasis (Internet). World Health Organization. World Health Organization: WHO, 2023. Available from: https://www.who.int/news-room/fact-sheets/detail/leishmaniasis
6. Burza S, Croft SL, Boelaert M. Leishmaniasis – Authors' reply. The Lancet. 2019 Mar 2;393(10174):872–3.
7. Montaner-Angoiti E, Llobat L. Is leishmaniasis the new emerging zoonosis in the world? Veterinary research communications. 2023 Dec;47(4):1777–99.
8. Ponte-Sucre A, Diaz E, Padrón-Nieves M. Drug Resistance in *Leishmania* Parasites. Springer, 2013.
9. Mann S, Frasca K, Scherrer S, Henao-Martínez AF, Newman S, Ramanan P, Suarez JA. A review of leishmaniasis: current knowledge and future directions. Current tropical medicine reports. 2021;8(2):121–32.
10. Killick-Kendrick R. The biology and control of phlebotomine sand flies. Clinics in dermatology. 1999;17:279–89.

11. Lewis DJ, Young DG, Fairchild GB, Minter DM. Proposals for a stable classification of the phlebotomine sandflies (Diptera: Psychodidae). Systematic entomology. 1977;2(4):319–32.
12. Maroli M, Feliciangeli MD, Bichaud L, Charrel RN, Gradoni L. Phlebotomine sandflies and the spreading of leishmaniases and other diseases of public health concern. Medical and veterinary entomology. 2013 Jun;27(2):123–47.
13. Leng YJ. A preliminary survey of phlebotomine sandflies in limestone caves of Sichuan and Guizhou Provinces, south-west China, and description and discussion of a primitive new genus *Chinius*. Annals of tropical medicine & parasitology. 1987;81(3):311–7.
14. World Health Organization. (2010) Control of the Leishmaniasis. Report of a Meeting of the WHO Expert Committee on the Control of Leishmaniases, 22–26 March 2010, WHO Technical Report Series. WHO.
15. Lewis DJ. A taxonomic review of the genus *Phlebotomus* (Diptera: Psychodidae). Bulletin of the British Museum (Natural History) (Entomology). 1982;45:121–209.
16. Lewis DJ, Young DG, Fairchild GB, Minter DM. Proposals for a stable classification of the phlebotomine sandflies (Diptera: Psychodidae). Systematic entomology. 1977;2(4):319–32.
17. Young DG, Duncan MA. (1994) Guide to the identification and geographic distribution of *Lutzomyia* sandflies in Mexico, the West Indies, Central and South America (Diptera: Psychodidae). Memoirs of the American Entomological Institute, 54, 1–881.
18. Galati EAB. (2003) Classificação de phlebotominae. Flebotomíneos do Brasil (ed. Rangel EF & Lainson R), pp. 23–51.
19. ECDC. Phlebotomine sand flies—Factsheet for experts. https://www.ecdc.europa.eu/en/disease-vectors/facts/phlebotomine-sand-flies (2014).
20. Benkova I, Volf P. Effect of temperature on metabolism of *Phlebotomus papatasi* (Diptera: Psychodidae). Journal of medical entomology. 2007;44(1):150–4.
21. Erguler K, Pontiki I, Zittis G, Proestos Y, Christodoulou V, Tsirigotakis N, Antoniou M, Kasap OE, Alten B, Lelieveld J. A climate-driven and field data-assimilated population dynamics model of sand flies. Scientific reports. 2019; 9(1):2469.
22. Kasap OE, Alten B. Laboratory estimation of degree-day developmental requirements of *Phlebotomus papatasi* (Diptera: Psychodidae). Journal of vector ecology: Journal of the Society for Vector Ecology. 2005;30(2):328–33.
23. Akhoundi, M. et al. A historical overview of the classification, evolution, and dispersion of leishmania parasites and sandflies. PLoS neglected tropical diseases. 2016;10:e0004349.
24. Volf P, Volfova V. Establishment and maintenance of sandfly colonies. Journal of Vector Ecology. 2011;36:S1–S9.
25. Feliciangeli MD. Natural breeding places of phlebotomine sandflies. Medical and Veterinary Entomology, 2004;18:71–80.
26. Molyneux DH, Ashford RW. The Biology of Trypanosoma and *Leishmania*, Parasites of Man and Domestic Animals. Taylor & Francis, 1983.
27. Killick-Kendrick R, Wilkes TJ, Bailly M, Bailly I, Righton LA. Preliminary field observations on the flight speed of a phlebotomine sandfly. Transactions of the Royal Society of Tropical Medicine and Hygiene. 1986;80(1):138–42.
28. Rioux JA, Killick-Kendrick R, Leaney AJ, Turner DP, Bailly M, Young CJ. Ecologie des Leishmanioses dans le sud de la France. 12. Dispersion horizontale de *Phlebotomus ariasi* Tonnoir, 1921. Expériences préliminaires. Annales de Parasitologie humaine et comparée. 1979;54(6):673–82.
29. Killick-Kendrick R, Rioux JA, Ratify M, Guy MW, Wilkes TJ, Guy FM, Davidson I, Knechtli R, Ward RD, Guilvard E, Perieres J. Ecology of leishmaniasis in the south of France. 20. Dispersal of *Phlebotomus ariasi* Tonnoir, 1921

as a factor in the spread of visceral leishmaniasis in the Cévennes. Annales de parasitologie humaine et comparee. 1984;59(6):555–72.
30. Oliveira SG, Bottecchia M, Bauzer LG, Souza NA, Ward RD, Kyriacou CP, Peixoto AA. Courtship song genes and speciation in sand flies. Memórias do Instituto Oswaldo Cruz. 2001;96:403–5.
31. Lazar LTY, Abbas KS. Morphology, life cycle, pathogenesis and virulence factors of genus *Leshmania*: a review. Plant archives. 2020;20(2):4057–60.
32. Bates PA. Transmission of Leishmania metacyclic promastigotes by phlebotomine sand flies. International journal for parasitology. 2007;37(10):1097–106.
33. Cecílio P, Cordeiro-da-Silva A, Oliveira F. Sand flies: basic information on the vectors of leishmaniasis and their interactions with *Leishmania* parasites. Communications biology. 2022;5(1):305.
34. Serafim, TD, et al. Sequential blood meals promote *Leishmania* replication and reverse metacyclogenesis augmenting vector infectivity. Nature microbiology. 2018;3:548–55.
35. Conceição-Silva F, Morgado FN. *Leishmania* spp-host interaction: there is always an onset, but is there an end? Frontiers in cellular and infection microbiology. 2019;9:330.
36. Moradin N, Descoteaux A. *Leishmania* promastigotes: building a safe niche within macrophages. Frontiers in cellular and infection microbiology. 2012;2:121.
37. Killick-Kendrick R. *Leishmania*. World Class Parasites (ed Farrell JP). Springer, 2001.
38. Service M. Medical Entomology for Students. Cambridge University Press, 2012.
39. Lawyer P, Killick-Kendrick M, Rowland T, Rowton E, Volf P. Laboratory colonization and mass rearing of phlebotomine sand flies (Diptera, Psychodidae). Parasite. 2017;24:42.
40. Volf P, Volfova V. Establishment and maintenance of sand fly colonies. Journal of Vector Ecology. 2011;36:S1–9.
41. Rutledge LC, Gupta RK. Medical and Veterinary Entomology (eds Mullen G & Durden L). Academic Press, 2002.
42. Lane RP. Medical Insects and Arachnids. Springer, 1993.
43. Dvorak V, Shaw J, Volf P. The Leishmaniases: Old Neglecte Tropical Diseases (eds Bruschi F & Gradoni F), 31–77. Springer International Publishing, 2018.
44. Maroli M, Feliciangeli MD, Bichaud L, Charrel RN, Gradoni L. Phlebotomine sandflies and the spreading of leishmaniases and other diseases of public health concern. Medical and veterinary entomology. 2013;27(2):123–47.
45. Mihalca AD, Cazan CD, Sulesco T, Dumitrache MO. A historical review on vector distribution and epidemiology of human and animal leishmanioses in Eastern Europe. Research in veterinary science. 2019 Apr;123:185–91.
46. Maroli M, Rossi L, Baldelli R, Capelli G, Ferroglio E, Genchi C, Gramiccia M, Mortarino M, Pietrobelli M, Gradoni L. The northward spread of leishmaniasis in Italy: evidence from retrospective and ongoing studies on the canine reservoir and phlebotomine vectors. Tropical medicine & international health. 2008 Feb;13(2):256–64.
47. Dereure J, Vanwambeke SO, Malé P, Martinez S, Pratlong F, Balard Y, Dedet JP. The potential effects of global warming on changes in canine leishmaniasis in a focus outside the classical area of the disease in southern France. Vector-borne and zoonotic diseases. 2009 Dec;9(6):687–94.
48. Baldelli R, Battelli G, Maroli M, Mollicone E, Gudi A, Stegagno G, Tasini G. A new stable focus of canine leishmaniasis in northern Italy. Parassitologia. 2001 Dec;43(4):151–3.
49. Gálvez R, Descalzo MA, Guerrero I, Miró G, Molina R. Mapping the current distribution and predicted spread of the leishmaniosis sand fly vector in the Madrid region (Spain) based on environmental variables and expected climate change. Vector-borne and zoonotic diseases. 2011 Jul;11(7):799–806.

50. Martín-Sánchez J, Morales-Yuste M, Acedo-Sánchez C, Barón S, Díaz V, Morillas-Márquez F. Canine leishmaniasis in southeastern Spain. Emerging infectious diseases. 2009 May;15(5):795.
51. Shaw SE, Langton DA, Hillman TJ. Canine leishmaniosis in the United Kingdom: a zoonotic disease waiting for a vector? Veterinary parasitology. 2009 Aug 26; 163(4):281–5.
52. Menn B, Lorentz S, Naucke TJ. Imported and travelling dogs as carriers of canine vector-borne pathogens in Germany. Parasites & vectors. 2010 Apr 8;3:34.
53. Otranto D, Dantas-Torres F, Mihalca AD, Traub RJ, Lappin M, Baneth G. Zoonotic parasites of sheltered and stray dogs in the era of the global economic and political crisis. Trends in parasitology. 2017 Oct;33(10):813–25.
54. Svobodova V, Svoboda M, Friedlaenderova L, Drahotsky P, Bohacova E, Baneth G. Canine leishmaniosis in three consecutive generations of dogs in Czech Republic. Veterinary Parasitology. 2017 Apr 15;237:122–4.
55. Toepp AJ, Schaut RG, Scott BD, Mathur D, Berens AJ, Petersen CA. *Leishmania* incidence and prevalence in U.S. hunting hounds maintained via vertical transmission. Veterinary parasitology. 2017;10:75–81.
56. Ready PD. Leishmaniasis emergence in Europe. Eurosurveillance. 2010 Mar 11; 15(10):29–39.
57. Ravel C, Cortes S, Pratlong F, Morio F, Dedet J-P, Campino L. First report of genetic hybrids between two very divergent *Leishmania* species: *Leishmania infantum* and *Leishmania major*. International journal for parasitology. 2006 Nov;36(13):1383–8.
58. Volf P, Benkova I, Myskova J, Sadlova J, Campino L, Ravel C. Increased transmission potential of *Leishmania major/Leishmania infantum* hybrids. International journal for parasitology. 2007 May;37(6):589–93.
59. Campino L, Cortes S, Dionisio L, Neto L, Afonso MO, Maia C. The first detection of *Leishmania major* in naturally infected *Sergentomyia minuta* in Portugal. Memorias do Instituto Oswaldo Cruz. 2013 Jun;108(4):516–8.
60. Bongiorno G, Di Muccio T, Bianchi R, Gramiccia M, Gradoni L. Laboratory transmission of an Asian strain of *Leishmania tropica* by the bite of the southern European sand fly *Phlebotomus perniciosus*. International journal for parasitology. 2019 May;49(6):417–21.
61. Alkan C, Bichaud L, de Lamballerie X, Alten B, Gould EA, Charrel RN. Sandfly-borne phleboviruses of Eurasia and Africa: epidemiology, genetic diversity, geographic range, control measures. Antiviral research. 2013 Oct;100(1):54–74.
62. Antoniou M, Gramiccia M, Molina R, Dvorak V, Volf P. The role of indigenous phlebotomine sand flies and mammals in the spreading of leishmaniasis agents in the Mediterranean region. Eurosurveillance. 2013 Jul 25;18(30):54–61.
63. Depaquit J, Ferte H, Leger N, Lefranc F, Alves-Pires C, Hanafi H, et al. ITS 2 sequences heterogeneity in *Phlebotomus sergenti* and *Phlebotomus similis* (Diptera, Psychodidae): possible consequences in their ability to transmit *Leishmania tropica*. International journal for parasitology. 2002 Aug;32(9):1123–31.
64. Christodoulou V, Antoniou M, Ntais P, Messaritakis I, Ivovic V, Dedet J-P, et al. Re-emergence of visceral and cutaneous leishmaniasis in the Greek island of Crete. Vector-borne and zoonotic diseases. 2012 Mar;12(3):214–22.
65. Ozbilgin A, Harman M, Karakus M, Bart A, Toz S, Kurt O, et al. Leishmaniasis in Turkey: visceral and cutaneous leishmaniasis caused by *Leishmania donovani* in Turkey. Acta Tropica. 2017 Sep;173:90–6.
66. Seblova V, Myskova J, Hlavacova J, Votypka J, Antoniou M, Volf P. Natural hybrid of *Leishmania infantum/L. donovani*: development in *Phlebotomus tobbi*, *P. perniciosus* and *Lutzomyia longipalpis* and comparison with non-hybrid strains differing in tissue tropism. Parasites & vectors. 2015 Nov 25;8:605.
67. Gonzalez E, Jimenez M, Hernandez S, Martin-Martin I, Molina R. Phlebotomine sandfly survey in the focus of leishmaniasis in Madrid, Spain (2012–2014): seasonal

dynamics, *Leishmania infantum* infection rates and blood meal preferences. Parasites & vectors. 2017 Aug 1;10:368.
68. Tomassone L, Berriatua E, De Sousa R, Duscher GG, Mihalca AD, Silaghi C, et al. Neglected vector-borne zoonoses in Europe: into the wild. Veterinary parasitology. 2018 Feb 15;251:17–26.
69. Risueno J, Ortuno M, Perez-Cutillas P, Goyena E, Maia C, Cortes S, et al. Epidemiological and genetic studies suggest a common *Leishmania infantum* transmission cycle in wildlife, dogs and humans associated to vector abundance in south east Spain. Veterinary parasitology. 2018 Aug 15;259:61–7.
70. Alvar J, Canavate C, Gutierrez Solar B, Jimenez M, Laguna F, Lopez Velez R, et al. *Leishmania* and human immunodeficiency virus coinfection: the first 10 years. Clinical microbiology reviews. 1997 Apr;10(2):298–319.
71. Botet Fregola J, Portus Vinyeta M. Leishmaniasis in peninsular Spain. A historical-bibliographic review (1912–1985). Revista de sanidad e higiene publica. 1993;67(4):255–66.

Chapter 3

Pathogenesis of leishmaniasis

Mechanisms of infection and disease progression

Shobha Kumari[#] *and Vikash Kumar*[#]

Department of Biochemistry, ICMR-Rajendra Memorial Research Institute of Medical Sciences, Patna, India

Department of Microbiology, University of Calcutta, Kolkata, India

*Ashish Kumar**

Department of Biochemistry, ICMR-Rajendra Memorial Research Institute of Medical Sciences, Patna, India

3.1 INTRODUCTION

Leishmaniasis is a complex vector-borne disease caused by protozoan parasites of the genus *Leishmania*. It is transmitted by the bite of infected female Phlebotomine sandflies to humans (Paramanik, 2023). Depending on the infecting *Leishmania* species and the immunological condition of the host, the disease, which affects millions of people worldwide, can take many different clinical manifestations, from self-healing cutaneous ulcers to fatal visceral leishmaniasis (VL). Despite decades of research, leishmaniasis continues to be a significant public health concern in tropical and subtropical regions, with over 90 countries reporting endemic transmission (Rocha et al., 2022). As per the World Health Organization (WHO) report, approximately 700,000–1 million new cases occur annually (Ahmadi et al., 2024). The persistent burden of leishmaniasis is exacerbated by limited treatment options, emerging drug resistance, and lack of an effective human vaccine.

We need to understand how the disease affects the body, meaning its pathogenesis, which is essential to developing effective targeted strategies for diagnosis, treatment, and its prevention (Ponte-Sucre et al., 2017; Srivastava et al., 2016). The disease pathogenesis is primarily driven by the complex interactions between the *Leishmania* and the host immune system, beginning at the site of the sandfly bite. Beneath the skin sandfly injects the motile promastigote form where parasites are phagocytosed by host macrophages. Inside the

* Corresponding author.

[#] These authors contributed equally.

DOI: 10.1201/9781003596455-3

macrophages it transforms into the amastigote form, capable of intracellular survival and replication. This ability to avoid the host's immune responses is a key feature of *Leishmania* infection and is essential for the parasite's long-term presence and preference for certain tissues (tissue tropism) (Silva-Moreira et al., 2025).

One key aspect of *Leishmania* infection is its ability to affect the host's immune signaling pathways, especially those that activate macrophages and produce cytokines. The outcome of the infection often depends on the balance between Th1-type and Th2-type immune responses. A protective Th1 response includes the production of interferon-gamma (IFN-γ), tumor necrosis factor-alpha (TNF-α), and nitric oxide (NO), which activate macrophages to eliminate the parasites (Kedzierski and Evans, 2014; Paton et al., 2025). In contrast, a dominant Th2 response is linked to interleukin-4 (IL-4), IL-10, and transforming growth factor-beta (TGF-β), promoting parasite survival and disease progression. Furthermore, regulatory T cells (Tregs) and myeloid-derived suppressor cells (MDSCs) play a growing role in immune suppression during chronic leishmaniasis (Dayakar et al., 2019; Paton et al., 2025). The pathogenic mechanisms and harmful effects of *Leishmania* are also influenced by differences among species in their virulence factors, tissue tropism, and immune evasion strategies. For example, *Leishmania donovani* and *Leishmania infantum* mainly target the spleen, liver, and bone marrow, which leads to VL.

Whereas *Leishmania major* and *Leishmania tropica* typically cause localized cutaneous lesions. *Leishmania braziliensis*, responsible for mucocutaneous leishmaniasis, induces chronic inflammation and destructive mucosal pathology (Ghosh et al., 2023). These clinical differences reflect variations in parasite surface molecules, such as lipophosphoglycan (LPG) and the gp63 metalloprotease, which interact with host receptors and influence immune cell function (Elmahallawy and Alkhaldi, 2021).

The host's genetic makeup, nutritional state, and co-infection with other pathogens like HIV are also crucial to disease progression. The combined presence of leishmaniasis and HIV in many regions creates a dual health challenge. Co-infected individuals often show unusual symptoms, poor treatment responses, and high relapse rates (Lindoso et al., 2018). New evidence indicates that post-kala-azar dermal leishmaniasis (PKDL), which occurs after treatment for VL, may serve as a potential reservoir for disease spread, complicating elimination efforts (Kumari et al., 2025).

This chapter aims to give an overview of leishmaniasis pathogenesis, focusing on how the infection occurs and how the disease progresses. We discuss the parasite's ways of avoiding the immune system, the role of host immune responses in determining disease outcomes, and the molecular and cellular processes related to parasite persistence. By elucidating these complex biological processes, we hope to contribute to the foundation for developing more effective therapeutic interventions and vaccines against this devastating disease.

3.2 BRIEF DISCUSSION ON TRANSMISSION AND EARLY EVENTS IN INFECTION

Leishmania has a digenetic (two-host) life cycle, alternating between the female Phlebotomine sandfly and mammalian hosts, mainly humans. The cycle starts when an infected sandfly bites a host, injecting metacyclic promastigotes and saliva that modulate immune responses and improve parasite survival (Paramanik, 2023).

At the infection site, neutrophils are the first to respond. Instead of destroying the parasites, they serve as temporary reservoirs. After undergoing apoptosis, these neutrophils are consumed by macrophages—a "Trojan horse" mechanism that facilitates parasite entry (Carreira and da Silva, 2021). Inside macrophages, promastigotes transform into amastigotes, adapting to acidic and warm conditions and initiating chronic infection.

Promastigotes avoid destruction by changing the phagolysosomal environment, allowing for their growth inside cells. When macrophages break apart, they release amastigotes, which then infect nearby phagocytes and maintain the infection.

The cycle continues when a non-infected sandfly feeds on the amastigotes from an infected host. These transform into procyclic promastigotes in the sandfly's midgut, multiply, and move to the foregut to become metacyclic promastigotes, the stage that can infect new hosts.

3.3 CELLULAR ENTRY AND ESTABLISHMENT OF INFECTION

Building on the early host–parasite interactions described previously, this section explores the detailed cellular mechanisms exploited by *Leishmania* to establish infection and persist within the host.

The main targets of *Leishmania* are specialized phagocytes like macrophages, dendritic cells (DCs), and neutrophils. These cells serve as the first line of immune defense and also act as reservoirs that allow the parasite to survive and multiply. *Leishmania*'s ability to manipulate host immune cells is a key feature of its pathogenicity.

Another important factor affecting the early stages of parasite entry is the complement system, which is a major part of the innate immune response (Costa-da-Silva et al., 2022; Yasmin et al., 2022). Once *Leishmania* promastigotes enter the host, they immediately encounter serum complement proteins. This activates the complement system, leading to C3b being deposited on the parasite surface. Normally, C3b promotes phagocytosis by immune cells and initiates the formation of the membrane attack complex (MAC), which can destroy invading pathogens. However, *Leishmania* avoids this destruction through surface molecules like Gp63, which splits C3b into iC3b (Costa-da-Silva et al., 2022). Unlike C3b, iC3b does not support MAC formation but retains opsonic activity and binds to complement receptor 3 (CR3) on phagocytic cells. This interaction allows the parasite to enter macrophages through a non-inflammatory route, termed "silent phagocytosis", thereby avoiding the activation of strong innate immune responses. This exploitation of the complement system not only aids parasite survival but also facilitates intracellular establishment.

3.3.1 Role of host cells in parasite entry

Neutrophils, macrophages, and DCs are three critical cell types that play significant roles during the entry of the *Leishmania* parasite (Figure 3.1).

3.3.1.1 Neutrophils as Trojan horses in early infection

Neutrophils are among the first immune cells to be drawn to the site of *Leishmania* infection after a sandfly bite. Their primary role is to engulf the metacyclic promastigotes, kicking off an immune response. However, instead of destroying the parasites, these neutrophils act like "Trojan horses" in the early infection stages (Carreira and da Silva, 2021; Passelli et al., 2021). The promastigotes can hide inside the neutrophils without causing a significant inflammatory response. A key part of *Leishmania*'s strategy to evade the immune system is its ability to influence neutrophil apoptosis. After engulfing the parasites, the infected neutrophils undergo a controlled apoptosis rather than a typical inflammatory cell death. These apoptotic cells are then cleaned up by macrophages through a process known as "silent phagocytosis" (Martínez-López et al., 2018; Regli et al., 2017). This type of clearance generally leads to an anti-inflammatory response, allowing *Leishmania* to enter macrophages without provoking a strong immune reaction. This tactic not

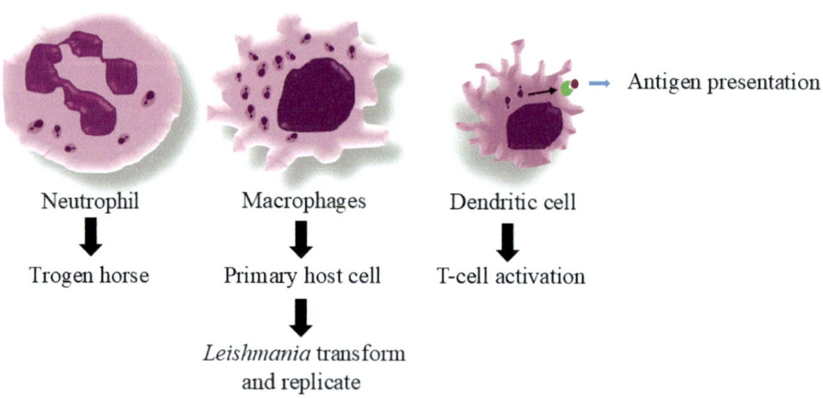

Figure 3.1 Role of host cells in parasite entry.

only helps the parasite survive but also creates a safe environment for its development within the host.

3.3.1.2 Macrophages: the primary host cells for *Leishmania*

Macrophages play a crucial role as the primary host cells for *Leishmania*, aiding in the parasite's transformation and replication process. When promastigotes enter macrophages, they typically get trapped within phagosomes that would normally fuse with lysosomes for degradation (Martínez-López et al., 2018). However, *Leishmania* has developed strategies to alter macrophage signaling, effectively blocking phagosomal maturation. Inside these macrophages, promastigotes transform into amastigotes, which are the non-flagellated and replicative forms of the parasite. This transformation is key to establishing the infection. The intracellular amastigotes reside in modified phagolysosomal compartments, allowing them to replicate and evade destruction by the immune system (Murray and Wynn, 2011). Eventually, when the infected macrophages burst, they release amastigotes into the surrounding environment, which allows them to target new macrophages and perpetuate the infection cycle.

3.3.1.3 *Dendritic cells (DCs)*

DCs play a vital role in the body's response to *Leishmania* infection. They can phagocytize *Leishmania* promastigotes, but their function goes beyond just phagocytosis; they also help modulate the immune system. DCs are vital for presenting antigens and activating T-cell responses, which are key in clearing the parasite from the body (Kautz-Neu et al., 2012). However, *Leishmania* has developed mechanisms to interfere with the maturation and function of DCs, which weakens their ability to stimulate T cells. This impaired adaptive immune response not only allows the parasite to persist but also contributes to the progression of the disease (Liu and Uzonna, 2012).

3.3.2 Molecular mechanisms of phagosome maturation arrest

One of the main strategies used by *Leishmania* to successfully establish infection is through disrupting the normal maturation process of phagosomes inside host macrophages. Under typical conditions, once a pathogen is engulfed, the phagosome undergoes several changes. It becomes more acidic, fuses with lysosomes, and activates enzymes that help break down the invader (Winberg et al., 2007). However, *Leishmania* promastigotes can block this process. They release specific virulence factors that interfere with the host's signaling pathways, allowing the parasite to survive inside the cells (Figure 3.2).

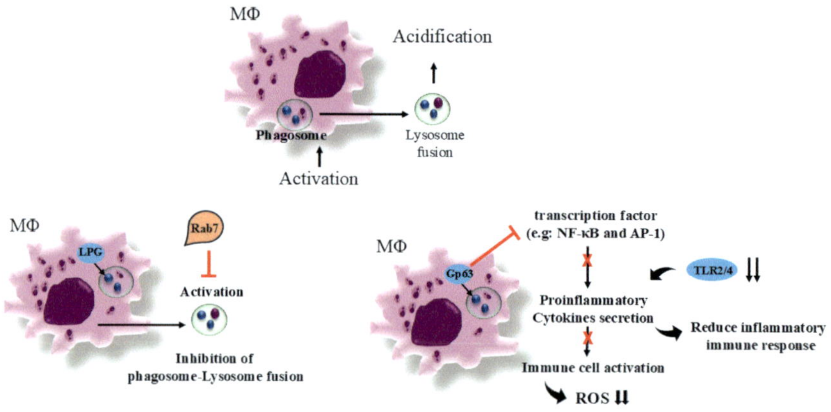

Figure 3.2 Mechanisms of phagosome maturation arrest: alteration in normal phagosome formation and involvement of Gp63 that cleaves host transcription factor ultimately leads to reduction in pro-inflammatory immune response.

3.3.2.1 Lipophosphoglycan (LPG) and inhibition of phagosome-lysosome fusion

One of the major virulence factors secreted by *Leishmania* is LPG, a surface glycoconjugate that plays a crucial role in immune evasion. LPG interferes with the normal maturation of phagosomes by modulating the function of Rab GTPases, which are key regulators of vesicular trafficking (Winberg et al., 2007). Specifically, LPG disrupts the transition of early phagosomes into late phagosomes by impairing the recruitment of Rab7, a key protein required for phagosome-lysosome fusion. As a result, the parasite-containing phagosomes remain in an immature state, preventing the activation of lysosomal enzymes that would otherwise degrade the parasite (Moradin and Descoteaux, 2012; Winberg et al., 2007) (Figure 3.2).

3.3.2.2 Glycoprotein gp63: a multifunctional immune evasion molecule

Another critical virulence factor produced by *Leishmania* is glycoprotein gp63, a zinc-dependent metalloprotease that modulates host immune signaling (Gupta et al., 2022). Gp63 exerts its immunomodulatory effects through multiple mechanisms (Gupta et al., 2022; Olivier et al., 2012):

- **Cleavage of Host Transcription Factors:** Gp63 cleaves key transcription factors such as NF-κB and AP-1, which are involved in the activation of pro-inflammatory responses. This prevents the upregulation of cytokines like TNF-α and IL-12, which are essential for macrophage activation and parasite clearance.

- **Degradation of Signaling Proteins:** Gp63 also targets host protein kinases, disrupting key signaling cascades that regulate immune cell activation. This further weakens the macrophage's ability to mount an effective response against the parasite.
- **Inhibition of Reactive Oxygen Species (ROS) Production:** Macrophages typically produce ROS and NO as part of their antimicrobial defense. Gp63 inhibits the production of these toxic molecules, thereby enhancing parasite survival (Winberg et al., 2007).
- **Subversion of Toll-like Receptor (TLR) Signaling:** TLRs play a critical role in pathogen recognition and innate immune activation. *Leishmania* selectively modulates TLR signaling to suppress immune responses while maintaining an environment conducive to its survival (Bamigbola and Ali, 2022). For instance, *Leishmania* can downregulate TLR2 and TLR4 signaling pathways, reducing the production of pro-inflammatory cytokines and preventing the effective activation of macrophages.

3.4 DISEASE-SPECIFIC PATHOGENESIS

Why different species of the same genus (*Leishmania*) cause different organ-based manifestations is a fundamental question in the pathogenesis of leishmaniasis. The divergence in disease outcome is not random but is shaped by a complex interplay of parasite factors, host immune responses, and vector influences (Ait Maatallah et al., 2022).

The possible reasons why different *Leishmania* species within the same genus cause different types of leishmaniasis (cutaneous, mucocutaneous, visceral, etc.) may be as follows:

3.4.1 Tissue tropism of *Leishmania* species

Different *Leishmania* species have evolved specific tissue preferences:

- **Dermotropic species** (e.g., *L. major*, *L. tropica*) preferentially infect and persist in **dermal macrophages.**
- **Viscerotropic species** (e.g., *L. donovani*, *L. infantum*) are adept at infecting **reticuloendothelial organs** such as the liver, spleen, and bone marrow.
- **Mucotropic species** (e.g., *L. braziliensis*) disseminate from skin to **mucosal tissues,** especially of the nasopharynx.

This tropism is governed by the following:

- Expression of surface molecules (e.g., LPG variants)
- Ability to survive in different tissue microenvironments
- Evasion or manipulation of local immune responses

Table 3.1 Species-specific virulent factors

Virulence factor	*Role*	*Species expression*
Lipophosphoglycan (LPG)	Affects complement resistance, macrophage entry, and phagolysosomal maturation	Highly variable among species
A2 protein family	Promotes survival in visceral organs	Prominent in *Leishmania donovani, Leishmania infantum*
GP63 (leishmanolysin)	Degrades host signaling proteins, enhances parasite entry	Common, but sequence/function differs
Cysteine proteases	Help escape lysosomal degradation	Vary in expression and activity

3.4.2 Species-specific virulence factors

Different *Leishmania* species express distinct virulence factors, which determine their ability to survive in different host tissues (Al-Khalaifah, 2022) listed in Table 3.1.

For example, A2 proteins are critical for visceralization, and their absence or downregulation in dermotropic species limits them to the skin (McCall et al., 2013).

3.4.3 Host immune response bias

The quality and direction of the host immune response after infection influences the disease form (Dayakar et al., 2019; McCall et al., 2013):

- Th1-dominant responses (IFN-γ, IL-12) → Control of parasite, usually localized CL
- Th2-skewed responses (IL-4, IL-10) → Immune suppression, favoring VL
- In MCL, hyperactive cytotoxic CD8+ T cells and high TNF-α cause tissue destruction despite low parasite loads

Each *Leishmania* species induces a slightly different immune profile:

- *L. donovani* and *L. infantum* actively promote IL-10 production and impair DC function to spread systemically.
- *L. braziliensis* elicits a strong Th1 and CD8+ response that causes mucosal immunopathology.
- *Leishmania mexicana* species tend to suppress Th1 activation, leading to DCL.

3.4.4 Genetic and epigenetic adaptations

Some *Leishmania* species possess unique gene arrangements, copy number variations, or epigenetic traits that give them survival advantages in specific environments:

- *L. donovani* has expanded gene families like A2, nucleoside transporters, and heat shock proteins that help resist high temperatures and oxidative stress in viscera (McCall et al., 2013; Requena et al., 2015).
- Gene amplification or deletion in surface antigens may also alter macrophage recognition and immune evasion strategies.

3.4.5 Vector–parasite interactions

The species of sandfly vector also contributes to the outcome:

- Sandfly saliva contains immunomodulatory molecules (e.g., maxadilan in *Lutzomyia longipalpis*), which enhance infection (Brodie et al., 2007).
- Different *Leishmania* species have co-evolved with specific sandfly species, influencing the initial immune modulation in the skin.

For example:

- *L. major* is transmitted by *Phlebotomus papatasi* (Old World CL)
- *L. infantum* by *Phlebotomus perniciosus* (VL in Mediterranean)
- The salivary proteins of these vectors differentially prime host immunity at the bite site

3.4.6 Parasite metabolic flexibility

Species like *L. donovani* and *L. infantum* can metabolically adapt to survive in the hostile visceral environment:

- Efficient antioxidant machinery (e.g., tryparedoxin peroxidase, ascorbate peroxidase) (Das et al., 2018; Kumari et al., 2024; Santos et al., 2024).
- Ability to acquire nutrients (e.g., iron, heme) in nutrient-deprived tissues (Goto et al., 2024).
- Differential expression of stress response proteins, enabling survival under hypoxia and low pH.

3.4.7 Intracellular niche preference and replication strategy

- Cutaneous species replicate slowly and maintain low burden, favoring persistence in macrophages of skin (Naderer et al., 2015).
- Visceral species show higher replication in liver and spleen macrophages, leading to organomegaly and systemic symptoms (Costa et al., 2023).
- Mucocutaneous species tend to disseminate hematogenously or via lymphatics from the primary cutaneous site (Abadías-Granado et al., 2021) (Table 3.2).

Table 3.2 Disease-specific intracellular niche preference and replication strategy

Factor	*Cutaneous species*	*Visceral species*	*Mucocutaneous species*
Tissue tropism	Skin	Liver, spleen, marrow	Skin to mucosa
Virulence factors	Moderate (GP63, LPG)	Strong (A2, LPG variants)	High inflammatory triggers
Immune response	Th1-biased or mixed	IL-10-rich, suppressed Th1	Overactive Th1/CD8+
Metabolic adaptation	Limited	High oxidative stress resistance	Intermediate
Vector influence	*Phlebotomus* spp.	*Phlebotomus* spp.	*Lutzomyia* spp.

3.5 MOLECULAR AND SIGNALING PATHWAYS IN IMMUNOPATHOGENESIS OF LEISHMANIASIS

The pathogenesis of leishmaniasis is deeply rooted in the parasite's remarkable ability to manipulate host immune signaling pathways. These molecular interactions determine whether the infection will be cleared or persist, and ultimately whether the disease will manifest as a localized, self-limiting cutaneous form or progress into systemic visceral or destructive mucocutaneous leishmaniasis. Once *Leishmania* promastigotes are phagocytosed by macrophages, they transform into amastigotes and initiate complex interactions with host receptors and intracellular signaling cascades, shaping the immune response in ways that favor their survival (Figure 3.3).

3.5.1 MyD88-dependent TLR pathway

One of the earliest signaling events triggered by *Leishmania* infection involves the activation of the TLR signaling cascade. TLRs such as TLR2, TLR4, and TLR9 recognize various *Leishmania*-derived pathogen-associated molecular patterns (PAMPs), including LPG, glycoprotein 63 (GP63), and unmethylated CpG DNA (Faria et al., 2012). Engagement of these receptors activates the adaptor molecule MyD88 (myeloid differentiation primary response 88), which in turn recruits downstream signaling molecules like IL-1 receptor-associated kinase (IRAK) and TNF receptor-associated factor 6 (TRAF6) (Hawn et al., 2002). This cascade culminates in the activation of transcription factors such as nuclear factor kappa-light-chain-enhancer of activated B cells (NF-κB) and activator protein 1 (AP-1), which translocate into the nucleus to promote the expression of pro-inflammatory cytokines, including IL-12, TNF-α, and IL-6. However, *Leishmania* has evolved sophisticated mechanisms to attenuate this response. For instance, LPG interferes with TLR2 and TLR4 signaling, suppressing NF-κB activation, while GP63 cleaves key signaling intermediates like MyD88 and IRAK1, thereby dampening the inflammatory response. Additionally, in *L. donovani* infections,

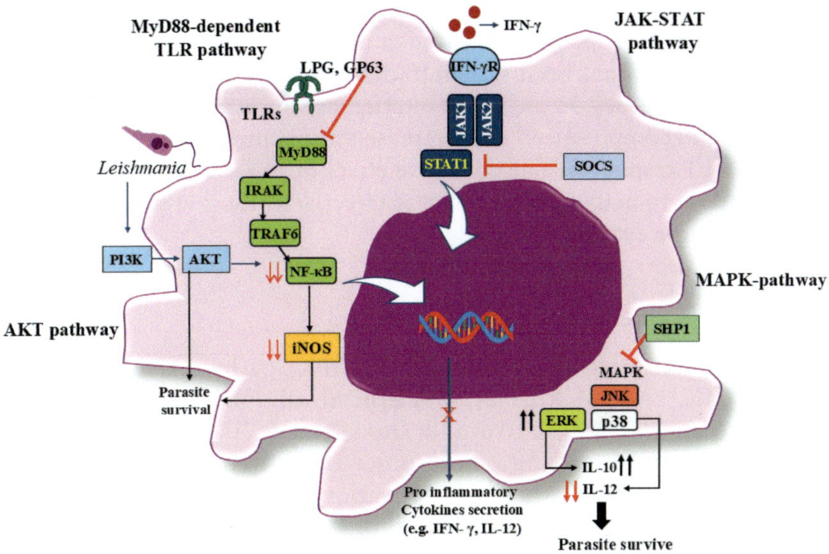

Figure 3.3 Phagocytosis of *Leishmania* by macrophages initiates complex interactions with host surface receptors and intracellular signaling cascades. This interaction modulates the host immune responses in ways that favor their survival for disease establishment.

the activation of TLR9 is inhibited, preventing effective interferon responses. These modulatory effects reduce IL-12 production, impairing Th1 polarization and facilitating parasite persistence, particularly in VL (Dayakar et al., 2019). Intriguingly, while the absence of MyD88 increases susceptibility to both cutaneous and VL due to impaired IL-12 signaling, uncontrolled or hyperactivated TLR signaling, especially in mucocutaneous cases, can contribute to tissue-destructive immunopathology.

3.5.2 JAK-STAT signaling pathway

Another critical pathway modulated by *Leishmania* is the Janus kinase-signal transducer and activator of transcription (JAK-STAT) signaling axis. IFN-γ, a key cytokine in the control of intracellular parasites, signals through its receptor by activating JAK1 and JAK2, which subsequently phosphorylate STAT1. Phosphorylated STAT1 dimerizes and enters the nucleus, initiating transcription of several pro-inflammatory genes, including inducible NO synthase (iNOS), which catalyzes NO production—a potent leishmanicidal molecule (Gurjar et al., 2022). Other STAT1-regulated genes include those encoding MHC class II molecules and various chemokines necessary for adaptive immune activation. However, *Leishmania* parasites actively suppress this axis to evade immune clearance. In *L. donovani* and *L. infantum* infections, the parasite induces the expression of suppressor of cytokine signaling (SOCS) proteins, particularly SOCS1 and SOCS3, which inhibit STAT1 phosphorylation and block JAK activity (Bichiou et al., 2021; Jafarzadeh et al., 2024). Furthermore, GP63 has been shown to directly degrade STAT1, reducing its transcriptional efficacy. In DCs, this interference leads to a reduction in IL-12 production and compromised antigen presentation, impairing the initiation of protective Th1 responses. The collective effect of JAK-STAT pathway inhibition is a failure to activate macrophages effectively, leading to chronic infection and immune evasion.

3.5.3 MAPK pathways (ERK, JNK, p38)

The mitogen-activated protein kinase (MAPK) pathways, which include ERK1/2, JNK, and p38 kinases, play crucial roles in managing the host's response to *Leishmania* (Privé and Descoteaux, 2000). When infection occurs, host pattern recognition receptors and cytokine receptors activate MAPK cascades. This leads to the activation of various transcription factors that are responsible for producing pro-inflammatory genes. The p38 and JNK pathways mainly help induce IL-12, TNF-α, and iNOS, which promote microbicidal activity (Junghae and Raynes, 2002). In contrast, the ERK1/2 pathway supports cell survival and the production of anti-inflammatory cytokines, especially IL-10 (Costa-da-Silva et al., 2022). *Leishmania* has figured out how to use these signaling pathways to its benefit (Figure 3.3).

GP63, a metalloprotease that *L. major* expresses in large amounts, removes phosphate groups from and inactivates MAPKs like ERK, JNK, and p38. This process reduces pro-inflammatory responses. During *L. donovani* infection, the parasite triggers host phosphatases such as SHP-1 and MAPK phosphatase-1 (MKP-1). These specifically deactivate p38 and JNK signaling, limiting IL-12 production (Junghae and Raynes, 2002). On the other hand, *Leishmania* may activate the ERK1/2 pathway to boost IL-10 expression, creating an environment that suppresses the immune response and helps it survive. This careful manipulation of MAPK pathways allows the parasite to weaken protective immune responses while avoiding excessive inflammation that could threaten its intracellular niche.

3.5.4 PI3K-Akt pathway

The phosphoinositide 3-kinase (PI3K)-Akt pathway is usually linked to cell survival and metabolism. However, it is now also seen as playing a role in the long-term survival of *Leishmania* within cells. When *Leishmania* infects macrophages, it activates PI3K and its downstream protein Akt (Kima, 2016). This action promotes cell survival and prevents the death of infected cells. By keeping the host cell alive longer, the parasite secures a safe environment for replication. Moreover, Akt signaling decreases NF-κB activation and lowers iNOS expression (Calegari-Silva et al., 2015). This response reduces the macrophage's ability to kill microbes. This strategy is especially clear in species that cause VL, where the long-term survival of the parasite is crucial for causing disease.

3.5.5 NF-κB pathway

The NF-κB pathway plays a key role in managing inflammatory responses. It gets activated through TLRs and cytokine receptors. In uninfected macrophages, activating NF-κB leads to the production of IL-12, TNF-α, and other cytokines necessary for a Th1 response. However, *Leishmania* can disrupt this pathway through various methods. For instance, in infections caused by *Leishmania*, the parasite stops the breakdown of IκB, which keeps NF-κB in the cytoplasm and prevents it from entering the nucleus (Reinhard et al., 2012). Additionally, *Leishmania* may increase host phosphatases that block parts of the IKK complex, effectively turning off the pathway. These strategies result in low IL-12 production and weak cell-mediated immunity, allowing the infection to persist and the disease to progress.

3.5.6 Host phosphatases and negative regulators

Adding another layer of complexity, *Leishmania* exploits host-derived phosphatases and negative regulatory proteins to suppress immune signaling. SHP-1, a tyrosine phosphatase activated during *Leishmania* infection, is a strong inhibitor of immune activation (Forget et al., 2006). It dephosphorylates JAK2, STAT1, and p38 MAPK, stopping several signaling pathways crucial for fighting the parasite. Other negative regulators, like PTEN and MKP-1, also increase during infection and contribute to the suppression of Akt and MAPK signaling (Kirk et al., 2020). These molecules act as critical checkpoints exploited by the parasite to maintain chronic infection.

Overall, the immune response during leishmaniasis is shaped by the parasite's ability to suppress pro-inflammatory pathways and promote anti-inflammatory signals. The blocking of TLR/MyD88 signaling, interference with the IFN-γ/STAT1 axis, disruption of MAPK and NF-κB pathways, and activation of host phosphatases all work together to weaken cellular immunity and help the parasite survive. These mechanisms influence the chronicity and severity of the disease while also providing insights into how different species affect clinical outcomes.

3.6 PATHOGENESIS OF LEISHMANIASIS: SCIENTIFIC RELEVANCE AND GLOBAL HEALTH IMPLICATIONS

The study of "Pathogenesis of Leishmaniasis: Mechanisms of Infection and Disease Progression" is critically important for several compelling scientific, clinical, and public health reasons.

3.6.1 Understanding disease diversity and species-specific manifestations

Leishmaniasis is not a uniform disease—it manifests in cutaneous, mucocutaneous, and visceral forms, depending largely on the infecting *Leishmania* species and host response. By studying the pathogenesis, researchers can understand why different species cause organ-specific manifestations and how they establish infection in distinct tissue environments (e.g., skin vs. visceral organs). This is crucial for developing targeted interventions.

3.6.2 Identifying immune evasion mechanisms

Leishmania has evolved complex strategies to evade host immunity, including modulation of TLR signaling, suppression of antigen presentation, inhibition of pro-inflammatory responses, and induction of regulatory cytokines like IL-10 and TGF-β. Dissecting these molecular mechanisms helps in understanding how chronic infection is maintained and why conventional immunity fails in certain forms, particularly VL.

3.6.3 Guiding therapeutic development

Current treatments for leishmaniasis are limited, often toxic, and prone to resistance. Pathogenesis studies identify key molecular targets and host signaling pathways (e.g., JAK-STAT, MyD88, MAPK) manipulated by the parasite. These insights enable the design of novel therapeutics that either inhibit parasite survival mechanisms or restore host immune responses.

3.6.4 Informing vaccine design

Effective vaccines against leishmaniasis remain elusive. Understanding which antigenic molecules, host immune pathways, and cytokine profiles correlate with protective immunity can inform vaccine development strategies, especially by identifying dominant Th1 responses, memory T cells, or the role of DCs in shaping adaptive immunity.

3.6.5 Addressing disease relapse and post-treatment complications

Relapse of VL and progression to PKDL remain major obstacles in elimination programs. By studying disease progression at the molecular level, especially in relapse-prone patients or those with co-infections like HIV, researchers can develop biomarkers for early detection and predictive tools for treatment monitoring.

3.6.6 Supporting precision medicine and biomarker discovery

Molecular understanding of host–pathogen interactions enables the identification of host susceptibility genes, cytokine profiles, and transcriptional signatures that predict disease outcome or treatment response. This supports the vision of precision medicine in leishmaniasis—tailoring treatment based on individual immune profiles and parasite species.

3.7 CONCLUSION

This chapter has explored the disease-causing mechanisms of leishmaniasis. It explains how *Leishmania* parasites interact with host cells to cause infection and disease progression. By looking into the molecular and immune events—from how the parasite enters and infects macrophages to how it avoids the immune system and persists—we gain insights into how this parasite survives in the human body and manipulates the immune system to survive.

Understanding how the disease develops is not just academic; it has real practical benefits. It helps clarify the range of symptoms, from skin forms to internal types, and informs the creation of targeted treatments and tests. More importantly, recognizing how the parasite takes advantage of specific immune pathways, such as the IL-10, STAT, and MyD88 signaling pathways, opens opportunities for therapies that can adjust these responses to help clear the parasite.

Moreover, insights into disease progression inform strategies for vaccine development by identifying critical antigens and immune mechanisms needed for protection. Studying pathogenesis also improves our understanding of complications such as PKDL, relapse, and drug resistance, which remain major hurdles in disease control.

In conclusion, understanding the development of leishmaniasis provides a basis for new approaches in diagnosis, treatment, and prevention. Ongoing research into these mechanisms is crucial for breaking the transmission cycle and ultimately controlling this neglected tropical disease.

ACKNOWLEDGMENT

Authors are thankful to the ICMR-RMRIMS, Patna, for the necessary research support.

REFERENCES

Abadías-Granado, I., Diago, A., Cerro, P.A., Palma-Ruiz, A.M. and Gilaberte, Y."Cutaneous and mucocutaneous leishmaniasis." *Actas Dermo Sifiliográficas (English Edition)* 112, no. 7 (2021): 601–618.

Ahmadi, Somayyeh, Maryam Hataminejad, Bahman Rahimi Esboei, Seyed Abdollah Hosseini, and Mahdi Fakhar. "An update on *Leishmania martiniquensis* infections: Transmission, clinical characteristics, and treatment." *Parasite Epidemiology and Control*, no. 27 (2024): e00386.

Ait Maatallah, Ikram, Khadija Akarid, and Meryem Lemrani. "Tissue tropism: Is it an intrinsic characteristic of Leishmania species?." *Acta Tropica* 232 (2022): 106512.

Al-Khalaifah, Hanan S. "Major molecular factors related to leishmania pathogenicity." *Frontiers in Immunology* 13 (2022): 847797.

Bamigbola, Ifeoluwa E., and Selman Ali. "Paradoxical immune response in leishmaniasis: The role of toll-like receptors in disease progression." *Parasite Immunology* 44, no. 4–5 (2022): e12910.

Bichiou, Haifa, Cyrine Bouabid, Imen Rabhi, and Lamia Guizani-Tabbane. "Transcription factors interplay orchestrates the immune-metabolic response of Leishmania infected macrophages." *Frontiers in Cellular and Infection Microbiology* 11 (2021): 660415.

Brodie, Tess M., Matthew C. Smith, Robin V. Morris, and Richard G. Titus. "Immunomodulatory effects of the *Lutzomyia longipalpis* salivary gland protein maxadilan on mouse macrophages." *Infection and Immunity* 75, no. 5 (2007): 2359–2365.

Calegari-Silva, Teresa C., Áislan C. Vivarini, Marina Miqueline, Guilherme R. R. M. Dos Santos, Karina Luiza Teixeira, Alessandra Mattos Saliba, Simone Nunes de Carvalho, Laís de Carvalho, and Ulisses G. Lopes. "The human parasite *Leishmania amazonensis* downregulates iNOS expression via NF-κB p50/p50 homodimer: Role of the PI3K/Akt pathway." *Open Biology* 5, no. 9 (2015): 150118.

Carreira, João Carlos Araujo, and Alba Valéria Machado da Silva. "The role of neutrophils in the interaction with *Leishmania*: Far beyond a simple Trojan horse?." *Open Journal of Animal Sciences* 11, no. 3 (2021): 399–421.

Costa, Carlos H. N., Kwang-Poo Chang, Dorcas L. Costa, and Francisco Valmor M. Cunha. "From infection to death: An overview of the pathogenesis of visceral leishmaniasis." *Pathogens* 12, no. 7 (2023): 969.

Costa-da-Silva, Ana Caroline, Danielle de Oliveira Nascimento, Jesuino RM Ferreira, Kamila Guimarães-Pinto, Leonardo Freire-de-Lima, Alexandre Morrot, Debora Decote-Ricardo, Alessandra Almeida Filardy, and Celio Geraldo Freire-de-Lima. "Immune responses in leishmaniasis: An overview." *Tropical Medicine and Infectious Disease* 7, no. 4 (2022): 54.

Das, Sanchita, Sagnik Giri, Shyam Sundar, and Chandrima Shaha. "Functional involvement of *Leishmania donovani* Tryparedoxin peroxidases during infection and drug treatment." *Antimicrobial Agents and Chemotherapy* 62, no. 1 (2018): e00806-17.

Dayakar, Alti, Sambamurthy Chandrasekaran, Suresh V. Kuchipudi, and Suresh K. Kalangi. "Cytokines: Key determinants of resistance or disease progression in visceral leishmaniasis: Opportunities for novel diagnostics and immunotherapy." *Frontiers in Immunology* 10 (2019): 670.

Elmahallawy, Ehab Kotb, and Abdulsalam A.M. Alkhaldi. "Insights into Leishmania molecules and their potential contribution to the virulence of the parasite." *Veterinary Sciences* 8, no. 2 (2021): 33.

Faria, Marilia S., Flavia C.G. Reis, and Ana Paula C.A. Lima. "Toll-like receptors in *Leishmania* infections: Guardians or promoters?." *Journal of Parasitology Research* 2012, no. 1 (2012): 930257.

Forget, Genevieve, David J. Gregory, Lorie A. Whitcombe, and Martin Olivier. "Role of host protein tyrosine phosphatase SHP-1 in *Leishmania donovani*-induced inhibition of nitric oxide production." *Infection and immunity* 74, no. 11 (2006): 6272–6279.

Ghosh, Sanhita, Supriya Nath, Kamalika Roy, Suman Karmakar, and Chiranjib Pal. "Leishmaniasis: Tissue tropism in relation to the species diversity." In *Pathobiology of Parasitic Protozoa: Dynamics and Dimensions*, pp. 133–153. Singapore: Springer Nature Singapore, 2023.

Goto, Yasuyuki, Tatsumi Ito, Souradeepa Ghosh, and Budhaditya Mukherjee. "Access and utilization of host-derived iron by Leishmania parasites." *The Journal of Biochemistry* 175, no. 1 (2024): 17–24.

Gupta, Anand Kumar, Sonali Das, Mohd Kamran, Sarfaraz Ahmad Ejazi, and Nahid Ali. "The pathogenicity and virulence of Leishmania-interplay of virulence factors with host defenses." *Virulence* 13, no. 1 (2022): 903–935.

Gurjar, Dhiraj, Sananda Kumar Patra, Neelam Bodhale, Nibedita Lenka, and Bhaskar Saha. "Leishmania intercepts IFN-γR signaling at multiple levels in macrophages." *Cytokine* 157 (2022): 155956.

Hawn, Thomas R., Adrian Ozinsky, David M. Underhill, Frederick S. Buckner, Shizuo Akira, and Alan Aderem. "*Leishmania major* activates IL-1α expression in macrophages through A MyD88-dependent pathway." *Microbes and Infection* 4, no. 8 (2002): 763–771.

Jafarzadeh, Abdollah, Dhiraj Gurjar, Neelam Bodhale, Sara Jafarzadeh, Maryam Nemati, Iraj Sharifi, and Bhaskar Saha. "Aberrant expression of SOCS impairs the anti-leishmanial immune response." *Cytokine* 174 (2024): 156461.

Junghae, Muthoni, and John G. Raynes. "Activation of p38 mitogen-activated protein kinase attenuates *Leishmania donovani* infection in macrophages." *Infection and Immunity* 70, no. 9 (2002): 5026–5035.

Kautz-Neu, Kordula, Kirsten Schwonberg, Michael R. Fischer, Anja I. Schermann, and Esther von Stebut. "Dendritic cells in *Leishmania major* infections: Mechanisms of parasite uptake, cell activation and evidence for physiological relevance." *Medical Microbiology and Immunology* 201 (2012): 581–592.

Kedzierski, Lukasz, and Krystal J. Evans. "Immune responses during cutaneous and visceral leishmaniasis." *Parasitology* 141, no. 12 (2014): 1544–1562.

Kima, Peter E. "PI3K signaling in Leishmania infections." *Cellular immunology* 309 (2016): 19–22.

Kirk, Sean G., Lobelia Samavati, and Yusen Liu. "MAP kinase phosphatase-1, a gatekeeper of the acute innate immune response." *Life Sciences* 241 (2020): 117157.

Kumari, Shobha, Pradeep Das, and Ashish Kumar. "Molecular Mechanism of PKDL Development." In *Visceral Leishmaniasis and Post-kala-azar Dermal Leishmaniasis*, pp. 98–111. CRC Press, 2025.

Kumari, Shobha, Saravanan Vijaykumar, Vikash Kumar, Ravi Ranjan, Dayakar Alti, Veer Singh, Ghufran Ahmed, Ganesh Chandra Sahoo, Krishna Pandey, and Ashish Kumar. "In silico and in vitro evaluation of the immunogenic potential of *Leishmania donovani* ascorbate peroxidase and its derived peptides." *Acta Tropica* 260 (2024): 107381.

Lindoso, José Angelo Lauletta, Carlos Henrique Valente Moreira, Mirella Alves Cunha, and Igor Thiago Queiroz. "Visceral leishmaniasis and HIV coinfection: Current perspectives." *HIV/AIDS-Research and Palliative Care*, no. 10, (2018): 193–201.

Liu, Dong, and Jude E. Uzonna. "The early interaction of Leishmania with macrophages and dendritic cells and its influence on the host immune response." *Frontiers in Cellular and Infection Microbiology* 2 (2012): 83.

Martínez-López, María, Manuel Soto, Salvador Iborra, and David Sancho. "Leishmania hijacks myeloid cells for immune escape." *Frontiers in Microbiology* 9 (2018): 883.

McCall, Laura-Isobel, Wen-Wei Zhang, and Greg Matlashewski. "Determinants for the development of visceral leishmaniasis disease." *PLoS Pathogens* 9, no. 1 (2013): e1003053.

Moradin, Neda, and Albert Descoteaux. "Leishmania promastigotes: Building a safe niche within macrophages." *Frontiers in Cellular and Infection Microbiology* 2 (2012): 121.

Murray, Peter J., and Thomas A. Wynn. "Protective and pathogenic functions of macrophage subsets." *Nature Reviews Immunology* 11, no. 11 (2011): 723–737.

Naderer, Thomas, Joanne Heng, Eleanor C. Saunders, Joachim Kloehn, Thusitha W. Rupasinghe, Tracey J. Brown, and Malcolm J. McConville. "Intracellular survival of *Leishmania major* depends on uptake and degradation of extracellular matrix glycosaminoglycans by macrophages." *PLoS Pathogens* 11, no. 9 (2015): e1005136.

Olivier, Martin, Vanessa Diniz Atayde, Amandine Isnard, Kasra Hassani, and Marina Tiemi Shio. "Leishmania virulence factors: Focus on the metalloprotease GP63." *Microbes and Infection* 14, no. 15 (2012): 1377–1389.

Paramanik, Manas. "Leishmaniasis and its vectors." In *Latest Trends in Zoology and Entomology Sciences*, pp. 33–51. New Delhi: AkiNik, 2023.

Passelli, Katiuska, Oaklyne Billion, and Fabienne Tacchini-Cottier. "The impact of neutrophil recruitment to the skin on the pathology induced by Leishmania infection." *Frontiers in Immunology* 12 (2021): 649348.

Paton, Hanna, Prabuddha Sarkar, and Prajwal Gurung. "An overview of host immune responses against *Leishmania* spp. infections." *Human Molecular Genetics* 34, no. R1 (2025): ddaf043.

Ponte-Sucre, Alicia, Francisco Gamarro, Jean-Claude Dujardin, Michael P. Barrett, Rogelio López-Vélez, Raquel García-Hernández, Andrew W. Pountain, Roy Mwenechanya, and Barbara Papadopoulou. "Drug resistance and treatment failure in leishmaniasis: A 21st century challenge." *PLoS Neglected Tropical Diseases* 11, no. 12 (2017): e0006052.

Privé, Charles, and Albert Descoteaux. "*Leishmania donovani* promastigotes evade the activation of mitogen-activated protein kinases p38, c-Jun N-terminal kinase, and extracellular signal-regulated kinase-1/2 during infection of naive macrophages." *European Journal of Immunology* 30, no. 8 (2000): 2235–2244.

Regli, Ivo B., Katiuska Passelli, Benjamin P. Hurrell, and Fabienne Tacchini-Cottier. "Survival mechanisms used by some *Leishmania* species to escape neutrophil killing." *Frontiers in Immunology* 8 (2017): 1558.

Reinhard, Katharina, Magdalena Huber, Michael Lohoff, and Alexander Visekruna. "The role of NF-κB activation during protection against *Leishmania* infection." *International Journal of Medical Microbiology* 302, no. 4–5 (2012): 230–235.

Requena, Jose M., Ana M. Montalvo, and Jorge Fraga. "Molecular chaperones of Leishmania: Central players in many stress-related and -unrelated physiological processes." *BioMed Research International* 2015, no. 1 (2015): 301326.

Rocha, Rafael, André Pereira, and Carla Maia. "Non-endemic *Leishmaniases* reported globally in humans between 2000 and 2021—A comprehensive review." *Pathogens* 11, no. 8 (2022): 921.

Santos, Isabella Fernandes Martins, Douglas de Souza Moreira, Karla Ferreira Costa, Juliana Martins Ribeiro, Silvane Maria Fonseca Murta, and Ana Maria Murta Santi. "Ascorbate peroxidase modulation confirms key role in *Leishmania infantum* oxidative defence." *Parasites & Vectors* 17, no. 1 (2024): 472.

Silva-Moreira, Anna Luiza, Artur Metzker Serravite, Laura Valéria Rios-Barros, Juliana Perrone Bezerra de Menezes, Maria Fátima Horta, and Thiago Castro-Gomes. "New insights into the life cycle, host cell tropism, and infection amplification of *Leishmania* spp." *Infection and Immunity* 93, no. 7 (2025): e00123–e00125.

Srivastava, Saumya, Prem Shankar, Jyotsna Mishra, and Sarman Singh. "Possibilities and challenges for developing a successful vaccine for leishmaniasis." *Parasites & Vectors* 9 (2016): 1–15.

Winberg, Martin E., Birgitta Rasmusson, and Tommy Sundqvist. "*Leishmania donovani*: Inhibition of phagosomal maturation is rescued by nitric oxide in macrophages." *Experimental Parasitology* 117, no. 2 (2007): 165–170.

Yasmin, Hadida, Anureeta Adhikary, Mohammed N. Al-Ahdal, Syamal Roy, and Uday Kishore. "Host–pathogen interaction in leishmaniasis: Immune response and vaccination strategies." *Immuno* 2, no. 1 (2022): 218–254.

Chapter 4

Clinical manifestations of different forms of leishmaniasis

Major Madhukar

Department of Clinical Medicine, ICMR-Rajendra Memorial Research Institute of Medical Sciences, Patna, Bihar, India

*Dayakar Alti**

Department of Immunology, ICMR-Rajendra Memorial Research Institute of Medical Sciences, Patna, Bihar, India

Roshan Kamal Topno

Department of Clinical Medicine, ICMR-Rajendra Memorial Research Institute of Medical Sciences, Patna, Bihar, India

4.1 INTRODUCTION

Leishmaniases are a spectrum of diseases caused by intracellular parasites of the genus *Leishmania* and transmitted by female sandflies. It is prevalent across the globe in different clinical forms, as shown in Figure 4.1 (Abbasi et al., 2018), of which cutaneous leishmaniasis (CL) is the most common version of the leishmaniases and persists in three different subclinical forms: localized, disseminated, and anergic diffuse CL (ADCL). Since the pathogenesis of all these forms involves skin tissue, it is broadly defined as tegumentary leishmaniasis. The dissemination of leishmaniasis to mucosal membranes manifests as mucocutaneous leishmaniasis (MCL), which follows the CL. The simultaneous occurrence of CL and MCL is very rare (Scorza et al., 2017). The most fatal version of this disease is visceral leishmaniasis (VL), which affects systemic organs and tissues. In certain cases, VL is followed by post-kala-azar dermal leishmaniasis (PKDL) with skin manifestations (Zijlstra et al., 2000) resembling CL. PKDL and VL coexistence is rare, and if it occurs, it is clinically defined as para-kala-azar dermal leishmaniasis (ParaKDL) with a spectrum of both disease symptoms (Zijlstra et al., 2003). The clear distinction between typical and atypical manifestations in the different forms of leishmaniasis is imperative to distinguish from other diseases, e.g., lepromatous leprosy and psoriasis skin diseases (Zijlstra, 2014), systemic

* Corresponding author: Dr. Dayakar Alti, Scientist-C, Department of Immunology, ICMR-Rajendra Memorial Research Institute of Medical Sciences, Agamkuan, Patna, Bihar, India

DOI: 10.1201/9781003596455-4

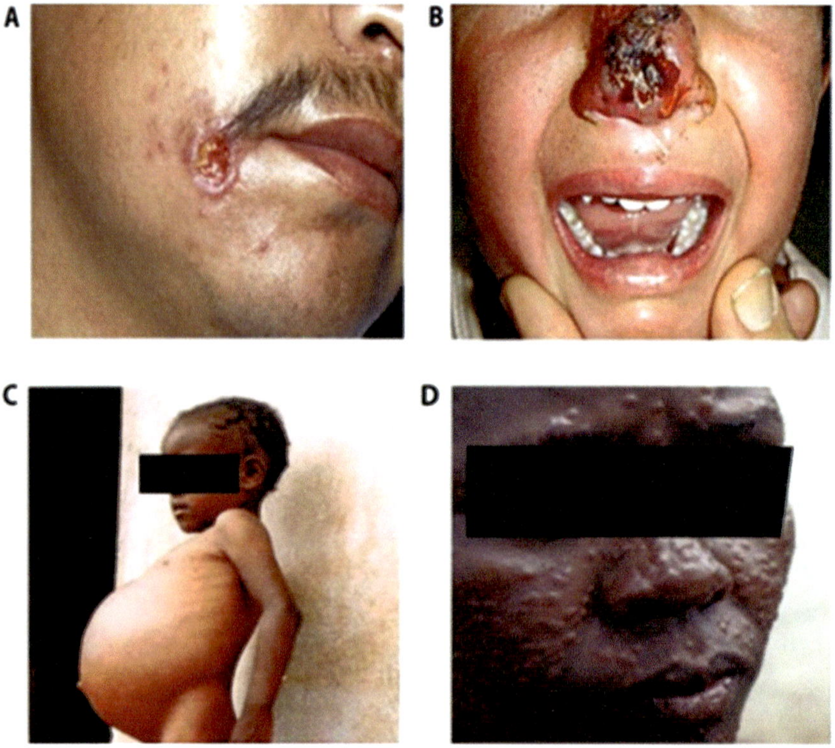

Figure 4.1 Different clinical forms of leishmaniasis (Abbasi et al., 2018). (A) Cutaneous leishmaniasis, (B) Mucocutaneous leishmaniasis, (C) Visceral leishmaniasis, and (D) Post-kala-azar dermal leishmanisis.

lupus erythematosus (SLE) (Bueno et al., 2019; Voulgarelis et al., 2003), autoimmune hepatitis, and rheumatoid arthritis (Tunccan et al., 2012). Physicians must consider differential diagnosis methods in order to avoid mismatched treatment regimens (Figure 4.1).

4.2 CLINICAL MANIFESTATIONS OF LEISHMANIASIS

4.2.1 Localized CL (LCL)

It is caused by the *Leishmania* spp. and includes *Leishmania major* (Pearson and de Queiroz Sousa, 1996), *Leishmania mexicana*, *Leishmania amazonensis*, *Leishmania (V.) braziliensis*, and *Leishmania (V.) panamanensis* (Grimaldi et al., 1989), *Leishmania tropica* (Kumar et al., 2007), *Leishmania infantum* (Oliveira Neto et al., 1986), *Leishmania donovani* (Karunaweera et al., 2003), and *Leishmania aethiopica* (Van Griensven et al., 2016). It is the abundantly occurring form of leishmaniasis that causes non-life-threatening manifestations limited to the skin. However, it is a serious threat to social stigma in patients. The incubation period is about two to eight weeks for LCL, which might change with the infected *Leishmania* species (El Hajj et al., 2004; Hochedez and Caumes, 2008; Schwartz et al., 2006). LCL is clinically polymorphic. Erythema is an early sign of sand fly bite. *Leishmania* deposition in the skin leads to the development of one or more nodular (lumps) or papular (bumps) phenotypic lesions (sores) that persist for months to years. These lesions are most apparent on the exposed body parts: face, arms, and legs (Figure 4.2A–F). During *L. (V.) braziliensis* infection, lymphadenopathy is the earliest manifestation, followed by lesion formation (Barral et al., 1995). The nodular or papular lesions may further develop into crateriform ulcers with discrete boundaries that appear as volcano-raised edges (Pearson and de Queiroz Sousa, 1996) covered by scabs or crust. The phenotype of ulcers is predominantly manifested by *Leishmania* species, e.g., dry nodules caused by *L. tropica* and wet lesions caused by *L. major* (Akilov et al., 2007). In general, LCL lesions are painless, but upon superficial infections of bacteria, e.g., *staphylococcus* spp. or fungus (Ziaei et al., 2008), they turn into fluid-discharging painful ulcers; otherwise, they are mostly self-healed (Reithinger et al., 2010). The resolution of lesions does not mean a sterile cure because the hypopigmented scars that remain on the skin may harbor parasites or their genetic material (Schubach et al., 1998a, 1998b). Subclinical LCL infections are widely reported with no scars or lesions on the skin but positive for delayed-type hypersensitivity (DTH) response (Ivonise et al., 2002) (Figure 4.2).

4.3 MUCOSAL CL (MCL)

It is a metastatic version of LCL, the *Viannia* subgenus *Leishmania* species; *L. (V.) braziliensis* (Costa et al., 1986), *L. (V.) panamanensis* (Osorio et al., 1998), *Leishmania (V.) guyanensis* (Santrich et al., 1990), and *L. amazonensis* (Costa et al., 1986; Lucas et al., 1998) are responsible for the occurrence of MCL in the New World countries. South America is at the top of the list with *L.* (*V.*) *braziliensis* infection causing MCL (Amato et al., 2008; Lucas et al., 1998; Saravia et al., 1998). About 1–10% of LCL cases progress to develop MCL in *L.* (*V.*) *braziliensis* endemic regions due to the dissemination of the parasite into mucosal membranes via blood or lymphatic stream. An individual's gender, age, nutritional status, diameter and profusion of lesions, body parts affected by lesions, and the treatment regimen influence the progression of LCL to MCL. Males with the progression of age, malnutrition, above-trunk lesions, and under-treatment are highly susceptible (Jones et al., 1987; Machado-Coelho et al., 2005; Marsden, 1986). MCL and LCL concurrent occurrence is very rare but after several months or years of cure, the latent parasites remain in the primary lesions of LCL patients (Jones et al., 1987; Machado-Coelho et al., 2005; Marsden, 1990) and may progress to

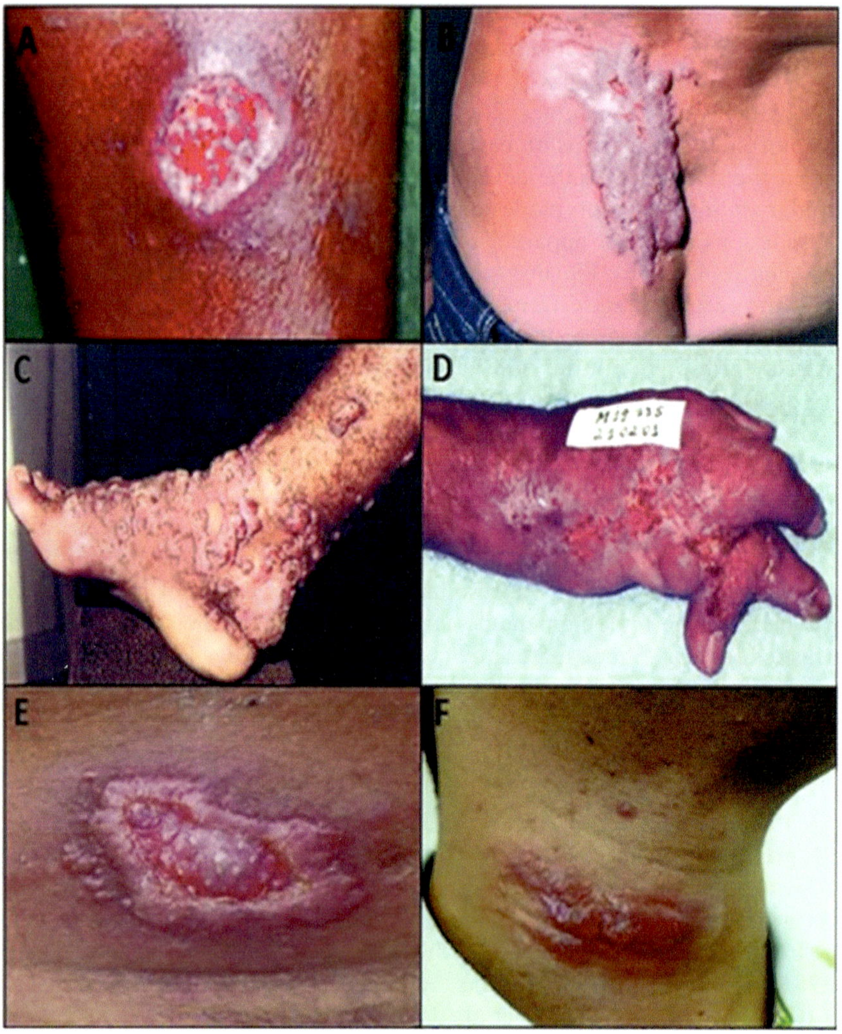

Figure 4.2 **Localized cutaneous leishmaniasis**: (A) ulcerated lesion; (B) keloid lesion; (C) verrucous lesion; (D) ulcerated lesion on the hand with bone deformity; (E) and (F) lesions forming plaque and ulcerative surface (Silveira et al., 2023).

develop metastatic lesions in nasopharyngeal regions (nose, larynx, trachea, soft palate, pharynx, lips, cheeks, etc.), and rarely genitalia (Figure 4.6D) (Pearson and de Queiroz Sousa, 1996). The destructive inflammation in these parts causes nasal bleeding and congestion, a tapir nose (damage to the nasal cartilage), severe disfigurement of facial tissues (Figure 4.3), and difficulty in speaking, breathing, swallowing, eating, etc. These patients ultimately face stigma and social exclusion, if left untreated. MCL therapy is complicated over the LCL and may require cosmetic surgeries and secondary treatments (Amato et al., 2007; Franke et al., 1990; Lessa et al., 2001) due to

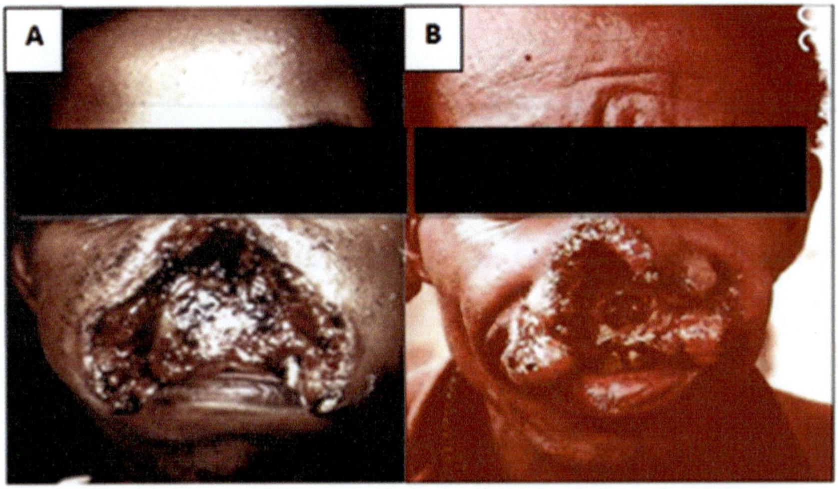

Figure 4.3 Severe clinical manifestations of MCL (Bowles et al., 2015; Mans et al., 2017). (A & B) Patients with tissue destructive mucosal lesions at nasopharyngeal regions.

superinfections (e.g., pneumonia and sepsis). MCL patients show a strong DTH response to *Leishmania* antigen (Schriefer et al., 2008) (Figure 4.3).

4.3.1 Disseminated CL (DCL)

After MCL, the disseminated CL (DCL) also emerged as a highly prevalent metastatic form of LCL, in some parts of the world, e.g., the state of Bahia, northeast Brazil, where DCL has surpassed MCL incidence (Schriefer et al., 2008). The DCL occurrence in the New World is similar to MCL due to the infection by the same *Leishmania* species: *L. (V.) braziliensis* (Carvalho et al., 1994), *L. (V.) panamanensis* (Hashiguchi et al., 2016), *L. (V.) guyanensis* (Couppie et al., 2004; Hashiguchi et al., 2016), and *L. amazonensis* (Carvalho et al., 1994). The distinct genetic polymorphism within the species of *L. (V.) braziliensis* introduces a new strain causing DCL in LCL endemic regions of Brazil (Queiroz et al., 2012). Unlike MCL, the diffusion of parasites from the initial lesion site is quite aggressive in DCL, and it takes only a few days or weeks (Carvalho et al., 1994; Turetz et al., 2002). It is characterized by numerous pleomorphic lesions, e.g., acneiform (Figure 4.4A–D), papular, or nodular (Figure 4.4E–H) (Hashiguchi et al., 2016; Turetz et al., 2002) in more than two anatomically distinct body parts involving mucus membranes (Costa et al., 1986; Goto and Lindoso, 2010). Moreover, the absence of lymphadenopathy in DCL emphasizes rapid hematogenous parasite dissemination. Similar to MCL, about 44% of DCL patients exhibit nasal mucosal lesions (Machado et al., 2011) and exhibit strong DTH responses (Turetz et al., 2002). Longer and multiple treatment schedules are recommended for DCL (Hashiguchi et al., 2016; Machado et al., 2011; Turetz et al., 2002) *vs.* LCL (Figure 4.4).

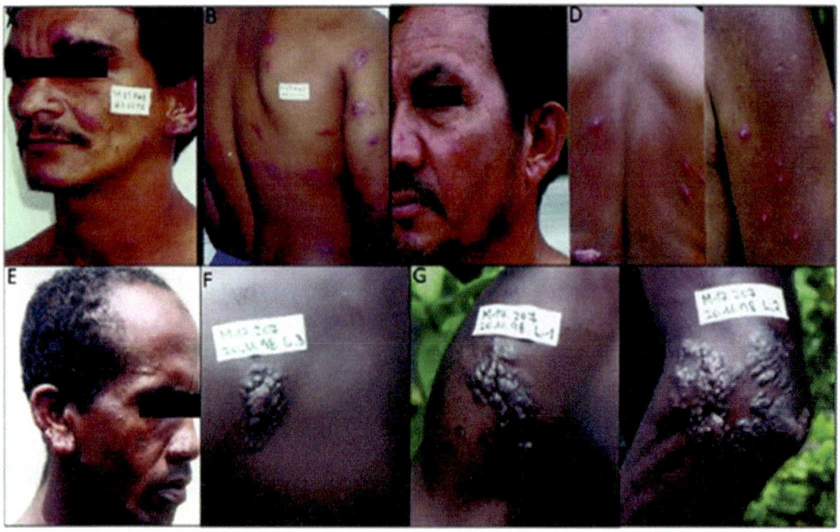

Figure 4.4 **Disseminated cutaneous leishmaniasis**: (A)–(D) acneiform ulcerative lesions; (E)–(H) plaque or nodular (non-ulcerated) lesions (Silveira et al., 2023).

4.3.2 Anergic diffuse CL (ADCL)

It is a severe form of DCL with a rare incidence. Initially, it was suspected that parasite strain variation may contribute to ADCL occurrence in endemic regions of other CL forms. However, the parasite species *Leishmania (V.) pifanoi* isolated from ADCL patients could develop classical LCL in healthy human volunteers (Convit et al., 1972). It suggests that the unusual occurrence of ADCL is due to something else beyond the parasite strain. The disease is prefixed as anergic due to the absence of skin DTH response and lymphocyte proliferation in response to the *Leishmania* antigens (Bomfim et al., 1996; Silveira et al., 2004), which contrasts with LCL, MCL, and DCL. ADCL is characterized by multiple, disseminated, non-ulcerative satellite lesions (nodular/papular) fused into plaques that cover a broad area of skin (Figure 4.5) (Goto and Lindoso, 2010). The lesions are characterized by the infiltration of abundant infected macrophages and a small number of lymphocytes, epidermal hyperplasia, and hyperkeratosis (Bryceson, 1969). *Leishmania*-specific anergic immune response in ADCL (Convit et al., 1962; Hashiguchi et al., 2016; Petersen et al., 1982) might be the consequence of uncontrolled amastigote growth in non-ulcerative lesions. ADCL infection is highly unresponsive to treatment and often exhibits relapse, and the lesions are not self-healing over time (Calvopina et al., 2006; Salaiza-Suazo et al., 1999; Zerpa et al., 2007) (Figure 4.5).

Atypical CL cases (Figure 4.6B–D) may be undistinguished from other skin diseases due to unusual lesions with zosteriform, erysiploid, lupoid,

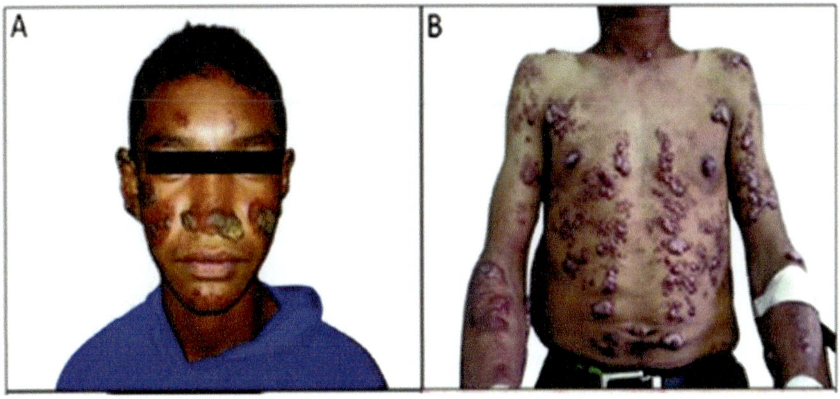

Figure 4.5 Anergic diffuse cutaneous leishmaniasis (Silveira et al., 2023). (A & B) Patients with infiltrating popular, nodular, and tuberous skin lesions on face and trunk regions.

sporotrichoid, nodular, and hyperkeratotic characteristics (Akilov et al., 2007). These unusual symptoms generally appear in HIV co-infection. Around 5% of LCL cases are relapsed to form a rare chronic disease called leishmaniasis recidivans (LR) in Iran (Momeni and Aminjavaheri, 1995), in which, the nascent lesions develop adjacent to the previously healed LCL scars (Berlin, 1940), which are painless, non-ulcerative (Figure 4.6A), and difficult to treat (Strick et al., 1983) (Figure 4.6).

4.3.3 Visceral leishmaniasis

It is a deadly version of the disease leishmaniasis, predominantly caused by the infection of *L. donovani* and *L. infantum/chagasi* (Aronson et al., 2016) and rarely caused by *L. mexicana* and *L. tropica* species (Barral et al., 1986; Magill et al., 1993; Monroy-Ostria et al., 2000; Sacks et al., 1995).

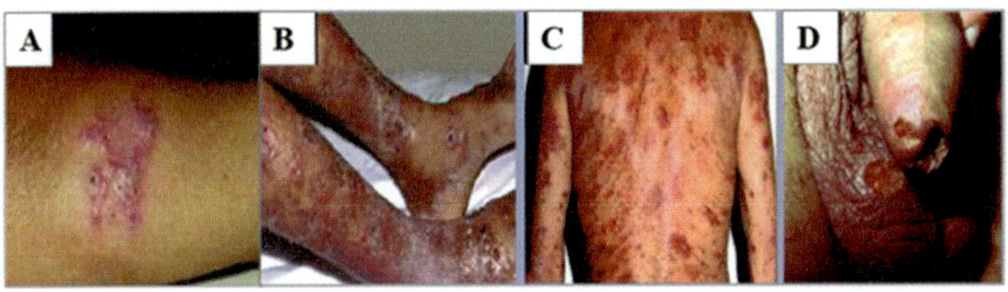

Figure 4.6 **Atypical clinical manifestations of tegumentary leishmaniasis**: (A) LR papules in the proximity of healed LCL lesion; (B)–(D) during HIV co-infection; ulcers on the legs and feet, erythematous plaques on the back, and ulcers on the scrotum and penis, respectively (Goto and Lindoso, 2010).

Asymptomatic (subclinical) infections are more prevalent (about 80–90%) in tropical and subtropical countries (Alencar, 1956; Jeronimo et al., 2011; Piyaraj et al., 2024; Singh et al., 2014), and their conversion into clinical disease is reflected by host genetic predisposition, parasite virulence, nutritional status, etc. Their conversion rate is higher at 1/6 in Brazil and 1/4 in Bangladesh than in Europe at 1/30 (Anstead et al., 2001; Badaro et al., 1986; Bern et al., 2007; Blackwell et al., 2009; Cerf et al., 1987; Karplus et al., 2002; Moral et al., 2002; Singh et al., 2014). VL is characterized by prolonged fever, chills, cold sweat, fatigue, malaise, adynamia, anemia, thrombocytopenia, leucopenia, weight loss, cachexia, lymphadenopathy, hepatosplenomegaly (Figure 4.1C), hyperpigmentation, hypergammaglobulinemia, jaundice, edema in lower limbs, etc. (World Health Organization, 2010). Clinical manifestations reported in VL patients infected by *L. donovani* or *L. infantum* are indistinguishable. Darkening of the skin and enlarged lymph nodes are common in Sudanese patients. The disease evolved in three different stages: acute, sub-acute, and chronic VL (Debono, 1947). Acute VL is quite often seen children under two years, manifested by prolonged high fever, moderate splenomegaly, and death within three to four months. The sub-acute infection lasts for 6–18 months with regular fever, progressive spleen enlargement, anemia, cachexia, occasionally diarrhea, and bronchitis. The opportunistic infections due to immunosuppression in VL may alter the course of the disease manifestations, and the chronic stage lasts for more than two years. During parasite invasion, manifestations like loss of appetite, pallor, hair loss, bleeding, hepatomegaly, apathy, and even sometimes seizures in children can be seen in VL patients (Pedrosa et al., 2013). In the later stages of VL, delayed puberty and amenorrhea can also occur. The DTH response in skin tests is negative in Sudanese patients and not executed in Indian patients (Zijlstra and El-Hassan, 1993). In most severe VL cases, rare clinical manifestations like disseminated intravascular coagulation (DIC) (De Santana Ferreira et al., 2021; Mishra et al., 2004), hemophagocytic lymphohistocytosis (HLH) (Daher et al., 2015; Rajagopala et al., 2008; Varma and Naseem, 2010), and renal impairment (Duarte et al., 2020; Libório et al., 2012) were associated with mortality.

4.3.4 Post-kala-azar dermal leishmaniasis

It is a clinical entity of VL succession in East Africa and the Indian subcontinent caused by *L. donovani* (Zijlstra et al., 2003). The New World *Leishmania* species (e.g., *L. infantum*) causing VL cases are unreported for PKDL conversion except in HIV co-infection (Lindoso et al., 2018). Instead, they are reported for causing non-ulcerative CL (Zhang et al., 2014). About 5–10% of Indian patients and 50–60% of Sudanese patients of VL develop PKDL (Zijlstra and El-Hassan, 1993). Interestingly, 15–20% of PKDL cases are developed without a previous history of VL, where the asymptomatic infections might progressed to PKDL (Ramesh and Mukherjee, 1995). PKDL

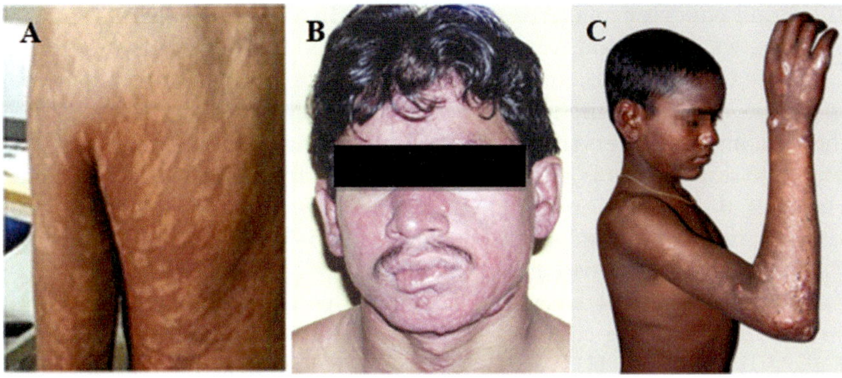

Figure 4.7 **Post-kala-azar dermal leishmaniasis**: (A) macular lesions (Duthie et al., 2019); (B) photosensitivity with papulo-nodular lesions; (C) hypopigmented macules and erythematous nodules (Das et al., 2014).

is manifested by monomorphic or polymorphic skin lesions with macular (Figure 4.7A) (hypopigmented patches), papular, or nodular phenotype (Figure 4.7B). Lesions are mostly apparent on the face and upper body; in severe cases, they may extend to lower limbs, mucosa of the tongue, buccal cavity, and genitalia (Kumar et al., 2021). Erythema and maculopapular plaques (Figure 4.7C) are common in Indian PKDL patients (Girgla et al., 1977; Ramesh and Mukherjee, 1995); in addition, lymphadenopathy (Kumar et al., 2021) and nodular lesions are frequent in African patients (Zijlstra et al., 2000). In general, PKDL lesions do not progress into ulcers (El Hassan et al., 1992); atypically, patients with mucosal lesions may develop ulcerative nodules (Das et al., 2014; Zijlstra et al., 2003). Three clinical grades are manifested in PKDL progression; grade I with scattered lesions on the face, with or without lesions on the upper chest and arms, grade II with dense lesions distributed on the face, chest, neck, back, and upper arms, and scattered lesions on forearms (Figure 4.7C) and legs, and grade III with dense lesions on whole body including mucosal tissues (Ganguly et al., 2010; Zijlstra et al., 2003). Photosensitivity (Figure 4.7B), verrucous (Jha et al., 2012), papillomatous, and hypertrophic lesions are atypical manifestations seen in some patients (Napier and Gupta, 1930, 1934). During post-kala-azar mucosal leishmaniasis (PKML), inflammatory manifestations like uveitis, blepharitis, conjunctivitis, laryngitis, and colitis were described in both Indian and Sudanese patients (Das et al., 2014; Zijlstra et al., 2003) (Figure 4.7).

4.3.5 Para kala-azar dermal leishmaniasis

It is a rare combination, i.e., concomitant occurrence of PKDL and VL manifestations with an intermediate pathophysiology characterized by fever, dermatoses, lymphadenopathy, hepatosplenomegaly, starvation (Zijlstra et al.,

2003), and pancytopenia (Alves et al., 2024). In addition to these, headache, dizziness, weakness, and weight loss are reported in some paraKDL patients (Lindoso et al., 2018). *L. donovani* and *L. infantum/chagasi* amastigotes persist in the skin and lymphoid tissues, e.g., lymph nodes and bone marrow. The combination of erythematous and maculopapular non-scaly and non-ulcerative skin lesions is described on the face, arms, and back of the chest (Hasan et al., 2023; Moniruzzaman et al., 2023). Skin biopsies are reported with inflammatory plasmacytoid structures (0.5–1.0 cm) and fibrosis of the papillary dermis (Lindoso et al., 2018). The spleen of paraKDL patients is of intermediate size between PKDL and VL patients (Kumar et al., 2021). ParaKDL patients are common (18%) in Sudan and rare in India (Zijlstra et al., 2017). During HIV co-infection in VL patients, the development of paraKDL manifestations with huge decorative nodular lesions mimics the atypical PKDL (Zijlstra, 2014).

4.4 DISCUSSION

An appropriate clinical illustration of leishmaniasis patients by physicians would lead epidemiologists to enumerate the actual disease prevalence, clinicians to set up new treatment guidelines, and entomologists to follow integrated vector control strategies. Since the asymptomatic cases are predominant and act as parasite reservoirs for leishmaniasis transmission (silent carriers), it is necessary to consider them a high priority for research. Developing a new set of rules for their prognosis/diagnosis and supportive therapy has to be implemented in endemic regions. Besides this, frequent surveillance for active parasitemia (at least once in six months) in endemic populations with the help of molecular diagnostic tools would provide insights for the prediction of symptomatic disease conversion and preparedness for clinical management. Identification of potential biomarkers (metabolites) specific to asymptomatic infection would be highly beneficial for non-invasive detection. Apart from asymptomatic cases, the occurrence of sporadic cases in non-endemic parts (Adhikari et al., 2010; Ejghal et al., 2022) and frequent relapses in endemic parts of the world are major hurdles to leishmaniasis elimination. Here, the identification of *Leishmania* species and strains would provide key insights to correlate the symptoms with the clinical form of the disease as well as drug resistance. Atypical clinical manifestations of leishmaniasis (Figure 4.6B–D) and PKDL development without a previous history of VL are sometimes misdiagnosed and lead to inappropriate treatment. CL (*L. major* infection) manifestations have been misdiagnosed as PKDL and vice versa. PKDL was misdiagnosed as other dermatoses and mistreated, resulting in exacerbation of parasite load in lesions and their visceralization in patients with co-morbidities (El Hassan et al., 2013). Therefore, parasite species identification using molecular techniques and employing differential diagnostic methods is highly requisite to improve the clinical judgment of the disease. Coexistence of two different clinical forms in the same patient, e.g., CL/MCL, ParaKDL, and co-infections, e.g., HIV and tuberculosis, must be carefully addressed by clinicians for their clinical appearance, lesion severity, and distribution pattern, previous history, geographical location, patient genetic background, nutri tional status, and *Leishmania* species, etc. In conclusion, a sound knowledge of disease prevalence, causative agents, pathogenesis, clinical manifestations, atypical symptoms, and differential diagnosis collectively helps physicians and patients to avoid misdiagnosis and inappropriate treatment.

4.5 CONCLUSION

Accurate clinical characterization of leishmaniasis is essential for guiding treatment, disease surveillance, and vector control strategies. Special attention must be given to asymptomatic carriers, atypical cases, and co-infections, which often complicate diagnosis and management. Regular molecular surveillance and species identification are critical for early detection, predicting disease progression, and preventing misdiagnosis. Establishing diagnostic biomarkers and improving differential diagnosis protocols can significantly enhance clinical outcomes. A comprehensive understanding of the disease's clinical spectrum, including rare forms like ParaKDL, and tailoring approaches based on geographic and patient-specific factors are vital steps toward effective control and eventual elimination of leishmaniasis

ACKNOWLEDGEMENT

The authors express their deepest gratitude to the ICMR-Rajendra Memorial Research Institute of Medical Sciences, Patna, for the invaluable support they provided throughout this study. The authors want to thank Anusandhan National Research Foundation (ANRF/ECRG/2024/000100/LS) for financial assistance.

REFERENCES

Abbasi M, Pal D, Mondal D, Anu Varghese B, Paul R, Singh S, Datta R. Carbonic anhydrase in acid acclimatization of *Leishmania*. 2018. doi: 10.13140/RG.2.2.15117.26085.

Adhikari L, Singh TS, Tsering D, Dhakal OP, Gupta A. Sporadic case of visceral leishmaniasis in Sikkim, India. Journal of Global Infectious Diseases. 2010 May 1; 2(2):196–7.

Akilov OE, Khachemoune A, Hasan T. Clinical manifestations and classification of Old World cutaneous leishmaniasis. International Journal of Dermatology. 2007 Feb 1;46(2):132–42.

Alencar JE. Leishmaniose visceral no Novo Mundo. Publicações Médicas. 1956; 27(196):1–2.

Alves NO, Oshiro JA, Silva YC, Pacher GC, Casaril AE, Rizk YS, Uehara SN, Paniago AM, Andrade IL, Arruda CC, Oliveira AG. VL-HIV co-infection with *Leishmania* containing skin lesions resembling para-kala-azar dermal leishmaniasis. PLOS Neglected Tropical Diseases. 2024 Aug 26;18(8):e0012438.

Amato VS, Tuon FF, Bacha HA, Neto VA, Nicodemo AC. Mucosal leishmaniasis: current scenario and prospects for treatment. Acta Tropica. 2008 Jan 1;105(1):1–9.

Amato VS, Tuon FF, Siqueira AM, Nicodemo AC, Neto VA. Treatment of mucosal leishmaniasis in Latin America: Systematic review. American Journal of Tropical Medicine and Hygiene. 2007 Aug 1;77(2):266–74.

Anstead GM, Chandrasekar B, Zhao W, Yang JU, Perez LE, Melby PC. Malnutrition alters the innate immune response and increases early visceralization following *Leishmania donovani* infection. Infection and Immunity. 2001 Aug 1;69(8): 4709–18.

Aronson N, Herwaldt BL, Libman M, Pearson R, Lopez-Velez R, Weina P, Carvalho EM, Ephros M, Jeronimo S, Magill A. Diagnosis and treatment of leishmaniasis: Clinical practice guidelines by the Infectious Diseases Society of America (IDSA) and the American Society of Tropical Medicine and Hygiene (ASTMH). Clinical Infectious Diseases. 2016 Dec 15;63(12):e202–64.

Badaro R, Jones TC, Lorenco R, Cerf BJ, Sampaio D, Carvalho EM, Rocha H, Teixeira R, Johnson Jr WD. A prospective study of visceral leishmaniasis in an endemic area of Brazil. Journal of Infectious Diseases. 1986 Oct 1;154(4): 639–49.

Barral A, Badaro R, Barral-Netto M, Grimaldi Jr G, Momem H, Carvalho EM. Isolation of *Leishmania mexicana amazonensis* from the bone marrow in a case of American visceral leishmaniasis. The American Journal of Tropical Medicine and Hygiene. 1986 Jul 1;35(4):732–4.

Barral A, Guerreiro J, Bomfim G, Correia D, Barral-Netto M, Carvalho EM. Lymphadenopathy as the first sign of human cutaneous infection by *Leishmania braziliensis*. The American Journal of Tropical Medicine and Hygiene. 1995 Sep 1; 53(3):256–9.

Berlin C. Leishmaniasis recidiva cutis; leishmanid. Archives of Dermatology and Syphilology. 1940 May 1;41(5):874–86.

Bern C, Haque R, Chowdhury R, Ali M, Kurkjian KM, Vaz L, Amann J, Wahed MA, Wagatsuma Y, Breiman RF, Williamson J. The epidemiology of visceral leishmaniasis and asymptomatic leishmanial infection in a highly endemic Bangladeshi village. American Journal of Tropical Medicine and Hygiene. 2007 May 1;76(5):909–14.

Blackwell JM, Fakiola M, Ibrahim ME, Jamieson SE, Jeronimo SB, Miller EN, Mishra A, Mohamed HS, Peacock CS, Raju M, Sundar S. Genetics and visceral leishmaniasis: Of mice and man. Parasite Immunology. 2009 May;31(5):254–66.

Bomfim G, Nascimento C, Costa J, Carvalho EM, Barral-Netto M, Barral A. Variation of cytokine patterns related to therapeutic response in diffuse cutaneous leishmaniasis. Experimental Parasitology. 1996 Nov 1;84(2):188–94.

Bowles DE, Britch S, Linthicum K, Johnson RN, Linton YM, White GB. Sand flies: Significance, surveillance, and control in contingency operations. Silver Spring: United States Department of Defense Armed Forces Pest Management Board. 2015:1–65.

Bryceson AD. Diffuse cutaneous leishmaniasis in Ethiopia I. The clinical and histological features of the disease. Transactions of the Royal Society of Tropical Medicine and Hygiene. 1969 Jan 1;63(6):708–37.

Bueno GC, Koerich AT, Burg LB, Kretzer SL, Moral JÂ, Pereira IA. Visceral leishmaniasis mimicking systemic lupus erythematosus. Revista da Sociedade Brasileira de Medicina Tropical. 2019 Feb 21;52:e20180208.

Calvopina M, Gomez EA, Sindermann H, Cooper PJ, Hashiguchi Y. Relapse of new world diffuse cutaneous leishmaniasis caused by *Leishmania* (*Leishmania*) *mexicana* after miltefosine treatment. American Journal of Tropical Medicine and Hygiene. 2006;75(6):1074–7.

Carvalho EM, Barral A, Costa JM, Bittencourt A, Marsden P. Clinical and immunopathological aspects of disseminated cutaneous leishmaniasis. Acta Tropica. 1994 Apr 1;56(4):315–25.

Cerf BJ, Jones TC, Badaro R, Sampaio D, Teixeira R, Johnson WD. Malnutrition as a risk factor for severe visceral leishmaniasis. The Journal of Infectious Diseases. 1987 Dec 1;156(6):1030–3.

Convit J, Kerdel-Vegas F, Gordon B. Disseminated anergic cutaneous leishmaniasis. British Journal of Dermatology. 1962 Apr 1;74(4):132–5.

Convit J, Pinardi ME, Rondon AJ. Diffuse cutaneous leishmaniasis: A disease due to an immunological defect of the host. Transactions of the Royal Society of Tropical Medicine and Hygiene. 1972 Jan 1;66(4):603–10.

Costa JM, Marsden PD, Llanos-Cuentas EA, Netto EM, Carvalho EM, Barral A, Rosa AC, Cuba CC, Magalhaes AV, Barreto AC. Disseminated cutaneous leishmaniasis in a field clinic in Bahia, Brazil: A report of eight cases. The Journal of Tropical Medicine and Hygiene. 1986 Dec 1;89(6):319–23.

Couppie P, Clyti E, Sainte-Marie D, Dedet JP, Carme B, Pradinaud R. Disseminated cutaneous leishmaniasis due to *Leishmania guyanensis*: Case of a patient with 425 lesions. The American Journal of Tropical Medicine and Hygiene. 2004 Nov 1;71(5):558–60.

Daher EF, Lima LL, Vieira AP, Nascimento LS, Soares DS, Abreu KL, Zuntini KL, Junior GB. Hemophagocytic syndrome in children with visceral leishmaniasis. The Pediatric Infectious Disease Journal. 2015 Dec 1;34(12):1311–4.

Das NK, Mukhopadhyay D, Datta PK, Chatterjee M. Post kala-azar dermal leishmaniasis: An update. Recent Advances in Dermatology. 2014 Feb 28;3:154–75.

de Santana Ferreira E, de Souza Júnior VR, de Oliveira JF, Costa MF, da Conceição de Barros Correia M, de Sá AF. Rare association of consumptive coagulopathy in visceral leishmaniasis: A case report. Tropical Doctor. 2021 Jan;51(1):120–2.

Debono JE. Kala-azar in infancy. Proceedings of the Royal Society of Medicine. 1947 Feb;40(4):155–9.

Duarte DB, Lacerda MC, Ribeiro YJ, Ribeiro MZ, Frederico MD, Oliveira MJ. Kidney biomarkers in tropical infections: An update. Pathogens and Global Health. 2020 Aug 17;114(6):302–8.

Duthie MS, Goto Y, Ghosh P, Mondal D. Impact of sequelae of visceral leishmaniasis and their contribution to ongoing transmission of *Leishmania donovani*. Pathogens and Disease. 2019 Aug;77(6):ftz057.

Ejghal R, Bekhti K, Lemrani M. The sporadic cutaneous leishmaniasis due to *Leishmania infantum* in Morocco: A presumably trend towards endemicity. Acta Tropica. 2022 Mar 1;227:106288.

El Hajj L, Thellier M, Carrière J, Bricaire F, Danis M, Caumes E. Localized cutaneous leishmaniasis imported into Paris: A review of 39 cases. International Journal of Dermatology. 2004 Feb;43(2):120–5.

El Hassan AM, Ghalib HW, Zijlstra EE, Eltoum IA, Satti M, Ali MS, Ali HM. Post kala-azar dermal leishmaniasis in the Sudan: Clinical features, pathology and treatment. Transactions of the Royal Society of Tropical Medicine and Hygiene. 1992 May 1;86(3):245–8.

El Hassan AM, Khalil EA, Elamin WM, El Hassan LA, Ahmed ME, Musa AM. Misdiagnosis and mistreatment of post-kala-azar dermal leishmaniasis. Case Reports in Medicine. 2013;2013(1):351579.

Franke ED, Wignall FS, Cruz ME, Rosales E, Tovar AA, Lucas CM, Llanos-Cuentas A, Berman JD. Efficacy and toxicity of sodium stibogluconate for mucosal leishmaniasis. Annals of Internal Medicine. 1990 Dec 15;113(12):934–40.

Ganguly S, Das NK, Barbhuiya JN, Chatterjee M. Post-kala-azar dermal leishmaniasis – An overview. International Journal of Dermatology. 2010 Aug;49(8):921–31.

Girgla HS, Marsden RA, Singh GM, Ryan TJ. Post-kala-azar dermal leishmaniasis. British Journal of Dermatology. 1977 Sep;97(3):307–11.

Goto H, Lindoso JA. Current diagnosis and treatment of cutaneous and mucocutaneous leishmaniasis. Expert Review of Anti-infective Therapy. 2010 Apr 1; 8(4):419–33.

Grimaldi Jr G, Tesh RB, McMahon-Pratt D. A review of the geographic distribution and epidemiology of leishmaniasis in the New World. The American Journal of Tropical Medicine and Hygiene. 1989 Dec 1;41(6):687–725.

Hasan MM, Proma SB, Hossain MS, Arifuzzaman M, Islam N, Siddique MA, Amiruzzaman. A case report on para-kala-azar dermal leishmaniasis: An unresolved mystery. BMC Infectious Diseases. 2023 Dec 18;23(1):885.

Hashiguchi Y, Gomez EL, Kato H, Martini LR, Velez LN, Uezato H. Diffuse and disseminated cutaneous leishmaniasis: Clinical cases experienced in Ecuador and a brief review. Tropical Medicine and Health. 2016 Dec;44:1–9.

Hochedez P, Caumes E. Common skin infections in travelers. Journal of Travel Medicine. 2008 Jul 1;15(4):252–62.

Ivonise F, Cibele A, Olívia B, Clarissa BA, Lucas PC, Roque PA, Edgar MC. Epidemiologic and immunologic findings for the subclinical form of *Leishmania braziliensis* infection. Clinical Infectious Diseases. 2002 Jun 1;34(11):e54–8.

Jeronimo SMB, de Queiroz Sousa A, Pearson RD. Leishmaniasis. In: Tropical Infectious Diseases: Principles, Pathogens and Practice, 3rd ed, Guerrant RL, Walker DH, Weller PF (Eds), Saunders Elsevier, Philadelphia 2011. p.696.

Jha AK, Mallik SK, Ganguly S, Jaykar KC. An unusual case of multiple erythematous nodules with ulcerative lesion. Indian Journal of Dermatology, Venereology and Leprology. 2012 Mar 1;78:229.

Jones TC, Johnson Jr WD, Barretto AC, Lago E, Badaro R, Cerf B, Reed SG, Netto EM, Tada MS, Franca F, Wiese K. Epidemiology of American cutaneous leishmaniasis due to *Leishmania braziliensis brasiliensis*. Journal of Infectious Diseases. 1987 Jul 1;156(1):73–83.

Karplus TM, Jeronimo SM, Chang H, Helms BK, Burns TL, Murray JC, Mitchell AA, Pugh EW, Braz RF, Bezerra FL, Wilson ME. Association between the tumor necrosis factor locus and the clinical outcome of *Leishmania chagasi* infection. Infection and Immunity. 2002 Dec;70(12):6919–25.

Karunaweera ND, Pratlong F, Siriwardane HV, Ihalamulla RL, Dedet JP. Sri Lankan cutaneous leishmaniasis is caused by *Leishmania donovani zymodeme* MON-37. Transactions of the Royal Society of Tropical Medicine and Hygiene. 2003 Jul 1;97(4):380–1.

Kumar P, Chatterjee M, Das NK. Post kala-azar dermal leishmaniasis: Clinical features and differential diagnosis. Indian Journal of Dermatology. 2021 Jan 1; 66(1):24–33.

Kumar R, Bumb RA, Ansari NA, Mehta RD, Salotra P. Cutaneous leishmaniasis caused by *Leishmania tropica* in Bikaner, India: Parasite identification and characterization using molecular and immunologic tools. The American Journal of Tropical Medicine and Hygiene. 2007 May 1;76(5):896–901.

Lessa HA, Machado P, Lima F, Cruz AA, Bacellar O, Guerreiro J, Carvalho EM. Successful treatment of refractory mucosal leishmaniasis with pentoxifylline plus antimony. The American Journal of Tropical Medicine and Hygiene. 2001 Aug 1;65(2):87–9.

Libório AB, Rocha NA, Oliveira MJ, Franco LF, Aguiar GB, Pimentel RS, Abreu KL, Silva Jr GB, Daher EF. Acute kidney injury in children with visceral leishmaniasis. The Pediatric Infectious Disease Journal. 2012 May 1;31(5):451–4.

Lindoso JA, Moreira CH, Celeste BJ, Oyafuso LK, Folegatti PM, Zijlstra EE. Para-kala-azar dermal leishmaniasis in a patient in Brazil: A case report. Revista da Sociedade Brasileira de Medicina Tropical. 2018;51(01):105–7.

Lucas CM, Franke ED, Cachay MI, Tejada A, Cruz ME, Kreutzer RD, Barker DC, McCann SH, Watts DM. Geographic distribution and clinical description of leishmaniasis cases in Peru. The American Journal of Tropical Medicine and Hygiene. 1998 Aug;59(2):312–7.

Machado PR, Rosa ME, Costa D, Mignac M, Silva JS, Schriefer A, Teixeira MM, Bacellar O, Carvalho EM. Reappraisal of the immunopathogenesis of disseminated leishmaniasis: In situ and systemic immune response. Transactions of the Royal Society of Tropical Medicine and Hygiene. 2011 Aug 1;105(8):438–44.

Machado-Coelho GL, Caiaffa WT, Genaro O, Magalhaes PA, Mayrink W. Risk factors for mucosal manifestation of American cutaneous leishmaniasis. Transactions of the Royal Society of Tropical Medicine and Hygiene. 2005 Jan 1; 99(1):55–61.

Magill AJ, Grogl M, Gasser Jr RA, Sun W, Oster CN. Visceral infection caused by *Leishmania tropica* in veterans of Operation Desert Storm. New England Journal of Medicine. 1993 May 13;328(19):1383–7.

Mans DR, Kent AD, Hu RV, Schallig HD. Epidemiological, biological and clinical aspects of leishmaniasis with special emphasis on Busi Yasi in Suriname. Journal of Clinical & Experimental Dermatology Research. 2017 Mar 20;8(2):e1000388.

Marsden PD. Mucocutaneous leishmaniasis. British Medical Journal. 1990 Sep 9; 301(6753):656.

Marsden PD. Mucosal leishmaniasis ("espundia" Escomel, 1911). Transactions of the Royal Society of Tropical Medicine and Hygiene. 1986 Jan 1;80(6):859–76.

Mishra P, Dixit A, Chatterjee T, Bhattacharya M, Bhattacharya J, Dutta P, Mahapatra M, Prasad Pati H, Choudhry VP, Saxena R. Disseminated intravascular coagulation as an unusual presentation of Kala-azar: Report of two cases. Scandinavian Journal of Infectious Diseases. 2004 Jul 1;36(6–7):519–21.

Momeni AZ, Aminjavaheri M. Treatment of recurrent cutaneous leishmaniasis. International Journal of Dermatology. 1995 Feb;34(2):129–33.

Moniruzzaman M, Sayeed SJ, Rahim MA, Hassan R, Rahman MM. Para kala-azar dermal leishmaniasis: A case report. Cureus. 2023 Jan;15(1):e33701.

Monroy-Ostria A, Hernandez-Montes O, Barker DC. Aetiology of visceral leishmaniasis in Mexico. Acta Tropica. 2000 Mar 25;75(2):155–61.

Moral L, Rubio EM, Moya M. A leishmanin skin test survey in the human population of l'Alacanti region (Spain): Implications for the epidemiology of *Leishmania infantum* infection in southern Europe. Transactions of the Royal Society of Tropical Medicine and Hygiene. 2002 Mar 1;96(2):129–32.

Napier LE, Gupta CD. A clinical study of post-kala-azar dermal leishmaniasis. The Indian Medical Gazette. 1930 May;65(5):249.

Napier LE, Gupta CD. Further clinical observations on post-kala-azar dermal leishmaniasis. The Indian Medical Gazette. 1934 Mar;69(3):121.

Oliveira Neto MP, Grimaldi Junior G, Momen H, Pacheco RS, Marzochi MC, McMahon Pratt D. Active cutaneous leishmaniasis in Brazil, induced by *Leishmania donovani chagasi*. Memórias do Instituto Oswaldo Cruz. 1986;81: 303–9.

Osorio LE, Castillo CM, Ochoa MT. Mucosal leishmaniasis due to *Leishmania (Viannia) panamensis* in Colombia: Clinical characteristics. The American Journal of Tropical Medicine and Hygiene. 1998 Jul;59(1):49–52.

Pearson RD, de Queiroz Sousa A. Clinical spectrum of Leishmaniasis. Clinical Infectious Diseases. 1996 Jan;22(1):1–1.

Pedrosa CM, Ximenes RA, Almeida WA, Rocha EM. Validity of the polymerase chain reaction in the diagnosis of clinically suspected cases of American visceral leishmaniasis. Brazilian Journal of Infectious Diseases. 2013;17:319–23.

Petersen EA, Neva FA, Oster CN, Diaz HB. Specific inhibition of lymphocyte-proliferation responses by adherent suppressor cells in diffuse cutaneous leishmaniasis. New England Journal of Medicine. 1982 Feb 18;306(7):387–92.

Piyaraj P, Bualert L, Kalrat A, Leelayoova S, Ruang-Areerate T, Theprin N, Naaglor T, Mungthin M. Asymptomatic *Leishmania* infection among blood donors in a Southern Province of Thailand. The American Journal of Tropical Medicine and Hygiene. 2024 Aug 13;111(4):804–13.

Queiroz A, Sousa R, Heine C, Cardoso M, Guimarães LH, Machado PR, Carvalho EM, Riley LW, Wilson ME, Schriefer A. Association between an emerging disseminated form of leishmaniasis and *Leishmania (Viannia) braziliensis* strain polymorphisms. Journal of Clinical Microbiology. 2012 Dec;50(12):4028–34.

Rajagopala S, Dutta U, Chandra KP, Bhatia P, Varma N, Kochhar R. Visceral leishmaniasis associated hemophagocytic lymphohistiocytosis – Case report and systematic review. Journal of Infection. 2008 May 1;56(5):381–8.

Ramesh V, Mukherjee A. Post-kala-azar dermal leishmaniasis. International Journal of Dermatology. 1995 Feb 1;34(2):85–91.

Reithinger R, Mohsen M, Leslie T. Risk factors for anthroponotic cutaneous leishmaniasis at the household level in Kabul, Afghanistan. PLoS Neglected Tropical Diseases. 2010 Mar 23;4(3):e639.

Sacks DL, Kenney RT, Neva FA, Kreutzer RD, Jaffe CL, Gupta AK, Sharma MC, Sinha SP, Saran R. Indian kala-azar caused by *Leishmania tropica*. The Lancet. 1995 Apr 15;345(8955):959–61.

Salaiza-Suazo N, Volkow P, Pérez Tamayo R, Moll H, Gillitzer R, Pérez-Torres A, Pérez-Montfort R, Domínguez JD, Velasco-Castrejón O, Crippa M, Becker I. Treatment of two patients with diffuse cutaneous leishmaniasis caused by *Leishmania mexicana* modifies the immunohistological profile but not the disease outcome. Tropical Medicine & International Health. 1999 Dec;4(12):801–11.

Santrich C, Segura I, Arias AL, Saravia NG. Mucosal disease caused by *Leishmania braziliensis guyanensis*. The American Journal of Tropical Medicine and Hygiene. 1990 Jan 1;42(1):51–5.

Saravia NG, Segura I, Holguin AF, Santrich C, Valderrama L, Ocampo C. Epidemiologic, genetic, and clinical associations among phenotypically distinct populations of *Leishmania (Viannia)* in Colombia. The American Journal of Tropical Medicine and Hygiene. 1998 Jul;59(1):86–94.

Schriefer A, Wilson ME, Carvalho EM. Recent developments leading toward a paradigm switch in the diagnostic and therapeutic approach to human leishmaniasis. Current Opinion in Infectious Diseases. 2008 Oct 1;21(5):483–8.

Schubach A, Haddad F, Neto MP, Degrave W, Pirmez C, Grimaldi Jr G, Fernandes O. Detection of *Leishmania* DNA by polymerase chain reaction in scars of treated human patients. Journal of Infectious Diseases. 1998a Sep 1;178(3):911–4.

Schubach A, Marzochi MC, Cuzzi-Maya T, Oliveira AV, Araujo ML, Oliveira AL, Pacheco RS, Momen H, Conceicao-Silva F, Coutinho SG, Marzochi KB. Cutaneous scars in American tegumentary leishmaniasis patients: A site of *Leishmania (Viannia) braziliensis* persistence and viability eleven years after antimonial therapy and clinical cure. The American Journal of Tropical Medicine and Hygiene. 1998b Jun;58(6):824–7.

Schwartz E, Hatz C, Blum J. New world cutaneous leishmaniasis in travellers. The Lancet Infectious Diseases. 2006 Jun 1;6(6):342–9.

Scorza BM, Carvalho EM, Wilson ME. Cutaneous manifestations of human and murine leishmaniasis. International Journal of Molecular Sciences. 2017 Jun 18; 18(6):1296.

Silveira FT, Campos MB, Müller SF, Ramos PK, Lima LV, dos Santos TV, Gomes CM, Laurenti MD, da Matta VL, Corbett CE. From biology to disease: Importance of Species-Specific *Leishmania* Antigens from the Subgenera *Viannia* (*L. braziliensis*) and *Leishmania* (*L. amazonensis*) in the Pathogenesis of American Cutaneous Leishmaniasis. In: Leishmania Parasites-Epidemiology, Immunopathology and Hosts. Almeida-Souza F, Cardoso FO, Abreu-Silva AL, Calabrese KS (Eds), IntechOpen, London 2023. https://doi.org/10.5772/intechopen.108967

Silveira FT, Lainson R, Corbett CE. Clinical and immunopathological spectrum of American cutaneous leishmaniasis with special reference to the disease in Amazonian Brazil: A review. Memórias do Instituto Oswaldo Cruz. 2004;99: 239–51.

Singh OP, Hasker E, Sacks D, Boelaert M, Sundar S. Asymptomatic *Leishmania* infection: A new challenge for *Leishmania* control. Clinical Infectious Diseases. 2014 May 15;58(10):1424–9.

Strick RA, Borok M, Gasiorowski HC. Recurrent cutaneous leishmaniasis. Journal of the American Academy of Dermatology. 1983 Sep 1;9(3):437–43.

Tunccan OG, Tufan A, Telli G, Akyürek N, Pamukçuoğlu M, Yılmaz G, Hızel K. Visceral leishmaniasis mimicking autoimmune hepatitis, primary biliary cirrhosis, and systemic lupus erythematosus overlap. The Korean Journal of Parasitology. 2012 Jun;50(2):133.

Turetz ML, Machado PR, Ko AI, Alves F, Bittencourt A, Almeida RP, Mobashery N, Johnson Jr WD, Carvalho EM. Disseminated leishmaniasis: A new and emerging form of leishmaniasis observed in northeastern Brazil. The Journal of Infectious Diseases. 2002 Dec 15;186(12):1829–34.

van Griensven J, Gadisa E, Aseffa A, Hailu A, Beshah AM, Diro E. Treatment of cutaneous leishmaniasis caused by *Leishmania aethiopica*: A systematic review. PLoS Neglected Tropical Diseases. 2016 Mar 3;10(3):e0004495.

Varma N, Naseem S. Hematologic changes in visceral leishmaniasis/kala azar. Indian Journal of Hematology and Blood Transfusion. 2010 Sep;26:78–82.

Voulgarelis M, Voulgari PV, Serelis J, Drosos AA, Skopouli FN. Visceral leishmaniasis resembling systemic lupus erythematosus. Clinical Rheumatology. 2003 Dec;22:452–5.

World Health Organization. Control of the Leishmaniases. World Health Organization technical report series 949, 2010. p.xii.

Zerpa O, Ulrich M, Blanco B, Polegre M, Avila A, Matos N, Mendoza I, Pratlong F, Ravel C, Convit J. Diffuse cutaneous leishmaniasis responds to miltefosine but then relapses. British Journal of Dermatology. 2007 Jun 1;156(6): 1328–35.

Zhang WW, Ramasamy G, McCall LI, Haydock A, Ranasinghe S, Abeygunasekara P, Sirimanna G, Wickremasinghe R, Myler P, Matlashewski G. Genetic analysis of *Leishmania donovani* tropism using a naturally attenuated cutaneous strain. PLoS Pathogens. 2014 Jul 3;10(7):e1004244.

Ziaei H, Sadeghian G, Hejazi SH. Distribution frequency of pathogenic bacteria isolated from cutaneus leishmaniasis lesions. The Korean Journal of Parasitology. 2008 Sep;46(3):191.

Zijlstra EE, Alves F, Rijal S, Arana B, Alvar J. Post-kala-azar dermal leishmaniasis in the Indian subcontinent: A threat to the South-East Asia Region Kala-azar Elimination Programme. PLoS Neglected Tropical Diseases. 2017 Nov 16;11(11): e0005877.

Zijlstra EE, El-Hassan AM. Leishmanin and tuberculin sensitivity in leishmaniasis in the Sudan, with special reference to kala-azar. Transactions of the Royal Society of Tropical Medicine and Hygiene. 1993 Jul 1;87(4):425–7.
Zijlstra EE, Khalil EA, Kager PA, El-Hassan AM. Post-kala-azar dermal leishmaniasis in the Sudan: Clinical presentation and differential diagnosis. British Journal of Dermatology. 2000 Jul 1;143(1):136–43.
Zijlstra EE, Musa AM, Khalil EA, El Hassan IM, El-Hassan AM. Post-kala-azar dermal leishmaniasis. The Lancet Infectious Diseases. 2003 Feb 1;3(2):87–98.
Zijlstra EE. PKDL and other dermal lesions in HIV co-infected patients with leishmaniasis: Review of clinical presentation in relation to immune responses. PLoS Neglected Tropical Diseases. 2014 Nov 20;8(11):e3258.

Chapter 5

Genetic and molecular basis: Diversity and virulence mechanisms of *Leishmania*

Shobha Kumari and Vikash Kumar

Department of Biochemistry, ICMR-Rajendra Memorial Research Institute of Medical Sciences, Agamkuan, Patna, India

Department of Microbiology, University of Calcutta, Kolkata, India

Ravi Ranjan

Department of Biochemistry, ICMR-Rajendra Memorial Research Institute of Medical Sciences, Agamkuan, Patna, India

Department of Biochemistry, University of Calcutta, Kolkata, India

*Pritam Narayan Singh and Dayakar Alti**

Department of Immunology, ICMR-Rajendra Memorial Research Institute of Medical Sciences, Agamkuan, Patna, India

5.1 INTRODUCTION

Leishmaniasis is a neglected tropical disease for which over a billion people are at risk; it is caused by digenetic protozoan *Leishmania* parasites. Leishmaniasis claims a large annual burden: an estimated 700,000 to 1 million new cases and up to 30,000 deaths. The disease is caused by parasites included in more than 20 species of *Leishmania* and occurs worldwide with the great majority occurring from tropical to subtropical climates (WHO, 2023). There are different clinical forms named cutaneous leishmaniasis (CL), mucocutaneous leishmaniasis (MCL), visceral leishmaniasis (VL) also known as kala-azar, and post-kala-azar dermal leishmaniasis (PKDL) which is the most severe and potentially fatal form (WHO, 2023). We highlight the genetic and molecular complexity of the parasite in our consideration of it as an organism whose diverse phenotypes are not simply a consequence of host or environmental variation. More than 20 *Leishmania* species are known as human pathogens. These comprise *Leishmania donovani*, *Leishmania*

* Corresponding author: Dr. Dayakar Alti, Scientist-C, dayakar.alti@nic.in, dayakar.pcu@gmail.com, 91-6305523657.

DOI: 10.1201/9781003596455-5

infantum, *Leishmania major*, *Leishmania tropica*, and *Leishmania braziliensis* and more so in case of New World Leishmaniasis, specifically Brazil where there are nine different species reported. This has facilitated the adaption of each species to unique sandfly vectors, mammalian hosts, and ecological niches across broad geographical regions (Mann et al., 2021). They also have a digenetic life cycle and similar basic morphological features but differ widely in their respective virulence, immunopathology, vector specificity, tissue tropism, chemotherapy response, and disease epidemiology (Ait Maatallah et al., 2022; Hanau et al., 2025).

Leishmania parasites are transmitted by phlebotomine sandflies and exhibit a remarkable ability to thrive and survive within two vastly different hosts: the insect vector and the mammalian phagocyte. *Leishmania* parasites exhibit extensive biological and clinical diversity, which is underpinned by their unique genomic and molecular characteristics. Unlike most eukaryotes, *Leishmania* parasites possess a complex genome architecture characterized by high levels of gene duplication, chromosomal aneuploidy, and polycistronic transcription (Black et al., 2023). This, coupled with its canonical transcriptional control mechanisms, allows the parasite an exceptional degree of genomic plasticity. That plasticity supports rapid response to selection, including drug treatment and immune responses from the host as well as environmental factors (Kamran et al., 2023; Black et al., 2023). Furthermore, the recent advents in molecular techniques such as next-generation sequencing (NGS), single-cell genomics, transcriptomics, and proteomic approaches have significantly improved our understanding of how genomic architecture and post-transcriptional regulation together contribute to functional diversity displayed by *Leishmania*.

Because understanding this diversity at the genetic and molecular level provides insight into the evolutionary success of a parasite, as vaccines are developed, there is an urgent need to grasp knowledge about such genotype–phenotype interactions in order to achieve globally effective or even durable vaccination/therapeutic strategies. Parasite genetics shape antigenicity and drug susceptibility profiles, as well as drive host–pathogen interactions. This is important for designing sustainable control measures. Examples include polymorphisms in drug transporter genes leading to resistance and changes in surface glyocproteins like GP63 and lipophosphoglycan (LPG) modulating immune evasion and virulence (Silva-Almeida et al., 2012; Moncada-Diaz et al., 2024). Hybridization, recombination, and clonal evolution have been increasingly considered a source of diversity both within species and between different .... Naturally occurring hybrids for the first time illustrated in nature producing new phenotypes and virulence profiles (Späth et al., 2025) provide compelling new reasons to reassess previous views about the mitotic propagation of *Leishmania*.

Here we summarise the molecular mechanisms associated with the diversity of *Leishmania* (Figure 5.1), including genome organization, mechanisms of genetic variation, and species-specific molecular signatures, and

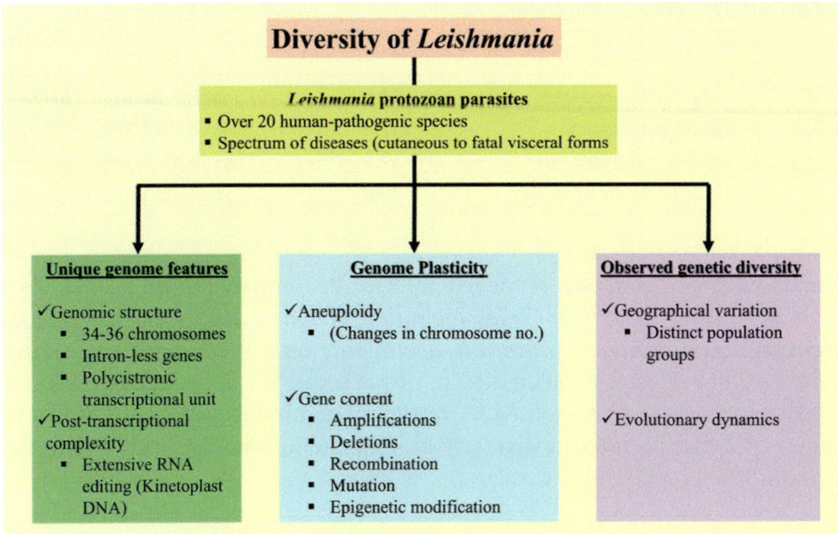

Figure 5.1 Factors influencing diversity in the genus *Leishmania*.

discuss their implications to parasite adaptability, pathogenesis, and control. Recent insights from large-scale genomic surveys and functional studies have advanced our understanding of the evolutionary and clinical relevance of genetic plasticity in this complex genus. We bring information on evolutionary genetics, molecular biology, and immune-parasitology together to give an overall vision of *Leishmania's* unique genetic characteristics that account for its remarkable biological versatility and widespread clinical spectrum. This knowledge is crucial for confronting the challenges of disease burden, drug resistance, in endemic regions and vaccine development.

5.2 GENOMIC ARCHITECTURE: THE FOUNDATION OF DIVERSITY

The genome of *Leishmania* is a paradigm of evolution providing stability and creating diversity at the same time. It is relatively small and around 32–35 Mb that separated into 34–36 chromosomes (Table 5.1) (Wincker et al., 1996). These chromosomes vary in size from about 0.3–2.7 Mb (Kazemi, 2011). Early sequencing projects of *L. major* (*Friedlin strain*), *L. infantum*, and *L. braziliensis* revealed a high level of conservation of gene order and content on chromosomes across different species, yet there are notable differences in gene content and repetitive elements (Ivens et al., 2005; Peacock et al., 2007). Also, to observe the diversity in the genus *Leishmania*, we performed a small exercise, where we selected a well-established marker, kinetoplast DNA (kDNA), to identify the presence in

Table 5.1 Diversity among common *Leishmania* spp. with different genome sizes

S. no.	*Leishmania species*	*Genome size (Mb)*	*Reference*
1.	*Leishmania donovani*	31.5	Lypaczewski et al. (2018)
2.	*Leishmania infantum*	32.8	Baig et al. (2009)
3.	*Leishmania braziliensis*	32.5	Lizzi et al. (2014)
4.	*Leishmania amazonensis*	30.8	Goes et al. (2023)

Leishmania. Further, we retrieved nucleotide sequences for kDNA from different species of *Leishmania*, including *chagasi*, *amazonensis*, *tropica*, *tarentolae*, *guliki*, *major*, *infantum*, *donovani*, *aethiopica*, and *panamensis*. These sequences were subjected to multiple sequence alignment (MSA) and a phylogenetic tree based on MSA results. The phylogenetic tree shows ten species of *Leishmania* and diversity among them, based on the kinetoplast nucleotide sequence (Figure 5.2).

5.2.1 Intron-less genes

In contrast to higher eukaryotes, *Leishmania* genes exhibit a remarkable absence of introns. The majority of protein-coding genes are present as uninterrupted open reading frames (ORFs), devoid of intervening sequences.

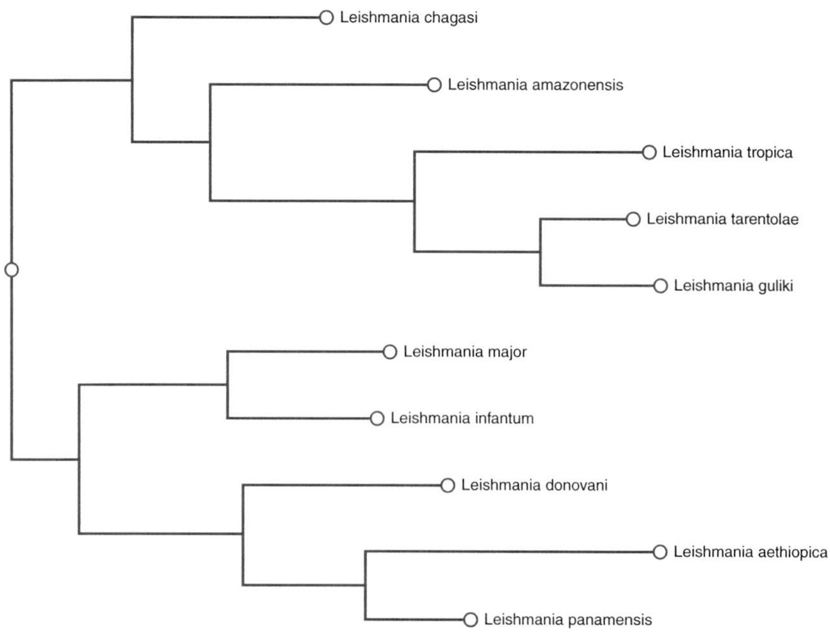

Figure 5.2 The phylogenetic tree representing the diversity among common *Leishmania* spp.

This intron-free organization facilitates faster gene transcription, simpler post-transcriptional regulation, and reduced genomic complexity (Fernandes et al., 1994). However, it also means that for regulating gene expression, *Leishmania* heavily relies on post-transcriptional and translational mechanisms, since classic transcriptional regulation by promoters and enhancers is largely absent (Müller, 2009).

5.2.2 Polycistronic transcription and post-transcriptional regulation

Unlike the majority of eukaryotes, *Leishmania* genes are structured into extensive, strand-specific polycistronic transcriptional units (PTUs). Genes within an array lack distinct promoters and undergo transcription. Consequently, post-transcriptional regulation, encompassing mRNA stability and translation efficiency, emerges as the primary control point for gene expression. These long transcripts are then processed to mature monocistronic mRNA by trans-splicing and polyadenylation. Trans-splicing introduces a spliced leader (SL) RNA at the 5′ end of each mRNA, while polyadenylation defines the 3′ end (Miller et al., 1986). This transcriptional organization affords flexibility to post-transcriptional mechanisms, such as RNA stability, translation efficiency, and RNA-binding proteins, thereby facilitating gene expression regulation (Clayton, 2019). In this context, polycistronic transcription feeds genetic diversity by uncoupling gene from regulation, which defines an opportunity to adapt through post-transcriptional modifications (Monnerat et al., 2004; Grünebast et al., 2025).

5.2.3 Post-transcriptional complexity

Mitochondrial DNA network found only in *Leishmania* is known as k (kinetoplast) DNA. It is made up of various minicircles and a small number of maxicircles (Clayton, 2003). RNA editing affects most transcripts encoded by kDNA minicircle. Minicircle-encoded short guide RNAs (gRNAs) dictates the site-specific insertion or deletion of uridylates into mitochondrial mRNAs. It results in the edition of new ORFs that are expressed due to alternative splicing, which were absent in the original DNA sequence. It is used to enhance the coding potential and Plasticity of the mitochondrial genes, by RNA editing. This allows the parasite to fitfully tune its production of ATP to different environments (sandfly vector, mammalian host). Conservation of these features identifies intron-light species, polycistronic transcription, genome flexibility, and complex post-transcriptional RNA editing as molecular hallmarks of this exciting group. Such a scheme lays the molecular foundation for *Leishmania's* genetic and graceful adaptiveness, as well as played an intrinsic role in its pathogenicity and epidemiological success (Shapira et al., 2001).

5.3 DOGMA OF *LEISHMANIA* GENETIC DIVERSITY

5.3.1 Genomic plasticity

One of the most striking features of *Leishmania* biology is its exceptional genomic plasticity. This property helps the parasite to easily acclimate with varied surroundings including insect vector and mammalian host (Bharadava et al., 2024). This plasticity is predominantly driven by several molecular mechanisms: aneuploidy and variation in copy number, gene amplification, gene deletion, genetic recombination, and mutation, each contributing significantly to the shaping of the parasite's genetic and phenotypic diversity.

5.3.2 Aneuploidy and copy number variation

Leishmania chromosomes are extensively polysomic and contribute greatly to genetic diversity. Aneuploidy results from cells often containing different numbers of copies of chromosomes. This is not a glitch but actually a defining mechanism in their means to survive (Mannaert et al., 2012). Distinct aneuploidy profiles are observed among various strains, even clones. Copy number variation (CNV) influences gene dosage and expression levels. More copies of chromosomes can, for instance, make drug resistance worse and help parasites survive stress or develop. This type of aneuploidy drives asexual reproduction on adaptive planes (Rogers et al., 2011; Mannaert et al., 2012).

5.3.3 Gene amplification

Gene amplification in *Leishmania* frequently entails the increase of the copy number of specific genes or chromosomal regions. A well-characterized instance is the amplification of the MRPA gene (multidrug resistance protein A), which encodes an ATP-binding cassette (ABC) transporter (Moreira et al., 2013). This gene was found to be amplified frequently in clinical antimonial resistance strains of *L. donovani*, leading and enabling the parasite to sequester toxic trivalent antimony out of the cell (Frézard et al., 2014). Preferential gene amplification was reported to activate aquaglyceroporin 1 (AQP1) for antimonial entry and drug sensitivity. Thereby resistance or deletion of AQP1 gene confers on resistance but amplification increases sensitivity to drug (Maharjan et al., 2008). *L. infantum* contains a number of amplification loci, for instance, H-locus (chromosome 23) whose circular DNS is being amplified upon drug exposure like methotrexate or antimonials (Chiquero et al., 1994). These amplicons are frequently overlap regions of redox-stable motifs and regions containing genes associated with metabolism or virulence, which could play a role in the parasite's coping mechanisms when exposed to stressful conditions.

5.3.4 Gene deletion

The deletion of specific genes is, like the other factors mentioned above, important for shaping the *Leishmania* genome. For example, deletions of the

A2 gene cluster results in inefficient mucocutaneous and visceralization with *L. donovani*. This results in a non-VL causing phenotype. Instead, it causes small skin lesions (Zhang and Matlashewski, 2001). Conversely, deletion of surface glycoconjugate genes encoding LPG biosynthetic machinery can modify parasite virulence and immune interaction with the host (Azevedo et al., 2020). There is a similar situation in which the gene PIFTC3 was found to be spontaneously deleted in lab-adapted strains of *Leishmania*. It affects the motility and infectivity (Feng et al., 2018). Thus, when we consider these examples, we can see that the loss of genes (deletions) might be a survival strategy or the result of relaxation in selective pressure within non-natural environments.

5.3.5 Genetic recombination

Although infrequent compared with clonal reproduction, genetic recombination has been observed both in experiments and in natural populations. Whole-genome sequencing and multilocus sequence typing (MLST) have confirmed the existence of interspecies hybrids such as *L. major* and *L. infantum* (these are present in areas of Middle East and Central Asia).

These hybrids often exhibit combined characteristics of both parental strains, including altered drug susceptibility and changes in tropism. Some hybrids even demonstrate novel organ-specific infection patterns (Ferreira, 2025). In sandfly vectors, genetic exchange through meiotic-like recombination has been experimentally demonstrated by co-infecting flies with two fluorescently labelled strains. The progeny exhibited recombinant chromosomes, indicating the inheritance of genetic material from both parents (Ferreira and Sacks, 2022). Such recombination events hold particular significance in the generation of mosaic aneuploidy and segmental chromosomal duplications, thereby expanding the genetic diversity of *Leishmania* (Ferreira et al., 2022).

5.3.6 Mutations

Mutations in *Leishmania* encompass both small-scale (point mutations) and large-scale chromosomal alterations. For instance, single-nucleotide mutations in the *Ld*MT gene (Miltefosine transporter) and its accessory subunit Ros3 result in impaired drug uptake and are frequently associated with clinical resistance to Miltefosine in *L. donovani* (Seifert et al., 2007; Espada et al., 2021). Another example is the mutation in the ubiquitin-fusion protein gene in *L. infantum*, which has been associated with in vitro adaptation and loss of virulence in vivo (Coutinho-Abreu et al., 2020). Certain mutations that occur in the genes coding for surface molecules like GP63 or hydrophilic acylated surface proteins (HASPs) can cause antigenic variation. This variation allows the parasite to escape the immune response and leads to a lifelong chronic infection (Doehl et al., 2017; Antonia et al., 2021). Curiously, these alterations are much more common in *Leishmania*, presumably because—like other *trypanosomatids*—it lacks conventional promoter-driven gene regulation, and selection operates predominantly post-transcriptionally and at the level of protein stability. These processes combine to render molecular diversity to *Leishmania*. Such diversity allows the parasite to adapt to different hosts, evade immune pressure, and escape drug treatment. This genetic plasticity emerges as a major conundrum in disease management.

5.3.7 Epigenetic modification

In addition to variations in the genetic factors mentioned above, the contribution of epigenetic changes is also an important aspect that causes heterogeneity at the molecular level in this parasite. Epigenetic changes include histones, chromatin, and regulation by RNA which affect gene expression without changing its primary sequence. It is known that *L. major* does not utilize classical promoters and transcription factors to any large extent but depends mostly upon control of transcription by histone acetylation and methylation (Thomas et al., 2009). In *Leishmania*, acetylation by histone H3 at strand switch regions (SSRs) has been linked to the initiation of transcription, as in *L. major* (Martínez-Calvillo et al., 2004). Similarly, SET-domain histone methylation in parasites is stage-specific as well and this occurs predominantly during the cellular reprogramming by which procyclic epimastigotes are converted to mammalian infectious amastigotes. This facilitates adaptation within the host (Clos et al., 2022). Chromatin remodelling proteins, such as ISWI, also contribute to nucleosome positioning and parasite survival, as observed in *L. braziliensis*. Additionally, while classical RNA interference (RNAi) is absent in most species, the presence of small RNAs and remnants of RNAi machinery in *L. braziliensis* and *Leishmania guyanensis* indicates a restricted role for RNA-mediated gene regulation. These epigenetic modifications, although reversible, contribute to phenotypic heterogeneity, enabling clonal populations to adapt to stress, evade immunity, or resist drugs. This adds a crucial, non-genetic layer to *Leishmania* adaptability (Martínez-Calvillo et al., 2018).

5.4 PRAGMATIC GENETIC DIVERSITY

The parasite *Leishmania* exhibits significant interspecific and intraspecific diversity. This way, the diversity is there between biochemically distinct strains within the same species. *Leishmania* spp. also show variation in chromosomal copy number, sometimes even within a clonal population. Natural interspecies hybrids have also been found, contributing new genotypes and potentially new disease phenotypes. These genetic variations impact the pathogenicity (different strains/species of *Leishmania* cause cutaneous, visceral, or mucocutaneous form of leishmaniasis), drug resistance (different genotypes/strains show variable sensitivity to the drugs), and diverse surface antigens like GP63 of LPG.

5.4.1 Geographical variation

The *Leishmania* species display strong dependency on the geographical structuring shaped by native vector species (Lutzomyia or Phlebotomus sandflies), ecological factors (climate, altitude, urbanization), and reservoir/hosts (Ranjan et al., 2024). For instance, the parasite *L. donovani* is mainly anthroponotic – spread of disease from human to human and spread over the Indian subcontinent and some Eastern African countries, while *L. infantum/Leishmania chagasi* causes VL in zoonotic forms in the Mediterranean and Latin America. Similarly, *L. tropica* and *L. major* mainly cause cutaneous form in the Old World foci (Hashim, 1997). This distribution highlights the disease's prevalence in diverse geographic regions, primarily in tropical and subtropical areas (Table 5.2).

5.4.2 Evolutionary dynamics

Leishmania's evolutionary success is attributed to several distinctive characteristics, including genome plasticity, sexual recombination, mutation, and vector–host coevolution (Patino et al., 2020). Frequent chromosomal CNVs (aneuploidy) and gene amplifications or deletions enable a flexible genome that can adapt swiftly to environmental stressors such as drug resistance (Moody, 2021). Besides expecting it to be clonally disseminated, sexual recombination is a process that had never been shown in its natural vector, a sand fly

Table 5.2 Species-specific diversity in different geographical areas

Disease/clinical form	*Leishmania species*	*Main endemic regions*
CL	*Leishmania major*	Middle East, Afghanistan, North Africa
	Leishmania tropica	Middle East, India, Pakistan, Afghanistan
	Leishmania aethiopica	Ethiopia, Kenya
	Leishmania mexicana	Mexico, Central America, Texas
	Leishmania amazonensis	Brazil, Peru, Bolivia
	Leishmania infantum (rare CL)	Mediterranean basin
	Leishmania donovani (rare CL)	Sudan, Sri Lanka, Kenya
	Leishmania arabica	Saudi Arabia, Yemen
	Leishmania killicki	Tunisia, Algeria
	Leishmania venezuelensis	Venezuela
	Leishmania garnhami	Ecuador
VL	*L. donovani*	India, Bangladesh, Nepal, Sudan, Ethiopia
	L. infantum/Leishmania chagasi	Mediterranean, Middle East, Brazil, Colombia
MCL	*Leishmania braziliensis*	Brazil, Peru, Bolivia, Ecuador
	Leishmania panamensis	Panama, Costa Rica, Colombia
	Leishmania guyanensis	Guyana, Suriname, French Guiana, N Brazil
	Leishmania peruviana	Andean Peru

(Ferreira et al., 2022). Continuous mutations and selection pressures generated by the system, including host immunity or chemotherapy, are constitutive elements of microevolution in parasite populations. Local evolutionary trajectories are determined by the coevolution of vectors and hosts, influenced by sandfly species-specific vectorial competences by adaptation+ of immune system gene regulation in host-derived cells and by ecological barriers (Poulin et al., 2024). These interactions are the dynamics on which *Leishmania* relies to sustain amorphous population structures favouring genetic diversity by which it can perpetuate bottlenecks. These properties of diversification make it highly versatile to respond to drug challenge through rapid development of drug resistance as well as change in virulence traits and adaptability on different ecological niches.

5.5 MOLECULAR MECHANISMS OF LEISHMANIA VIRULENCE

Leishmania species are an example of protozoa that live in two different organisms in their life cycle (the sand fly vector and mammalian hosts). In order to pathology and cause disease, these obligate intracellular parasites have adapted greatly due to the intricate life cycle that they employ. *Leishmania* is an obligate parasite, meaning that it requires its hosts for life and reproduction. To thrive, they have evolved a diverse arsenal of virulence mechanisms that enable them to establish infection, evade host immune responses, and ensure successful transmission. From the moment of transmission through the bite of an infected sand fly, *Leishmania* initiates a series of complex interactions with the host immune system (Figure 5.3). Neutrophils are the first responders in the skin. *Leishmania* promastigotes can induce neutrophils to form neutrophil extracellular traps (NETs), but many parasites escape NET-mediated killing by secreting nucleases (including salivary nucleases from sandflies). For example, *L. infantum* and *L. major* produce nucleases that degrade NET DNA, allowing parasites to avoid this trap. Even when taken up by neutrophils, *Leishmania* delays apoptosis and uses dying neutrophils as "Trojan horses" to infect macrophages (Gupta et al., 2022). Some species (e.g., *Leishmania amazonensis* and *L. donovani*) can survive within neutrophils by resisting granule peptides and oxidative bursts (Séguin et al., 2016). The key surface molecules of *Leishmania*, such as LPG and GP63, help the parasite to resist complement-mediated lysis and facilitate entry into host cells (Gupta et al., 2022). Once inside macrophages, *Leishmania* exhibits resistance to oxidative and nitrosative stress, interferes with phagosome maturation, manipulates host signalling pathways, and suppresses pro-inflammatory responses. Although infection induces a strong T helper (Th)1 immune response marked by interleukin (IL)-12 and interferon (IFN)-γ production, the parasite often persists due to its ability to neutralize key killing pathways (Martínez-López et al., 2018). Understanding these sophisticated molecular

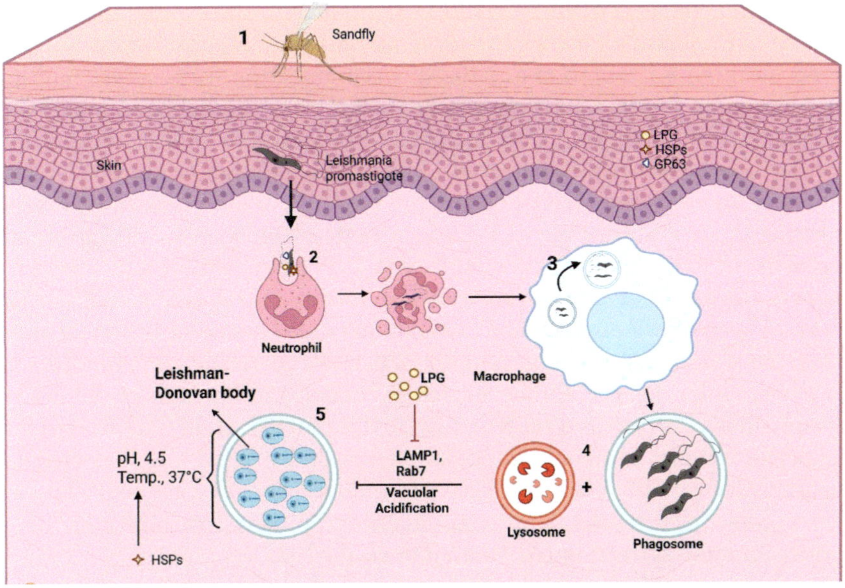

Figure 5.3 Entry and differentiation of *Leishmania* in host cells. (1) *Leishmania* promastigotes are introduced into the host skin during the blood meal of an infected sandfly. (2) These parasites are rapidly engulfed by neutrophils, where they temporarily reside without being destroyed. (3) Exploiting a "Trojan horse" strategy, the dying neutrophils are phagocytosed by macrophages, allowing *Leishmania* to enter a more permissive host cell while evading immune detection. (4) Promastigotes are sequestered within parasitophorous vacuoles in the cytoplasm of macrophages. Now, they are able to resist degradation by inhibiting the normal fusion of phagosomes with lysosomes. This is achieved mainly via the activity of parasite molecules such as lipophosphoglycan (LPG), which inhibit key host trafficking machinery components (e.g., LAMP1, Rab7) and block the vacuolar proton ATPase acidification (Martínez-López et al., 2018). (5) An acidic pH and slightly elevated temperature of the intracellular environment help in recruiting promastigotes (the infective form) into amastigotes (Mann et al., 2021; Gupta et al., 2022).

strategies is essential for developing effective vaccines, diagnostics, and novel therapeutic interventions against leishmaniasis.

This is supported by *Leishmania*-derived factors like LPG, heat shock proteins (HSPs), and GPI-anchored molecules which enable parasite to adapt and survive the harsh conditions present inside phagolysosomal compartment giving rise to reproducing body of *L. donovani* organisms for perpetuation of infection (Martínez-López et al., 2018; Gupta et al., 2022).

5.5.1 Surface molecules

Molecules anchored on the surface of the parasite by GPI Foremost among these is LPG, a large polymer that is tremendously expressed on the surface of the extracellular promastigote. LPG also binds to the sandfly midgut epithelium (a requirement for vector survival) and protects parasites during the blood meal (Al-Khalaifah et al., 2022). Upon infection, LPG serves multiple virulence roles: it acts as a complement receptor on macrophages, inactivates complement by promoting C3bi formation, and strongly inhibits host oxidative defences (Martínez-López et al., 2018; Al-Khalaifah et al., 2022). LPG blocks protein kinase C (PKC) dependent NADPH oxidase assembly, sharply reducing the respiratory burst. It also prevents phagosome acidification and fusion by delaying vacuolar-type ATPase (V-ATPase) assembly and reorganizing periphagosomal F-actin (Martínez-López et al., 2018; Gupta et al., 2022). Thus, LPG enables *Leishmania* to avoid being killed inside early phagosomes. When expressed on amastigotes, lower LPG levels correlate with reduced need for extracellular defence. Other surface glycoconjugates include glycoinositolphospholipids (GIPLs) and proteophosphoglycans (PPGs). PPGs, secreted into the parasitophorous vacuole, modulate macrophage signalling and promote parasite survival. Together, these glycoconjugates form a dense coat that subverts complement, masks parasite antigens, and interacts with host lectins to temper immune activation (Al-Khalaifah et al., 2022).

Leishmania expresses abundant proteases and enzymes that modulate host immunity. The zinc metalloprotease GP63, also called leishmanolysin, is the most prevalent surface protein ($\sim 5 \times 10^5$ copies per cell) (Al-Khalaifah et al., 2022). GP63 acts as an adhesin/opsonin by binding to host macrophages, and it cleaves complement C3b to inactive C3bi, further blocking lysis. GP63 degrades extracellular matrix components (collagen, fibronectin, laminin), facilitating tissue invasion (Wozencraft et al., 1987; Yao et al., 2003). Crucially, GP63 also targets host signalling molecules: it cleaves and activates host protein tyrosine phosphatases (e.g., SHP-1) and degrades transcription factors, undermining macrophage activation. Knockouts of GP63 in *Leishmania mexicana* severely attenuate virulence and increase complement sensitivity, highlighting its virulence importance (Wozencraft et al., 1987). Cysteine proteases (CPA, CPB, and CPC) are secreted by amastigotes; these degrade host antimicrobial peptides and can cleave NF-κB and other transcription factors, further blunting inflammatory cytokine production. Nucleoside diphosphate kinases and ecto-nucleotidases help raise extracellular adenosine, shifting macrophages towards an anti-inflammatory (IL-10) phenotype. Kinetoplastid membrane protein 11 (KMP-11) expressed on amastigotes upregulates host IL-10 and arginase while suppressing nitric oxide (NO) synthesis (Contreras et al., 2010).

5.5.2 Enzymes of key metabolic pathways

To survive the microbicidal milieu, amastigotes bolster their antioxidant systems. Trypanothione reductase, unique to trypanosomatids, maintains high levels of reduced trypanothione and is essential for neutralizing host-derived H_2O_2 and peroxynitrite (Krobitsch et al., 1999). Knockouts of this enzyme result in drastically reduced infectivity in macrophages. *Leishmania* also upregulates host enzymes, favouring its growth: it induces host arginase-1 to consume L-arginine (limiting NO production) and generate polyamines for parasite survival. Heat-shock proteins (HSP70 and HSP90) are upregulated during intracellular stress, helping parasites fold proteins in the acidic phagolysosome (Moczko et al., 1995; Krobitsch et al., 1999).

5.5.3 Secretory or excretory molecules

Leishmania parasites actively release exosomes and microvesicles packed with virulence-associated molecules such as elongation factor-1α (EF-1α), enolase, and various small RNAs, which collectively alter host immune responses. EF-1α, secreted specifically by *L. donovani*, interacts with and activates the host phosphatase SHP-1, leading to suppression of the JAK/STAT signalling cascade (Rawlings et al., 2004; Forget et al., 2006). This disruption impairs IFN-γ-driven NO production, a critical antimicrobial defence mechanism in macrophages. Additionally, *Leishmania* manipulates host ubiquitin signalling by inducing deubiquitinases like A20, which inhibit toll-like receptor (TLR)-mediated pathways by blocking the activation of NF-κB and downstream pro-inflammatory cytokines (Srivastav et al., 2012). Although some of these enzymes, say for example enolase, have now been identified as facilitating immune evasion by the parasite in addition to performing their traditional metabolic roles. The strategies synchronized prime the macrophages for parasite existence by minimizing inflammatory responses, deactivating microbicidal activity, and establishing a favourable cellular environment as strategic intervention site to stimulate persistent infections (Kar et al., 2010).

5.6 *LEISHMANIA* IMMUNE EVASION STRATEGIES

Similar to the many successful pathogens, *Leishmania* has developed effective strategies for evasion of host immune responses, thereby ensuring the survival within the infected host environment. Virulence factors of *Leishmania*, many of which have been discussed here (Section 5.3), fall in this category targeting the immune interrogation. The ability of the parasite to establish and maintain chronic infection depends largely on its capacity to evade immune recognition and clearance. Ultimately, the course of disease is dependent upon the complex interplay between an effective host immune response and the adaptive strategies evolved by *Leishmania* spp. These are some of the key strategies that the parasite uses to avoid immune targeting.

5.6.1 Resists oxidative burst

After ingestion, macrophages generate ROS (iNADPH oxidase) and NO (iNOS). The parasite blocks these defences at multiple points (Lodge et al.,

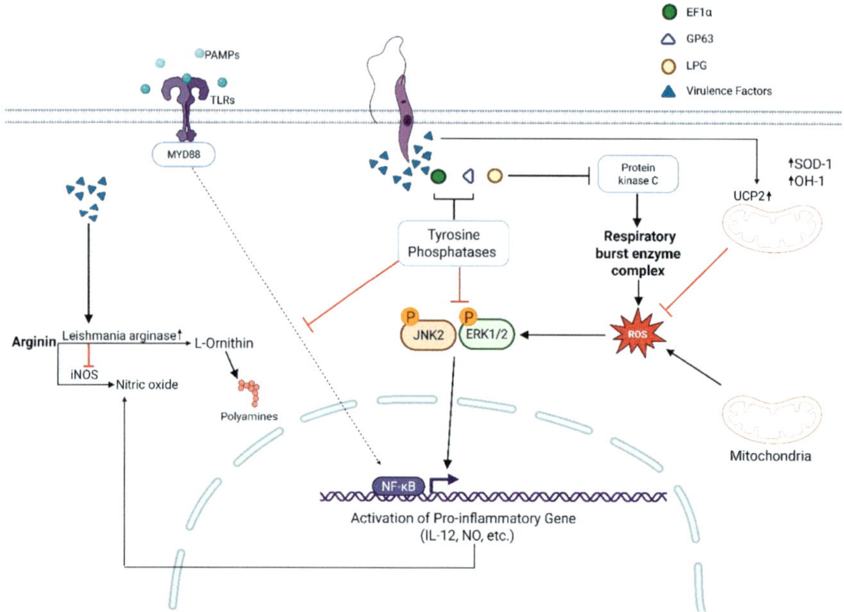

Figure 5.4 Mechanisms of suppression of host oxidative and nitrosative immune responses by *Leishmania*.

2006). LPG inhibits PKC, preventing assembly of the NADPH oxidase complex and blocking the respiratory burst. *Leishmania* upregulates host uncoupling protein 2 (UCP2), decreasing mitochondrial superoxide production. It also induces host antioxidant enzymes, heme oxygenase-1 (HO-1), and superoxide dismutase (SOD-1), to detoxify residual ROS (Gupta et al., 2022). For RNS, *Leishmania* triggers host phosphatases (SHP-1) via GP63 and EF-1α, which inhibit JAK/STAT and MAPK pathways needed for iNOS induction. Concurrently, parasite-derived arginases (and host arginase-1 induction) divert L-arginine away from NO synthesis to produce polyamines for parasite growth. These strategies together ensure that infected macrophages fail to mount an effective oxidative/nitrosative kill (Figure 5.4) (Lodge et al., 2006; Gupta et al., 2021).

5.6.2 Subverts phagosome maturation

Leishmania effectively subverts phagosome maturation to ensure its survival within host macrophages. One of the key strategies involves the inhibition of phagosome–lysosome fusion, a process primarily mediated by its surface molecule, LPG (Martínez-López et al., 2018; Gupta et al., 2022). Parasites coated with LPG prevent the normal acidification of the phagosomal vacuole, thereby avoiding activation of degradative enzymes. *Leishmania* also recruits host phosphoinositide 3-kinases (PI3Ks), which disrupt vesicular trafficking and delay fusion with lysosomes. Concurrently, the parasite stabilizes an actin cytoskeletal shell around the vacuole, further hindering its maturation into a degradative phagolysosome (Gupta et al., 2022). This delay creates a more favourable environment that allows promastigotes to differentiate into amastigotes under less acidic conditions (Zilberstein et al., 2021). In addition, *Leishmania* secretes acid phosphatases and reinforces LPG-mediated inhibition of the V-ATPase, a critical proton pump required for lowering phagosomal pH. By blocking V-ATPase function and acidification, the parasite successfully prevents enzymatic activation and creates a protected niche for intracellular replication (Marshansky and Futai, 2008).

5.6.3 Interferes with TLR and cytokine signalling

Leishmania blunts macrophage activation by crippling TLR pathways and downstreams MAP kinases. Generally, TLR engagement leads to MyD88/IRAK/TRAF6 signalling, ubiquitination of TRAF6, and activation of NF-κB, AP-1, and IRF transcription factors to induce pro-inflammatory genes (Kar et al., 2011). *Leishmania* upregulates host deubiquitinases (e.g., A20), which remove ubiquitin from TRAF6, blocking TLR signal propagation. It also exploits host phosphatases (SHP-1, MKP-1, and PP2A) to dephosphorylate and inactivate MAPKs (ERK, JNK, and p38), preventing AP-1 and IRF activation. Cysteine proteases from some species can further degrade existing MAPKs (ERK1/2, JNK) (Ukil et al., 2011). The net effect is failure to produce IL-12, tumour necrosis factor (TNF)-α, IL-1β, and other cytokines needed to drive a Th1 response (Ukil et al., 2011; Sharma et al., 2014).

5.6.4 Modulates apoptosis and autophagy

Leishmania delays the apoptosis of infected host cells. By inducing anti-apoptotic pathways (e.g., Bcl-2 upregulation via ERK) and inhibiting caspases, the parasite prolongs macrophage survival. Autophagy is subverted; the phagolysosomal niche co-opts autophagic membranes, allowing amastigote replication. Meanwhile, host antigen presentation is impaired: infected dendritic cells (DCs) produce less IL-12 and fail to prime T cells effectively (Martínez-López et al., 2018). Regulatory pathways are amplified, for instance, KMP-11 drives IL-10/transforming growth factor (TGF)-β production and arginase-1 activity, further suppressing NO (Ukil et al., 2005). Together, these tactics prevent effective innate clearance and skew the adaptive response towards a non-protective profile. Cytokine environment skewing by secreting or inducing IL-10 and IL-4, *Leishmania* promotes a Th2/regulatory milieu. In visceral disease, high IL-10 is associated with parasite persistence. Subclinical or diffuse cutaneous forms caused by *L. mexicana* and *L. amazonensis* often correlate with IL-10-dominated responses. Conversely, a strong Th1 response (IFN-γ, TNF-α, and IL-12) is needed to activate infected macrophages (Mukherjee et al., 2015). By dampening IL-12 and promoting IL-10/TGF-β, the parasite ensures that Th1 responses fail to develop fully. In patients, an overactive Th1 response can lead to localized pathology (e.g., mucosal disease from *L. braziliensis*), whereas a Th2 bias leads to uncontrolled parasitemia and dissemination (Mukherjee et al., 2015; Martínez-López et al., 2018).

5.6.5 Exploits host non-coding RNAs

To establish a successful infection, *Leishmania* species modulates the host microRNAs (miRNAs) expression profile in macrophages, which plays a complex role in acute inflammation, and creates an immunosuppressive microenvironment (Gregory and Olivier, 2005). The miRNAs such as let-7c, miR-9, miR-124, miR-146a/b, miR-181a, and miR-210 are associated with M2 macrophages, thereby actively involved in anti-inflammation and tissue-repair functions, and promote the parasite survival. In contrast, miR-21, miR-26a, miR-125a/b, miR-130a/b, and miR-720 are associated with M1 macrophages, which play a dominant role in pro-inflammation, thereby the parasite clearance (Li et al., 2018). *L. donovani* infection in macrophages and DCs exhibits notably low miR-155 levels. This miRNA plays a critical role in host defence by inhibiting the expression of arginase 2 in DCs and facilitating T cell activation, both contribute to parasite clearance. But *L. donovani* effectively suppresses the miR-155 expression to make the host environment favourable for its succession (Muxel et al., 2018). *L. amazonensis* also upregulates miR-30e, miR-294-3p, miR-302, and miR-721 to avoid NO-mediated killing in macrophages, and as is now well-established, these miRNAs target NOS2 mRNA which reduces overall NO production and supports parasite survival. Moreover, miR-294-3p also suppresses TNF-α secretion and escapes ROS-mediated death (Fernandes et al., 2019).

During *L. donovani* infection in human monocyte-derived macrophages (MDMs) and DCs, the upregulation of miR-21 and let-7a targeted the SMAD7 and SOCS4, respectively, which are key regulators of the JAK-STAT pathway, and the upregulation of miR-511 and miR-21 targeted the TLR 4 signalling and PU.1 transcription factor (Geraci et al., 2015). The protozoan *L. donovani* employs c-Myc as a type of mimic virulence factor to overcome global miRNA expression in human MDMs. This strategic exploitation identifies c-Myc as a new Trojan horse host factor, utilized by the parasite to increase its fitness within the host (Colineau et al., 2018). In macrophages, the hypoxia-induced HIF-1α evoked by *L. donovani* modulates miRNA-210 activity of targeting p50 subunit of NF-κB signalling, which in turn suppresses the production of pro-inflammatory cytokines IL-12a and TNF-α (Kumar et al., 2018). Moreover, miR-21 is a second important candidate for the IL-12a control. Both of these observations imply that *C. elegans* utilizes a previously unappreciated survival benefit conferred by hypoxia-induced miRNA activation to suppress host immune responses. In human U937 and THP-1 cells, *L. infantum* infection induces expression of miR-346 which inhibits the MHC I/II-mediated antigen presentation targeting TAP-1, RFX1, and BCAP31 (Diotallevi et al., 2018). *L. donovani* inhibits Th1 cell IFN-γ secretion via inducing miR-7a-1-3p, miR-690, miR-7017-5p, miR574-5p, and miR7235-5 p. Furthermore, the cytokine-induced signalling transducers STAT1 (targeted by miR-574) and STAT4 (targeted by miR-6994-5p) and direct-targets IFN-γ gene (targets of miR-6994-5p and miR-5128) that reduce $CD4^+$ T cell signalling through downregulation of IFN and IL 12 have been reported to favour immune evasion by *L. donovani* regulating different pathways in a synergistic manner (Kumar et al., 2020).

5.6.6 Hijacks host epigenetic landscape

Recent studies unveiled how *Leishmania* parasites modulate the host epigenetic landscape, especially histone dynamics at key inflammatory gene loci, to subvert immune responses and thrive in the host. This epigenetic manipulation is largely executed through various histone-modifying enzymes, commonly referred to as "writer" proteins. A compelling link between histone methylation patterns and macrophage polarization was established during *L. donovani* infection. The Smyd2-mediated H3K36 dimethylation of the TNF-α promoter and Ezh2-mediated H3K27 trimethylation of the NOS2 promoter suppress the TNF-α and iNOS expression, respectively, and the Ash11-mediated H3K4 trimethylation of the IL-10 promoter enhances IL-10 expression in the J774 cell line (Parmer et al., 2020). Similarly, *L. amazonensis* infection in macrophages promotes the formation of complexes between histone deacetylases (HDACs) and the p50/p50 homodimer, which induce the removal of H3K9Ac acetylation and CBP/p300 from the iNOS promoter and suppress iNOS expression. Activation of HDAC1 by *L. amazonensis* infection contributes to a hyporesponsive immune state (Calegari-Silva et al., 2018). Consistent with this, a genome-wide decrease in H3K27 acetylation at the promoter regions of macrophage defence genes was observed in *L. donovani*-infected THP-1 cells. Accordingly, HDAC1 is considered a potential target for therapeutics against leishmaniasis (Roy et al., 2020). To target the inflammasome activity, *Leishmania* hijacks the host histone H3 modifications as a survival strategy (Kamhawi and Serafim, 2020). For instance, *L. amazonensis* infection causes hypo-acetylation of H3K9/14 and hypo-trimethylation of H3K4, to suppress NLRP3 and NF-κB pathways (Lecoeur et al., 2020). Recently identified super enhancer (SE) elements, occupied by "writer" proteins at miRNA promoters, regulate their expression, as a novel mechanism. For example, establishing an initiating positively correlated SE at upstream of the miR-146a-5p promoter that is being occupied by bromodomain 4 (BRD 4) leads the *L. donovani*-infected macrophages to exhibit M2 phenotype exploiting immune evasion strategy (Das et al., 2021).

5.7 CONCLUSION REMARKS

Leishmania has developed clever ways to survive inside the human body by avoiding the immune system. It uses special molecules called virulence factors like LPG, GP63, and EF1-α to block the body's defence tools, reduce harmful molecules like ROS and RNS, and change the immune response to help the infection grow. It also tricks the host by using proteins like A20, HO-1, and SHP-1, though we don't fully understand how. More new factors, such as HSP78 and various enzymes, are being discovered. Studying these tricks is helping scientists develop better tests and possible treatments to fight leishmaniasis more effectively.

ACKNOWLEDGEMENT

The authors express their deepest gratitude to the ICMR-Rajendra Memorial Research Institute of Medical Sciences, Patna, for the invaluable support they provided throughout this study. The authors want to thank Anusandhan National Research Foundation (ANRF/ECRG/2024/000100/LS) for financial assistance.

REFERENCES

Ait Maatallah, Ikram, Khadija Akarid, and Meryem Lemrani. "Tissue tropism: Is it an intrinsic characteristic of *Leishmania* species?" *Acta Tropica* 232 (2022): 106512.

Al-Khalaifah, Hanan S. "Major molecular factors related to *Leishmania* pathogenicity." *Frontiers in Immunology* 13 (2022): 847797.

Antonia, Alejandro L., Alyson B. Barnes, Amelia T. Martin, Liuyang Wang, and Dennis C. Ko. "Variation in Leishmania chemokine suppression driven by diversification of the GP63 virulence factor." *PLoS Neglected Tropical Diseases* 15, no. 10 (2021): e0009224.

Azevedo, Lucas Gentil, Artur Trancoso Lopo de Queiroz, Aldina Barral, Luciane Amorim Santos, and Pablo Ivan Pereira Ramos. "Proteins involved in the biosynthesis of lipophosphoglycan in Leishmania: A comparative genomic and evolutionary analysis." *Parasites & Vectors* 13, no. 1 (2020): 44.

Baig, Saeeda, Anwar Siddiqui, Runu Chakravarty, and Tariq Moatter. "Hepatitis B virus subgenotypes D1 and D3 are prevalent in Pakistan." *BMC Research Notes* 2 (2009): 1–7.

Bharadava, Krupanshi, Tarun Kumar Upadhyay, Radhey Shyam Kaushal, Irfan Ahmad, Yasser Alraey, Samra Siddiqui, and Mohd Saeed. "Genomic insight of leishmania parasite: In-depth review of drug resistance mechanisms and genetic mutations." *ACS Omega* 9, no. 11 (2024): 12500–12514.

Black, Jennifer A., João Luís Reis-Cunha, Angela K. Cruz, and Luiz RO Tosi. "Life in plastic, it's fantastic! How Leishmania exploit genome instability to shape gene expression." *Frontiers in cellular and infection microbiology* 13 (2023): 1102462.

Calegari-Silva, Teresa C., Áislan C. Vivarini, Renata de MS Pereira, Karina L. Dias-Teixeira, Carolina T. Rath, Amanda S. S. Pacheco, Gabrielle B. L. Silva, et al. "*Leishmania amazonensis* downregulates macrophage iNOS expression via Histone Deacetylase 1 (HDAC1): A novel parasite evasion mechanism." *European Journal of Immunology* 48, no. 7 (2018): 1188–1198.

Chiquero, M. Jesus, Asusción Olmo, Pilar Navarro, Luis M. Ruiz-Perez, Santiago Castanys, Dolores Gonzalez-Pacanowska, and Francisco Gamarro. "Amplification of the H locus in *Leishmania infantum*." *Biochimica et Biophysica Acta (BBA)-Molecular Basis of Disease* 1227, no. 3 (1994): 188–194.

Clayton, Christine. "Regulation of gene expression in trypanosomatids: Living with polycistronic transcription." *Open Biology* 9, no. 6 (2019): 190072.

Clayton, Christine. "Post-transcriptional regulation." In *Molecular medical parasitology*, pp. 67–87. Academic Press, 2003.

Clos, Joachim, Janne Grünebast, and Myrine Holm. "Promastigote-to-amastigote conversion in Leishmania spp.—A molecular view." *Pathogens* 11, no. 9 (2022): 1052.

Colineau, Lucie, Ulrike Lambertz, Oriol Fornes, Wyeth W. Wasserman, and Neil E. Reiner. "c-Myc is a novel *Leishmania* virulence factor by proxy that targets the host miRNA system and is essential for survival in human macrophages." *Journal of Biological Chemistry* 293, no. 33 (2018): 12805–12819.

Contreras, Irazu, María Adelaida Gómez, Oliver Nguyen, Marina T. Shio, Robert W. McMaster, and Martin Olivier. "*Leishmania*-induced inactivation of the macrophage transcription factor AP-1 is mediated by the parasite metalloprotease GP63." *PLoS Pathogens* 6, no. 10 (2010): e1001148.

Costa-da-Silva, Ana Caroline, Danielle de Oliveira Nascimento, Jesuino R. M. Ferreira, Kamila Guimarães-Pinto, Leonardo Freire-de-Lima, Alexandre Morrot, Debora Decote-Ricardo, Alessandra Almeida Filardy, and Celio Geraldo Freire-de-Lima. "Immune responses in leishmaniasis: An overview." *Tropical Medicine and Infectious Disease* 7, no. 4 (2022): 54.

Coutinho-Abreu, Iliano V., Tiago D. Serafim, Claudio Meneses, Shaden Kamhawi, Fabiano Oliveira, and Jesus G. Valenzuela. "Distinct gene expression patterns in vector-residing *Leishmania infantum* identify parasite stage-enriched markers." *PLoS Neglected Tropical Diseases* 14, no. 3 (2020): e0008014.

Das, Sonali, Sohitri Mukherjee, and Nahid Ali. "Super enhancer-mediated transcription of miR146a-5p drives M2 polarization during *Leishmania donovani* infection." *PLoS Pathogens* 17, no. 2 (2021): e1009343.

Diotallevi, Aurora, Mauro De Santi, Gloria Buffi, Marcello Ceccarelli, Fabrizio Vitale, Luca Galluzzi, and Mauro Magnani. "*Leishmania* infection induces microRNA hsa-miR-346 in human cell line-derived macrophages." *Frontiers in Microbiology* 9 (2018): 1019.

Doehl, Johannes S. P., Jovana Sádlová, Hamide Aslan, Kateřina Pružinová, Sonia Metangmo, Jan Votýpka, Shaden Kamhawi, Petr Volf, and Deborah F. Smith. "*Leishmania* HASP and SHERP genes are required for in vivo differentiation, parasite transmission and virulence attenuation in the host." *PLoS Pathogens* 13, no. 1 (2017): e1006130.

Elmahallawy, Ehab Kotb, and Abdulsalam A. M. Alkhaldi. "Insights into *Leishmania* molecules and their potential contribution to the virulence of the parasite." *Veterinary Sciences* 8, no. 2 (2021): 33.

Espada, Caroline R., Andreia Albuquerque-Wendt, Valentin Hornillos, Eva Gluenz, Adriano C. Coelho, and Silvia R. B. Uliana. "Ros3 (Lem3p/CDC50) gene dosage is implicated in miltefosine susceptibility in *Leishmania (Viannia) braziliensis* clinical isolates and in *Leishmania (Leishmania) major*." *ACS Infectious Diseases* 7, no. 4 (2021): 849–858.

Feng, Xiuhong, Khoa D. Tran, Marco A. Sanchez, Hakima Al Mezewghi, and Scott M. Landfear. "Glucose transporters and virulence in *Leishmania mexicana*." *Msphere* 3, no. 4 (2018): 1–18.

Fernandes, Juliane Cristina Ribeiro, Juliana Ide Aoki, Stephanie Maia Acuña, Ricardo Andrade Zampieri, Regina P. Markus, Lucile Maria Floeter-Winter, and Sandra Marcia Muxel. "Melatonin and *Leishmania amazonensis* infection altered miR-294, miR-30e, and miR-302d impacting on Tnf, Mcp-1, and Nos 2 expression." *Frontiers in Cellular and Infection Microbiology* 9 (2019): 60.

Fernandes, Octavio, Vimal K. Murthy, Ursula Kurath, Wim M. Degrave, and David A. Campbell. "Mini-exon gene variation in human pathogenic *Leishmania* species." *Molecular and Biochemical Parasitology* 66, no. 2 (1994): 261–271.

Ferreira, Tiago R. "At the genetic crossroads of *Leishmania*: Emerging hybrids reshaping disease patterns." *PLoS Pathogens* 21, no. 6 (2025): e1013213.

Ferreira, Tiago R., and David L. Sacks. "Experimental hybridization in *Leishmania*: Tools for the study of genetic exchange." *Pathogens* 11, no. 5 (2022): 580.

Ferreira, Tiago R., Ehud Inbar, Jahangheer Shaik, Brendan M. Jeffrey, Kashinath Ghosh, Deborah E. Dobson, Stephen M. Beverley, and David Sacks. "Self-hybridization in *Leishmania major*." *MBio* 13, no. 6 (2022): e02858-22.

Forget, Genevieve, David J. Gregory, Lorie A. Whitcombe, and Martin Olivier. "Role of host protein tyrosine phosphatase SHP-1 in *Leishmania donovani*-induced inhibition of nitric oxide production." *Infection and Immunity* 74, no. 11 (2006): 6272–6279.

Frézard, Frédéric, Rubens Monte-Neto, and Priscila G. Reis. "Antimony transport mechanisms in resistant leishmania parasites." *Biophysical Reviews* 6, no. 1 (2014): 119–132.

Geraci, Nicholas S., John C. Tan, and Mary Ann McDowell. "Characterization of microRNA expression profiles in *Leishmania*-infected human phagocytes." *Parasite Immunology* 37, no. 1 (2015): 43–51.

Goes, Wanessa Moreira, Carlos Rodolpho Ferreira Brasil, João Luis Reis-Cunha, Anderson Coqueiro-dos-Santos, Viviane Grazielle-Silva, Júlia de Souza Reis, Tatiane Cristina Souto, et al. "Complete assembly, annotation of virulence genes and CRISPR editing of the genome of *Leishmania amazonensis* PH8 strain." *Genomics* 115, no. 5 (2023): 110661.

Gregory, David J. and Martin Olivier. "Subversion of host cell signalling by the protozoan parasite *Leishmania*." *Parasitology* 130, no. S1 (2005): S27–S35.

Grünebast, Janne, Stephan Lorenzen, and Joachim Clos. "Genome-wide quantification of polycistronic transcription in *Leishmania major*." *MBio* 16, no. 1 (2025): e02241–24.

Gupta, Anand Kumar, Shalini Roy, and Pijush K. Das. "Antileishmanial effect of the natural immunomodulator genipin through suppression of host negative regulatory protein UCP2." *Journal of Antimicrobial Chemotherapy* 76, no. 1 (2021): 135–145.

Gupta, Anand Kumar, Sonali Das, Mohd Kamran, Sarfaraz Ahmad Ejazi, and Nahid Ali. "The pathogenicity and virulence of *Leishmania*-interplay of virulence factors with host defenses." *Virulence* 13, no. 1 (2022): 903–935.

Hanau, Stefania, Martina Maritati, Carlo Contini, Alessandro Trentini, Maria Cristina Manfrinato, and Shawgi Hago Almugadam. "Commentary on the issue of *Leishmania* infection: Focus on some pathogenetic, clinical, and epidemiological aspects." *Veterinary Sciences* 12, no. 6 (2025): 536.

Hashim, Amna Osman Yousif and University of Khartoum, Khartoum (Sudan). *The Polymerase chain reaction as a tool of molecular diagnosis of Leishmania infection in the Sudan*. University of Khartoum, 1997.

Ivens, Alasdair C., Christopher S. Peacock, Elizabeth A. Worthey, Lee Murphy, Gautam Aggarwal, Matthew Berriman, Ellen Sisk, et al. "The genome of the kinetoplastid parasite, *Leishmania major*." *Science* 309, no. 5733 (2005): 436–442.

Kamhawi, Shaden, and Tiago D. Serafim. "*Leishmania*: A maestro in epigenetic manipulation of macrophage inflammasomes." *Trends in Parasitology* 36, no. 6 (2020): 498–501.

Kamran, Mohd, Rahul Bhattacharjee, Sonali Das, Sohitri Mukherjee, and Nahid Ali. "The paradigm of intracellular parasite survival and drug resistance in leishmanial parasite through genome plasticity and epigenetics: Perception and future perspective." *Frontiers in Cellular and Infection Microbiology* 13 (2023): 1001973.

Kar, Susanta, Anindita Ukil, and Pijush K. Das. "Cystatin cures visceral leishmaniasis by NF-κB-mediated proinflammatory response through co-ordination of TLR/MyD88 signaling with p105-Tpl2-ERK pathway." *European Journal of Immunology* 41, no. 1 (2011): 116–127.

Kar, Susanta, Anindita Ukil, Gunjan Sharma, and Pijush K. Das. "MAPK-directed phosphatases preferentially regulate pro-and anti-inflammatory cytokines in experimental visceral leishmaniasis: Involvement of distinct protein kinase C isoforms." *Journal of Leukocyte Biology* 88, no. 1 (2010): 9–20.

Kazemi, Bahram. "Genomic organization of *Leishmania* species." *Iranian Journal of Parasitology* 6, no. 3 (2011): 1.

Krobitsch, Sylvia, and Joachim Clos. "A novel role for 100 kD heat shock proteins in the parasite *Leishmania donovani*." *Cell Stress & Chaperones* 4, no. 3 (1999): 191–198.

Kumar, Vinod, Ajay Kumar, Sushmita Das, Ashish Kumar, Kumar Abhishek, Sudha Verma, Abhishek Mandal, Rakesh K. Singh, and Pradeep Das. "*Leishmania donovani* activates hypoxia inducible factor-1α and miR-210 for survival in macrophages by downregulation of NF-κB mediated pro-inflammatory immune response." *Frontiers in Microbiology* 9 (2018): 385.

Kumar, Vinod, Sushmita Das, Ajay Kumar, Neeraj Tiwari, Ashish Kumar, Kumar Abhishek, Abhishek Mandal, et al. "*Leishmania donovani* infection induce differential miRNA expression in CD4+ T cells." *Scientific Reports* 10, no. 1 (2020): 3523.

Lecoeur, Hervé, Eric Prina, Thibault Rosazza, Kossiwa Kokou, Paya N'Diaye, Nathalie Aulner, Hugo Varet, et al. "Targeting macrophage histone H3 modification as a leishmania strategy to dampen the NF-κB/NLRP3-mediated inflammatory response." *Cell Reports* 30, no. 6 (2020): 1870–1882.

Li, Heng, Ting Jiang, Meng-Qi Li, Xi-Long Zheng, and Guo-Jun Zhao. "Transcriptional regulation of macrophages polarization by MicroRNAs." *Frontiers in Immunology* 9 (2018): 1175.

Lizzi, Karina M., Whitney A. Qualls, Scott C. Brown, and John C. Beier. "Expanding integrated vector management to promote healthy environments." *Trends in Parasitology* 30, no. 8 (2014): 394–400.

Lodge, Robert, and Albert Descoteaux. "Phagocytosis of *Leishmania donovani amastigotes* is Rac1 dependent and occurs in the absence of NADPH oxidase activation." *European Journal of Immunology* 36, no. 10 (2006): 2735–2744.

Lypaczewski, Patrick, Johanna Hoshizaki, Wen-Wei Zhang, Laura-Isobel McCall, John Torcivia-Rodriguez, Vahan Simonyan, Amanpreet Kaur, Ken Dewar, and Greg Matlashewski. "A complete *Leishmania donovani* reference genome identifies novel genetic variations associated with virulence." *Scientific Reports* 8, no. 1 (2018): 16549.

Maharjan, Mahendra, Sushma Singh, Mitali Chatterjee, and Rentala Madhubala. "Role of aquaglyceroporin (AQP1) gene and drug uptake in antimony-resistant clinical isolates of *Leishmania donovani*." *The American Journal of Tropical Medicine and Hygiene* 79, no. 1 (2008): 69–75.

Mann, Sarah, Katherine Frasca, Sara Scherrer, Andrés F. Henao-Martínez, Sabrina Newman, Poornima Ramanan, and José A. Suarez. "A review of leishmaniasis: Current knowledge and future directions." *Current Tropical Medicine Reports* 8 (2021): 121–132.

Mannaert, An, Tim Downing, Hideo Imamura, and Jean-Claude Dujardin. "Adaptive mechanisms in pathogens: Universal aneuploidy in *Leishmania*." *Trends in Parasitology* 28, no. 9 (2012): 370–376.

Marshansky, Vladimir, and Masamitsu Futai. "The V-type H+-ATPase in vesicular trafficking: Targeting, regulation and function." *Current Opinion in Cell Biology* 20, no. 4 (2008): 415–426.
Martínez-Calvillo, Santiago, Dan Nguyen, Kenneth Stuart, and Peter J. Myler. "Transcription initiation and termination on *Leishmania major* chromosome 3." *Eukaryotic Cell* 3, no. 2 (2004): 506–517.
Martínez-Calvillo, Santiago, Gabriela Romero-Meza, Juan C. Vizuet-de-Rueda, Luis E. Florencio-Martínez, Rebeca Manning-Cela, and Tomás Nepomuceno-Mejía. "Epigenetic regulation of transcription in trypanosomatid protozoa." *Current Genomics* 19, no. 2 (2018): 140–149.
Martínez-López, María, Manuel Soto, Salvador Iborra, and David Sancho. "*Leishmania* hijacks myeloid cells for immune escape." *Frontiers in Microbiology* 9 (2018): 883.
Miller, Samuel I., Scott M. Landfear, and Dyann F. Wirth. "Cloning and characterization of a *Leishmania* gene encoding a RNA spliced leader sequence." *Nucleic Acids Research* 14, no. 18 (1986): 7341–7360.
Moczko, Martin, Birgit Schönfisch, Wolfgang Voos, Nikolaus Pfanner, and Joachim Rassow. "The mitochondrial ClpB homolog Hsp78 cooperates with matrix Hsp70 in maintenance of mitochondrial function." *Journal of Molecular Biology* 254, no. 4 (1995): 538–543.
Moncada-Diaz, Maria Juliana, Cristian Camilo Rodríguez-Almonacid, Eyson Quiceno-Giraldo, Francis T. H. Khuong, Carlos Muskus, and Zemfira N. Karamysheva. "Molecular mechanisms of drug resistance in *Leishmania* spp." *Pathogens* 13, no. 10 (2024): 835.
Monnerat, Séverine, Santiago Martinez-Calvillo, Elizabeth Worthey, Peter J. Myler, Kenneth D. Stuart, and Nicolas Fasel. "Genomic organization and gene expression in a chromosomal region of *Leishmania major*." *Molecular and Biochemical Parasitology* 134, no. 2 (2004): 233–243.
Moody, Rhiannon. "Genome plasticity and its role in leishmania adaptation and drug resistance." *Journal of Medical Research and Innovation* 5, no. 2 (2021): 15–22.
Moreira, Douglas S., Rubens L. Monte Neto, Juvana M. Andrade, Ana Maria M. Santi, Priscila G. Reis, Frédéric Frézard, and Silvane M. F. Murta. "Molecular characterization of the MRPA transporter and antimony uptake in four New World *Leishmania* spp. susceptible and resistant to antimony." *International Journal for Parasitology: Drugs and Drug Resistance* 3 (2013): 143–153.
Mukherjee, Budhaditya, Joydeep Paul, Sandip Mukherjee, Rupkatha Mukhopadhyay, Shantanabha Das, Kshudiram Naskar, Shyam Sundar, Jean-Claude Dujardin, Bhaskar Saha, and Syamal Roy. "Antimony-resistant *Leishmania donovani* exploits miR-466i to deactivate host MyD88 for regulating IL-10/IL-12 levels during early hours of infection." *The Journal of Immunology* 195, no. 6 (2015): 2731–2742.
Müller, Michaela Beatrice. Characterization of a new large family of extinct short retroposons in the protozoan parasite *Leishmania* and their role in post-transcriptional regulation of gene expression. (2009).
Muxel, Sandra M., Juliana I. Aoki, Juliane C. R. Fernandes, Maria F. Laranjeira-Silva, Ricardo A. Zampieri, Stephanie M. Acuña, Karl E. Müller, Rubia H. Vanderlinde, and Lucile M. Floeter-Winter. "Arginine and polyamines fate in *Leishmania* infection." *Frontiers in Microbiology* 8 (2018): 2682.
Parmar, Naveen, Pragya Chandrakar, and Susanta Kar. "*Leishmania donovani* subverts host immune response by epigenetic reprogramming of macrophage M (Lipopolysaccharides+ IFN-γ)/M (IL-10) polarization." *The Journal of Immunology* 204, no. 10 (2020): 2762–2778.
Patino, Luz H., Marina Muñoz, Lissa Cruz-Saavedra, Carlos Muskus, and Juan David Ramírez. "Genomic diversification, structural plasticity, and hybridization

in *Leishmania (Viannia) braziliensis*." *Frontiers in Cellular and Infection Microbiology* 10 (2020): 582192.
Peacock, Christopher S., Kathy Seeger, David Harris, Lee Murphy, Jeronimo C. Ruiz, Michael A. Quail, Nick Peters, et al. "Comparative genomic analysis of three *Leishmania* species that cause diverse human disease." *Nature Genetics* 39, no. 7 (2007): 839–847.
Poulin, Robert, Priscila M. Salloum, and Jerusha Bennett. "Evolution of parasites in the Anthropocene: New pressures, new adaptive directions." *Biological Reviews* 99, no. 6 (2024): 2234–2252.
Ranjan, Ravi, Shobha Kumari, and Ashish Kumar. "Geographical distribution of VL and PKDL in India." In *Visceral Leishmaniasis and Post-kala-azar Dermal Leishmaniasis*, pp. 18–26. CRC Press, 2025
Rawlings, Jason S., Kristin M. Rosler, and Douglas A. Harrison. "The JAK/STAT signaling pathway." *Journal of Cell Science* 117, no. 8 (2004): 1281–1283.
Rogers, Matthew B., James D. Hilley, Nicholas J. Dickens, Jon Wilkes, Paul A. Bates, Daniel P. Depledge, David Harris, et al. "Chromosome and gene copy number variation allow major structural change between species and strains of *Leishmania*." *Genome Research* 21, no. 12 (2011): 2129–2142.
Roy, Gargi, Harsimran Kaur Brar, Rohini Muthuswami, and Rentala Madhubala. "Epigenetic regulation of defense genes by histone deacetylase1 in human cell line-derived macrophages promotes intracellular survival of *Leishmania donovani*." *PLoS Neglected Tropical Diseases* 14, no. 4 (2020): e0008167.
Séguin, Olivier, and Albert Descoteaux. "*Leishmania*, the phagosome, and host responses: The journey of a parasite." *Cellular Immunology* 309 (2016): 1–6.
Seifert, Karin, F. Javier Pérez-Victoria, Marianne Stettler, María P. Sánchez-Cañete, Santiago Castanys, Francisco Gamarro, and Simon L. Croft. "Inactivation of the miltefosine transporter, LdMT, causes miltefosine resistance that is conferred to the amastigote stage of *Leishmania donovani* and persists in vivo." *International Journal of Antimicrobial Agents* 30, no. 3 (2007): 229–235.
Shapira, Michal, Alon Zilka, Srinivas Garlapati, Edit Dahan, Irit Dahan, and Victoria Yavesky. "Post transcriptional control of gene expression in *Leishmania*." *Medical Microbiology and Immunology* 190, no. 1 (2001): 23–26.
Sharma, Gunjan, Susanta Kar, Writoban Basu Ball, Kuntal Ghosh, and Pijush K. Das. "The curative effect of fucoidan on visceral leishmaniasis is mediated by activation of MAP kinases through specific protein kinase C isoforms." *Cellular & Molecular Immunology* 11, no. 3 (2014): 263–274.
Silva-Almeida, Mariana, Bernardo Acácio Santini Pereira, Michelle Lopes Ribeiro-Guimarães, and Carlos Roberto Alves. "Proteinases as virulence factors in *Leishmania* spp. infection in mammals." *Parasites & Vectors* 5 (2012): 1–10.
Späth, Gerald F., Laura Piel, and Pascale Pescher. "*Leishmania* genomic adaptation: More than just a 36-body problem." *Trends in Parasitology* 41, no. 6 (2025): 441–448.
Srivastav, Supriya, Susanta Kar, Ajit G. Chande, Robin Mukhopadhyaya, and Pijush K. Das. "*Leishmania donovani* exploits host deubiquitinating enzyme A20, a negative regulator of TLR signaling, to subvert host immune response." *The Journal of Immunology* 189, no. 2 (2012): 924–934.
Thomas, Sean, Amanda Green, Nancy R. Sturm, David A. Campbell, and Peter J. Myler. "Histone acetylations mark origins of polycistronic transcription in *Leishmania major*." *BMC Genomics* 10, no. 1 (2009): 152.
Ukil, Anindita, Aruna Biswas, Tapasi Das, and Pijush K. Das. "18β-glycyrrhetinic acid triggers curative Th1 response and nitric oxide up-regulation in experimental visceral leishmaniasis associated with the activation of NF-κB." *The Journal of Immunology* 175, no. 2 (2005): 1161–1169.

Ukil, Anindita, Susanta Kar, Supriya Srivastav, Kuntal Ghosh, and Pijush K. Das. "Curative effect of 18β-glycyrrhetinic acid in experimental visceral leishmaniasis depends on phosphatase-dependent modulation of cellular MAP kinases." *PLoS One* 6, no. 12 (2011): e29062.

World Health Organization: Leishmaniasis (2023). https://www.who.int/news-room/fact-sheets/detail/leishmaniasis.

Wincker, Patrick, Christophe Ravel, Christine Blaineau, Michel Pages, Yann Jauffret, Jean-Pierre Dedet, and Patrick Bastien. "The *Leishmania* genome comprises 36 chromosomes conserved across widely divergent human pathogenic species." *Nucleic Acids Research* 24, no. 9 (1996): 1688–1694.

Wozencraft, Anne O. and Jenefer M. Blackwell. "Increased infectivity of stationary-phase promastigotes of *Leishmania donovani*: Correlation with enhanced C3 binding capacity and CR3-mediated attachment to host macrophages." *Immunology* 60, no. 4 (1987): 559.

Yao, Chaoqun, John E. Donelson, and Mary E. Wilson. "The major surface protease (MSP or GP63) of *Leishmania* sp. biosynthesis, regulation of expression, and function." *Molecular and Biochemical Parasitology* 132, no. 1 (2003): 1–16.

Zhang, Wen-Wei, and Greg Matlashewski. "Characterization of the A2–A2rel gene cluster in *Leishmania donovani*: Involvement of A2 in visceralization during infection." *Molecular Microbiology* 39, no. 4 (2001): 935–948.

Zilberstein, Dan. "Lysosome sensing is a key mechanism in *Leishmania* intracellular development." *Frontiers in Microbiology* 12 (2021): 667807.

Chapter 6

Diagnostic methods for *Leishmania*

Traditional techniques and advances in molecular diagnosis

Anurag Kumar Kushwaha
Department of Medicine, Institute of Medical Sciences, Banaras Hindu University, Varanasi, India

Vishal Kumar Singh
Centre of Experimental Medicine & Surgery, Institute of Medical Sciences, Banaras Hindu University, Varanasi, India

Ashish Shukla, Sarita Singh, Priya Maurya, and Shyam Sundar
Department of Medicine, Institute of Medical Sciences, Banaras Hindu University, Varanasi, India

*Rajiv Kumar**
Centre of Experimental Medicine & Surgery, Institute of Medical Sciences, Banaras Hindu University, Varanasi, India

6.1 INTRODUCTION

Leishmaniasis, a vector-borne infectious disease caused by protozoan parasites of the genus *Leishmania*, poses a significant public health concern, especially in tropical and subtropical regions. It affects millions of people across tropical, subtropical, and Mediterranean regions, with more than 90% of visceral leishmaniasis (VL) cases reported from limited number of countries such as India, Bangladesh, Sudan, and Brazil (1, 2). Leishmaniasis presents in three primary clinical forms: cutaneous leishmaniasis (CL), which causes localized skin lesions; mucocutaneous leishmaniasis (MCL), characterized by destructive mucosal lesions; and VL, the most severe version that impacts internal organs, including the spleen, liver, and bone marrow, and is lethal if left untreated (3, 4). The global epidemiological landscape of leishmaniasis is complex, driven by multiple factors including environmental change,

* Corresponding author: rajiv.kumar@bhu.ac.in

DOI: 10.1201/9781003596455-6

urbanization, immunosuppression (notably HIV co-infection), and human migration (4–6). Accurate diagnosis is the cornerstone of effective disease management, enabling timely therapeutic intervention, surveillance, and vector control strategies (4, 7). However, diagnosis is often hampered by the overlapping clinical manifestations with other febrile illnesses, variations in host immune response, and limited access to healthcare infrastructure in endemic regions. As a result, there is a critical need for sensitive, specific, cost-effective, and field-deployable diagnostic tools (8). These disease manifestations necessitated accurate and timely diagnosis for effective management. Diagnostic methodologies have evolved from conventional microscopic and culture-based techniques to highly sensitive and specific molecular approaches, significantly improving disease detection and surveillance. Accurate and early diagnosis is important for effective case management, control strategies, and epidemiological surveillance. However, diagnosis of leishmaniasis still challenging due to the variable clinical presentations, co-infections, and the overlap of symptoms with different diseases such as tuberculosis, malaria, and fungal infections (9–11). Over the past century, diagnostic approaches have evolved from basic parasitological methods to sophisticated molecular techniques that offer greater sensitivity and specificity. This chapter provides an in-depth assessment of diagnostic methods for leishmaniasis, beginning with classical parasitological and serological techniques and advancing through molecular and emerging technologies. Each method is evaluated in terms of its biological principles, diagnostic performance, field applicability, advantages, and limitations. We also highlight the recent innovations that are transforming

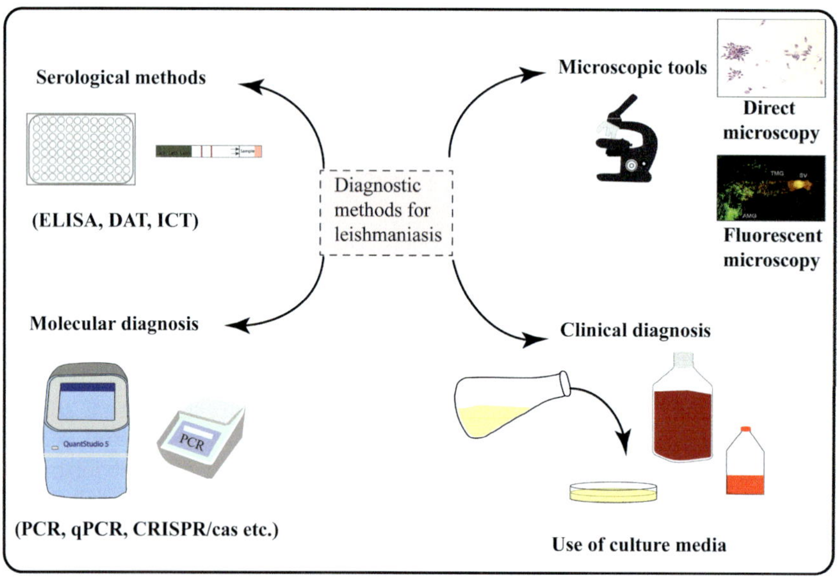

Figure 6.1 Diagrammatic representation of different diagnostic methods for *Leishmania*.

the diagnostic landscape, including Clustered Regularly Interspaced Short Palindromic Repeats (CRISPR)-based platforms and biosensors, and explore how these tools can be integrated into public health agencies to support disease elimination initiatives (Figure 6.1).

6.2 TRADITIONAL DIAGNOSTIC TECHNIQUES

Traditional diagnostic techniques for leishmaniasis encompass parasitological, cultural, and immunological approaches that have historically formed the backbone of case detection. These tools are critical in endemic regions due to their affordability and relative simplicity, even though they sometimes fall short in sensitivity or specificity compared to molecular approaches. Traditional diagnostic techniques form the foundation of leishmaniasis diagnosis, particularly in low-resource and endemic settings. These include microscopy for direct parasite visualization, culture methods for parasite isolation, and serological tests for detecting host immune responses. Despite the advent of advanced molecular diagnostics, these methods remain indispensable due to their cost-effectiveness and feasibility in the field.

6.2.1 Microscopy-based methods

The earliest diagnostic approach for *leishmaniasis* was based on microscopy, first introduced following the discovery of *Leishmania donovani* by Leishman and Donovan in 1903 (12). This procedure involves the direct visualization of *Leishmania* amastigotes in tissue smears stained with Giemsa, taken from bone marrow, spleen aspirates, lymph nodes, or skin lesions (13). Despite its simplicity and low cost, microscopy has low sensitivity, particularly in patients with low parasite loads or chronic infections (14). Microscopy continues to be one of the primary diagnostic tools for detecting *Leishmania* parasites, especially in resource-limited settings. Microscopy refers to the direct visualization of the parasite within host cells. Amastigotes (non-flagellated intracellular forms) of *Leishmania* are identified by their small, oval shape and presence of both a nucleus and a kinetoplast. The sensitivity of microscopy largely depends on the sample source, preparation, and expertise of the examiner. Common microscopy-based techniques include the following.

6.2.2 Light microscopy

Direct visualization of *Leishmania* amastigotes in Giemsa-stained smears obtained from skin lesions, bone marrow, spleen aspirates, or lymph nodes. Light microscopy using Giemsa-stained smears is considered the gold standard for parasitological confirmation. Giemsa staining enhances the visualization of intracellular amastigotes based on their characteristic morphology—round to oval-shaped bodies with a distinct kinetoplast and nucleus. For VL, specimens are typically obtained from bone marrow, splenic aspirates, or lymph nodes, while in CL and post-kala-azar dermal leishmaniasis (PKDL), skin lesion scrapings or biopsies are used. Amastigotes appear as round or oval organisms, 2–4 μm in diameter, with a nucleus and a rod-shaped kinetoplast. Splenic aspiration offers the highest sensitivity (93–99%) but carries a risk of hemorrhage. Bone marrow aspirates are safer and commonly used, with sensitivities ranging from 60 to 85%. Skin lesion smears for CL demonstrate variable sensitivity (50–70%), largely dependent on lesion chronicity and technician skill (14).

6.2.3 Fluorescence microscopy

Fluorescence microscopy is an enhanced optical diagnostic technique that employs fluorescent dyes to detect *Leishmania* parasites in host tissue with greater sensitivity than traditional staining methods. In this approach, DNA-binding fluorochromes such as acridine orange, DAPI (4′,6-diamidino-2-phenylindole), or SYBR Green stain nucleic acids in both the nucleus and kinetoplast of the parasite caused them to fluoresce under specific wavelengths of light (15). This results in improved contrast, especially in specimens with low parasite burden or obscured morphology. In the context of leishmaniasis diagnosis, fluorescence microscopy has been applied to tissue smears from lesions, bone marrow, and splenic aspirates. Amastigotes typically appear as small, oval-shaped bright green or blue structures against a dark background, allowing easier differentiation from surrounding cellular material. Studies have reported increased sensitivity with this method. Fluorescence microscopy using acridine orange could detect parasites in up to 95% of confirmed cases of CL, surpassing conventional Giemsa staining (16). This technique is particularly valuable in diagnosing chronic or atypical cases where parasite load is low, and light microscopy may yield false-negative results. Moreover, it reduces observer fatigue and error by providing clearer visualization, which is particularly beneficial for less experienced microscopists.

6.2.4 Culture-based methods

Culturing *Leishmania* parasites in specialized media allows parasite isolation, species identification, and drug susceptibility testing. Despite its high specificity, culture methods are time-consuming, requiring weeks for parasite growth. Examples include the following:

- **Novy-MacNeal-Nicolle (NNN) Medium:** A biphasic medium supporting the growth of promastigotes from clinical samples.
- **Schneider's Drosophila Medium:** Used for axenic culture of *Leishmania* promastigotes under controlled laboratory conditions.
- **Blood Agar-Based Culture:** Enhances parasite isolation and is often combined with other culture media to improve sensitivity.

6.2.5 Serological tests

Serological assays detect circulating anti-*Leishmania* antibodies or parasite-specific antigens in patient serum that are developed in response to *Leishmania* infection. These methods are frequently used for VL diagnosis but have limitations in distinguishing active from past infections. Serological tests were introduced to address the limitations of microscopy and culture, especially for VL. The rK39 antigen-based rapid diagnostic test (RDT) emerged as a breakthrough, providing rapid, point-of-care diagnosis with acceptable sensitivity and specificity in endemic regions (17). Other serological assays, including enzyme-linked immunosorbent assays (ELISAs) and direct agglutination tests (DATs), have also been widely applied. However, these methods detect antibodies rather than parasites, limiting their ability to distinguish between active and past infections (18). Due to their simplicity, rapid turnaround, and low invasiveness, serological tests are frequently used in field settings and surveillance programs.

One of the most widely used tools in endemic regions is the rK39 immunochromatographic test (ICT), a lateral flow assay that detects antibodies against the recombinant kinesin-related protein (rK39) derived from *Leishmania chagasi*. Its simplicity—requiring only a drop of blood and providing results within 15–20 minutes makes it especially suitable for field deployment (19, 20). Studies from the Indian subcontinent report sensitivity ranging from 85 to 98%, and specificity between 75 and 95% (17). However, in East Africa, the test's performance is reduced, with sensitivities as low as 71%, likely due to genetic heterogeneity among circulating *Leishmania* strains.

Another established method is the DAT, which uses stained and fixed whole *Leishmania* promastigotes as antigen. This test involves serial dilutions of patient serum, leading to visible agglutination if specific antibodies are present. DAT has been shown to possess high diagnostic accuracy, with sensitivity rates of 90–95% and specificity between 80 and 98% (21). While more labor-intensive than rK39, DAT can be used in mass screening efforts and has a long shelf life without requiring a cold chain (22).

ELISA is a laboratory-based technique that quantitatively measures anti-*Leishmania* antibodies using either crude antigens or recombinant proteins (e.g., rK26, rKE16, rA2). This method allows for high-throughput testing and is often used in surveillance and seroprevalence studies. Depending on the antigen and regional epidemiology, ELISA exhibits sensitivity between 85 and 97% and specificity ranging from 80 to 95% (18, 22). New-generation ELISAs incorporating peptide epitopes or cocktail antigens are being explored to enhance sensitivity and minimize cross-reactivity with other pathogens such as *Trypanosoma cruzi*.

A key limitation of all serological methods is their inability to differentiate between current and past infections, as antibodies may persist for years after successful treatment. This limitation severely hampers their use for post-treatment monitoring or confirming relapse. Moreover, these assays perform poorly in CL and PKDL, where antibody titers are generally lower or

Table 6.1 Comparison of traditional diagnostic techniques for *Leishmaniasis*

Diagnostic method	*Type*	*Sensitivity (%)*	*Specificity (%)*	*Time to result*	*Sample type*
Light microscopy	Parasitological	53–99	~100	1–2 hrs	Spleen, BM, skin
Fluorescence microscopy	Parasitological	60–95	~100	1–2 hrs	Lesions, BM
Culture (NNN/ Schneider)	Parasitological	40–70	~100	7–30 days	BM, skin, spleen
rK39 RDT	Serological	85–98	75–95	<30 min	Blood, serum
DAT	Serological	90–95	80–98	2–6 hrs	Serum
ELISA	Serological	85–97	80–95	3–5 hrs	Serum, plasma

undetectable. Additionally, in immunocompromised patients, especially those co-infected with HIV, serological responses are often blunted, resulting in false-negative results. Ongoing research is focused on improving test specificity and developing multi-antigen assays capable of distinguishing between *Leishmania* species. Studies are also investigating the use of IgG subclass profiling, avidity testing, and biomarkers such as circulating immune complexes or cytokine levels to enhance diagnostic utility. Efforts are underway to refine ELISA platforms for use with dried blood spots, which would facilitate sample transport and storage in remote regions (Table 6.1).

6.2.6 Advances in molecular diagnosis

Molecular methods have revolutionized *Leishmania* diagnosis by offering unparalleled sensitivity and specificity. These approaches detect *Leishmania* DNA or RNA in clinical specimens, enabling species identification and parasite quantification. The advent of molecular diagnostics has dramatically transformed the landscape of *Leishmania* detection, particularly with the introduction of polymerase chain reaction (PCR) (23). PCR-based methods target conserved regions of the *Leishmania* genome, including kinetoplast DNA (kDNA), ribosomal RNA genes (rRNA), and species-specific minicircles, achieving unparalleled sensitivity and specificity (24). Quantitative PCR (qPCR) further enhances these capabilities by enabling parasite quantification, useful for monitoring treatment response and relapse risk (25, 26). Major techniques include the following.

6.2.7 Polymerase chain reaction (PCR)

Amplifies *Leishmania* DNA from various clinical samples, allowing species differentiation and low parasite burden detection. PCR is widely regarded as the most sensitive technique for *Leishmania* detection. It involves the amplification of target DNA sequences using primers specific to conserved or species-specific genetic elements of the parasite. There are a lot of parameters that influence the performance of PCR, including the method of nucleic acid extraction that is used, the nature of material, the copy number of the gene target, and the design of the primers that target these genes. Commonly used targets include kDNA, small subunit ribosomal RNA (SSU rRNA), internal transcribed spacer (ITS) regions, and heat shock protein 70 (HSP70) (27). PCR methods can be performed on a variety of clinical samples, including blood, bone marrow, spleen aspirates, and skin lesion biopsies (14). The choice of target and primer design influences the assay's sensitivity and specificity. For instance, kDNA-based PCR, due to the high copy number of minicircles (10,000–20,000 copies/cell), allows for extremely low detection thresholds, making it suitable for early diagnosis and screening (27, 28).

6.2.8 Real-time PCR (qPCR)

Provides quantitative assessment of parasite load, aiding in treatment monitoring and prognosis. qPCR, also known as real-time PCR, builds upon conventional PCR by incorporating fluorescent probes or dyes that allow for real-time monitoring of the amplification process. It provides quantitative data on parasite load, which is valuable for treatment monitoring and prognosis. The commonly used chemistries include TaqMan probes, SYBR Green, and molecular beacons. qPCR has been employed for species-specific detection, monitoring therapeutic response, and studying parasite kinetics (3). It exhibits excellent diagnostic accuracy (>95% sensitivity and >98% specificity) and has been successfully validated in multiple studies across endemic regions. However, its use is limited by cost, technical complexity, and infrastructure requirements (29).

6.2.9 Loop-mediated isothermal amplification (LAMP)

Loop-mediated isothermal amplification (LAMP) offer simplified, field-friendly alternatives to PCR, requiring minimal equipment and delivering results within an hour (30). LAMP has been validated for VL and CL diagnosis, demonstrating robust performance even with minimally invasive samples like skin scrapings or blood spots or non-invasive specimens such as conjunctival or oral swabs, hence enhancing the safety and ease of diagnosis (31). LAMP is a highly sensitive and rapid molecular method that amplifies DNA under isothermal conditions using a set of four to six primers and a DNA polymerase with strand displacement activity. LAMP assays have demonstrated sensitivity and specificity comparable to PCR for VL and CL diagnosis provides increased sensitivity (80–100%) and specificity (94–100%), frequently equivalent to or exceeding PCR performance (32). The approach is efficient, economical, and appropriate for field application as it functions at a stable temperature without requiring complicated machinery. It eliminates the need for thermal cyclers and can produce results within 30–60 minutes (31, 33). Target sequences commonly used include the 18S rRNA and kDNA (32). The technique can be adapted for colorimetric detection, turbidity measurement, or fluorescence-based readouts, making it highly suitable for field application. The development of lyophilized LAMP reagents and simple heating platforms further enhances its portability (32).

6.2.10 Reverse transcription PCR (RT-PCR)

Reverse transcription PCR (RT-PCR) enables RNA-based detection for studying gene expression in *Leishmania* research. It involves the conversion of parasite RNA into complementary DNA (cDNA), followed by amplification using PCR. It is particularly useful for detecting RNA transcripts of viable parasites, thereby distinguishing active infections from residual DNA. RT-PCR is valuable in research settings for studying gene expression, drug resistance markers, and parasite viability post-treatment (34).

6.2.11 Next-generation sequencing (NGS)

Next-Generation Sequencing (NGS) refers to high-throughput sequencing technologies capable of analyzing entire genomes or transcriptomes. In *leishmaniasis*, NGS enables detailed phylogenetic studies, identification of mixed infections, resistance gene profiling, and discovery of novel biomarkers and population genetics (35). Approaches include whole-genome sequencing (WGS), amplicon sequencing, and metagenomics. Metagenomic approaches also enable simultaneous detection of *Leishmania* alongside co-infecting pathogens, enhancing the differential diagnosis of febrile syndromes in endemic areas (36). Despite its comprehensive capabilities, NGS is currently confined to research and reference laboratories due to its high cost, data analysis complexity, and infrastructure demands. Nonetheless, as the cost of sequencing continues to decline, its integration into diagnostic pipelines for surveillance and outbreak investigation is increasingly feasible (Table 6.2).

6.2.12 Emerging diagnostic technologies

While conventional molecular methods like PCR and qPCR have improved the sensitivity and specificity of leishmaniasis diagnosis, novel diagnostic platforms are being developed to overcome challenges related to cost, portability, ease of use, and rapid field deployment. These technologies aim to decentralize diagnostics and enable point-of-care testing in endemic areas lacking sophisticated laboratory infrastructure. Among the most promising innovations are CRISPR-based assays, biosensor technologies, and integrated digital health tools.

6.2.12.1 CRISPR-based diagnostics

CRISPR-Cas systems, widely recognized for their gene editing capabilities, have recently emerged as powerful tools for molecular diagnostics. Platforms

Table 6.2 Comparison of molecular diagnostic techniques for *Leishmaniasis*

Method	*Target gene(s)*	*Sensitivity (%)*	*Specificity (%)*	*Turnaround time*	*Equipment required*	*Field suitability*	*Notes*
PCR	kDNA, SSU rRNA, ITS	90–99	>95	4–6 hrs	Thermocycler	Low–moderate	Gold standard for DNA detection
qPCR	kDNA, HSP70, 18S	>95	>98	2–4 hrs	Real-time PCR system	Low	Allows quantification
LAMP	kDNA, 18S rRNA	85–98	>95	30–60 mins	Water bath/heat block	High	Suitable for field settings
RT-PCR	mRNA, rRNA	80–95	>95	4–6 hrs	Thermocycler, RT kit	Low	Detects viable parasites
NGS	Whole genome, amplicons	>99	>99	1–3 days	Sequencer, bioinformatics	Very low	For surveillance, species typing

like SHERLOCK (Specific High-sensitivity Enzymatic Reporter unLOCKing) and DETECTR (DNA Endonuclease Targeted CRISPR Trans Reporter) utilize the sequence-specific recognition and collateral cleavage activity of Cas enzymes (e.g., Cas12a, Cas13a) to detect nucleic acids with high sensitivity and specificity (37, 38). CRISPR-Cas systems are emerging as precise and rapid molecular diagnostic tools for *Leishmania* detection. CRISPR-based assays offer high specificity by targeting conserved genetic sequences, with potential applications in portable, field-deployable diagnostic platforms in which entire reaction can be completed in under an hour without the need for thermal cycling. Recent advancements in CRISPR-Cas diagnostics hold promise for ultra-sensitive detection of *Leishmania* DNA in clinical and environmental samples. By leveraging the sequence-specific cleavage activity of Cas nucleases, CRISPR-based assays offer highly specific detection with potential for integration into point-of-care platforms (37). A study by Fozouni et al. demonstrated that Cas13-based CRISPR diagnostics could achieve single-copy sensitivity for RNA viruses (39). For *Leishmania*, CRISPR-Cas12a assays targeting kDNA have shown detection limits of 10–20 copies per reaction, with diagnostic sensitivity comparable to or exceeding that of qPCR (40). Preliminary validations using clinical samples from VL patients have shown promising results, although more field evaluations are needed.

6.2.12.2 Biosensors

Biosensors are pivotal in the emerging field of nanodiagnostics, which includes the detection of DNA/RNA, proteins, and volatile chemical molecules. These devices, in essence, metamorphose a biological signal into an electrical signal through a recognition mechanism; they are noted for being cost-effective and portable, with verified high sensitivity and specificity in their abilities. Biosensors utilize biorecognition elements (e.g., DNA probes, antibodies) for real-time detection of *Leishmania*-specific biomolecules, offering promising applications in point-of-care diagnostics (41). Biosensors leverage multiple signal production techniques, such as electrochemiluminescence, optical signals, or surface plasmon resonance (SPR) (42). Genosensors, which are able to identify DNA (or RNA), are the most common type of biosensor used in the diagnostic process for *Leishmania*. Additional recognition components may include antibodies, antigens, or novel aptamer-based sensors. Biosensors and microarrays are being developed to detect *Leishmania* antigens, nucleic acids, or metabolites in a multiplex format, facilitating rapid screening and epidemiological surveillance (43).

6.2.12.2.1 Types of biosensors in Leishmaniasis

Electrochemical Biosensors: Utilize electrodes modified with nanomaterials such as gold nanoparticles or graphene oxide to detect DNA or antigens via impedance or current change.

Optical Biosensors: Based on fluorescence or plasmonic resonance for detecting biomolecular interactions in real-time. SPR sensors have been employed for antigen-antibody interactions.

Lateral Flow Biosensors: Paper-based platforms combining colorimetric detection with nanomaterials (e.g., gold nanorods) for enhanced signal visualization.

The technological development of biosensors for *Leishmania* diagnostics remains in its early stages and necessitates additional investigation to enhance efficacy and standardization for the disease. Despite this, recent studies on their application in NTDs have suggested the possibility of their improved performance, as well as a reduction in interactions with substances that interfere with their operation, good stability, and the miniaturization of products, which enables them to be portable (44). Although biosensor technology is still in the early stages with reference to the detection of leishmaniasis, it has the potential to be an efficient answer to the problem of providing a detection method that is not only quick and portable but also cost-effective (45).

6.2.12.3 Point-of-care testing (POCT)

The term "point of care testing" (POCT) refers to laboratory investigations that are carried out in close proximity to the patient or using diagnostic equipment that is very simple to operate. These tests are completed at the bedside in a very short amount of time. On a global scale, the utilization of the POCT has been progressively growing over the past few years (46, 47). Advancements in portable molecular diagnostic devices, including miniaturized PCR and biosensor-based assays, are transforming field diagnosis of leishmaniasis. POCT systems provide rapid, on-site results with minimal infrastructure requirements, improving accessibility in endemic areas (48). Microscopical examination of the amastigote form and culture of the parasite are the two primary methodologies that are utilized in the conventional examinations. Both diagnostics tests are associated with the constraint of low sensitivity and necessitate considerable effort. A significant majority of the serological tests that were accessible in the past were depends on ELISA. The diagnosis of VL and CL can now be accomplished with a variety of POCTs that target *Leishmania* antigens such as rK9, rK16, rK26, rK28, or rk39 (49, 50). The antigen known as rK39, a recombinant sequence of 39 amino acids derived from *L. chagasi*, is the one that is utilized the most frequently (20). Unlike the culture methods, which can take up to three weeks to produce findings, the ICTs that contain the rK39 antigen produce results promptly within 20 minutes (47). Nevertheless, these tests are not able to diagnose VL in patients who are immunocompromised, and the number of false positive results is higher in those who come from areas where the disease is endemic. In order to circumvent these limitations, tests based on rK28 (recombinant synthetic gene of kinesin genes of *L. donovani*) have been investigated and found to have sensitivity and specificity that are comparable to those based on rk39. These tests are suitable for use in areas where the disease is not endemic. The IT-LEISH® (DiaMed AG, Switzerland—now Biorad, France), Kalazar Detect® (InBios International, USA), and Onsite *Leishmania* Ab Rapid Test (CTK Biotech, USA) are some of the other RDTs that are currently available for commercial use for VL. Additionally, there are a number of prototypes available in a variety of formats. According to recent findings the majority of these diagnostic kits have a sensitivity that ranges from 93 to 100% and a specificity that ranges from 93 to 97% (50).

Commercially available POCT for CL is referred to as the CL Detect Rapid Test. Antigen detection in amastigotes originating from skin lesions is carried out using a qualitative membrane-based examination. The sensitivity is greater than 95%; nevertheless, the test has a restriction in that it cannot be performed on serum samples. As a result, if blood is present during the collection of the skin sample, it may interfere with the performance of the test (50). Some of the drawbacks of serology-based RDTs include the possibility of cross-reactivity and false positive results. The LAMP assay is a unique molecular approach that was developed by Eiken Chemical Company in collaboration with FIND and other partners (30). A detection limit of around 1000 copies per genome is utilized by this assay, which targets 16S RNA as its target. There is no cold chain required because it is in a ready-to-use format. Using ultraviolet or blue light to illuminate the targeted region can result in an improvement in the signals that are produced. Both sensitivity and specificity range from 80 to 100% and 94 to 100%, respectively. Due to the fact that the test is focused on visual evaluation, it is potentially prone to inaccuracy (30). A number of the non-invasive assays that facilitate the use of the latex agglutination test for the diagnosis of VL in urine samples are currently approaching the conclusion of their development. In order to render the heat-sensitive material inactive, the urine samples are first subjected to a pre-treatment, and then they are treated with a latex-sensitized antibody that is directed against the *L. donovani* antigen.

6.3 CURRENT CHALLENGES AND FUTURE PERSPECTIVES

Despite these technological advancements, major challenges persist. Molecular diagnostics remain expensive and require well-equipped laboratories and trained personnel, limiting their accessibility in many endemic areas (40). Moreover, the diversity of *Leishmania* species and the varying performance of diagnostic tools across different epidemiological settings complicate the development of universal diagnostic platforms (51). Efforts to develop standardized protocols, enhance diagnostic accuracy for asymptomatic infections, and combine serological and molecular markers for comprehensive diagnostics are ongoing (18). Looking ahead, integrating artificial intelligence (AI) for image analysis in microscopy, developing wearable biosensors for real-time surveillance, and leveraging multi-omics approaches for biomarker discovery could revolutionize leishmaniasis diagnosis. Combined with enhanced genomic surveillance and mobile health (mHealth) platforms, these innovations have the potential to reshape control programs, bringing leishmaniasis diagnosis closer to the point of need (52).

6.4 CONCLUSION

Traditional diagnostic methods, such as microscopy, culture, and serological assays, continue to play a crucial role in the diagnosis of leishmaniasis, particularly in resource-constrained settings where their simplicity and lower cost are advantageous. These methods have provided the foundational understanding of the disease for over a century. However, their inherent limitations in sensitivity, specificity, and the ability to differentiate active from past infections highlight the need for more advanced solutions. The advent of molecular diagnostics has dramatically improved the sensitivity, specificity, and rapidity of *Leishmania* detection, fundamentally shaping the future of leishmaniasis diagnostics. Techniques like PCR, qPCR, and LAMP offer highly precise detection of parasite nucleic acids, enabling accurate diagnosis even with low parasite burdens and facilitating parasite quantification for treatment monitoring. Emerging technologies such as NGS, CRISPR-based diagnostics, biosensors, and advancements in point-of-care testing are further pushing the boundaries of accuracy, accessibility, and multiplexing capabilities.

Despite these remarkable scientific and technological progress, significant challenges persist, primarily related to the cost of advanced molecular techniques, the requirement for sophisticated infrastructure, the need for specialized human resources, and the complexities posed by the genetic diversity of *Leishmania* species. Looking forward, the strategic integration of advanced molecular approaches with field-deployable assays, coupled with cutting-edge innovations such as AI-driven image analysis, wearable biosensors, multi-omics biomarker discovery, and robust genomic surveillance tools, holds the promise of revolutionizing leishmaniasis control. The integration of advanced molecular approaches, field-deployable assays, and genomic surveillance tools will enhance disease monitoring, facilitate targeted treatment strategies, and contribute to the global efforts to control leishmaniasis. These integrated strategies, supported by mHealth platforms, will not only enhance disease monitoring but also facilitate more targeted treatment strategies, ultimately contributing significantly to the global efforts to control and potentially eliminate leishmaniasis. The ongoing research and development in diagnostic technologies are crucial steps toward a future where leishmaniasis diagnosis is accessible, rapid, and accurate for all affected populations.

ACKNOWLEDGMENT

Authors are thankful to the Banaras Hindu University, Varanasi, for providing research support.

REFERENCES

1. Alvar J, Vélez ID, Bern C, Herrero M, Desjeux P, Cano J, et al. Leishmaniasis worldwide and global estimates of its incidence. PLoS One. 2012;7(5):e35671.
2. World Health Organization. WHO fact sheets. 1998. https://doi.org/10.1371/journal.pone.0035671.
3. Singh OP, Tiwary P, Kushwaha AK, Singh SK, Singh DK, Lawyer P, et al. Xenodiagnosis to evaluate the infectiousness of humans to sandflies in an area endemic for visceral leishmaniasis in Bihar, India: A transmission-dynamics study. Lancet Microbe. 2021;2(1):e23–31.
4. Kushwaha AK, Scorza BM, Singh OP, Rowton E, Lawyer P, Sundar S, et al. Domestic mammals as reservoirs for *Leishmania donovani* on the Indian subcontinent: Possibility and consequences on elimination. Transbound Emerg Dis. 2022;69(2):268–77.
5. Cloots K, Marino P, Burza S, Gill N, Boelaert M, Hasker E. Visceral Leishmaniasis-HIV coinfection as a predictor of increased Leishmania transmission at the village level in Bihar, India. Front Cell Infect Microbiol. 2021;11:604117.
6. Singh OP, Chaubey R, Kushwaha AK, Fay MP, Sacks D, Sundar S. Visceral leishmaniasis-human immunodeficiency virus-coinfected patients are highly infectious to sandflies in an endemic area in India. J Infect Dis. 2024;229(6):1909–12.
7. Kushwaha AK, Shukla A, Scorza B, Chaubey R, Maurya DK, Rai TK, et al. Dogs as reservoirs for *Leishmania donovani*, Bihar, India, 2018–2022. Emerg Infect Dis J. 2024;30(12):2604.
8. Sundar S, Jitendra S, Kumar SV, Neha A, Kumar R. Current and emerging therapies for the treatment of leishmaniasis. Expert Opin Orphan Drugs. 2024;12(1):19–32.
9. Desjeux P. Leishmaniasis: Current situation and new perspectives. Comp Immunol Microbiol Infect Dis. 2004;27(5):305–18.
10. Grimaldi Jr G, Teva A, Porrozzi R, Pinto MA, Marchevsky RS, Rocha MGL, et al. Clinical and parasitological protection in a *Leishmania infantum* -macaque model vaccinated with adenovirus and the recombinant A2 antigen. PLoS Negl Trop Dis. 2014;8(6):e2853.
11. Ornellas-Garcia U, Cuervo P, Ribeiro-Gomes FL. Malaria and leishmaniasis: Updates on co-infection. Front Immunol. 2023;14:1122411.
12. Herwaldt BL. Leishmaniasis. Lancet. 1999;354(9185):1191–9.
13. Zijlstra E, Ali MS, El-Hassan A, El-Toum IA, Satti M, Ghalib H, et al. Kala-azar: A comparative study of parasitological methods and the direct agglutination test in diagnosis. Trans R Soc Trop Med Hyg. 1992;86(5):505–7.
14. Sundar S, Rai M. Laboratory diagnosis of visceral leishmaniasis. Clin Diagn Lab Immunol. 2002;9(5):951–8.
15. Singh N, Dube A. Fluorescent leishmania: Application to anti-leishmanial drug testing. Am J Trop Med Hyg. 2004;71(4):400–2.
16. de Paiva-Cavalcanti M, de Morais RCS, Pessoa-e-Silva R, Trajano-Silva LAM, Gonçalves-de-Albuquerque SdC, Tavares DdHC, et al. Leishmaniases diagnosis: An update on the use of immunological and molecular tools. Cell Biosci. 2015;5(1):31.
17. Boelaert M, El-Safi S, Hailu A, Mukhtar M, Rijal S, Sundar S, et al. Diagnostic tests for kala-azar: A multi-centre study of the freeze-dried DAT, rK39 strip test

and KAtex in East Africa and the Indian subcontinent. Trans R Soc Trop Med Hyg 2008;102(1):32–40.

18. Chappuis F, Sundar S, Hailu A, Ghalib H, Rijal S, Peeling RW, et al. Visceral leishmaniasis: what are the needs for diagnosis, treatment and control? Nat Rev Microbiol. 2007;5(11):873–82.
19. Kumar D, Khanal B, Tiwary P, Mudavath SL, Tiwary NK, Singh R, et al. Comparative evaluation of blood and serum samples in rapid immunochromatographic tests for visceral leishmaniasis. J Clin Microbiol. 2013;51(12):3955–9.
20. Cloots K, Singh OP, Singh AK, Kushwaha AK, Malaviya P, Kansal S, et al. Diagnosis of visceral leishmaniasis in an elimination setting: A validation study of the diagnostic algorithm in India. Diagnostics. 2022;12(3):670.
21. Roberts T, Keddie SH, Rattanavong S, Gomez SR, Bradley J, Keogh RH, et al. Accuracy of the direct agglutination test for diagnosis of visceral leishmaniasis: A systematic review and meta-analysis. BMC Infect Dis. 2023;23(1):782.
22. Srivastava P, Gidwani K, Picado A, Van der Auwera G, Tiwary P, Ostyn B, et al. Molecular and serological markers of *Leishmania donovani* infection in healthy individuals from endemic areas of Bihar, India. Trop Med Int Health. 2013;18(5):548–54.
23. Rodríguez-Cortés A, Ojeda A, Francino O, López-Fuertes L, Timón M, Alberola J. Leishmania infection: laboratory diagnosing in the absence of a "gold standard". Am J Trop Med Hyg. 2010;82(2):251–6.
24. Deborggraeve S, Boelaert M, Rijal S, De Doncker S, Dujardin JC, Herdewijn P, et al. Diagnostic accuracy of a new Leishmania PCR for clinical visceral Leishmaniasis in Nepal and its role in diagnosis of disease. Trop Med Int Health 2008;13(11):1378–83.
25. Galluzzi L, Ceccarelli M, Diotallevi A, Menotta M, Magnani M. Real-time PCR applications for diagnosis of leishmaniasis. Parasit Vectors. 2018;11(1):273.
26. Molina I, Fisa R, Riera C, Falcó V, Elizalde A, Salvador F, et al. Ultrasensitive real-time PCR for the clinical management of visceral leishmaniasis in HIV-infected patients. Am Soc Trop Med Hyg. 2013;89(1):105–10.
27. Sundar S, Singh OP. Molecular diagnosis of visceral leishmaniasis. Mol Diagn Ther. 2018;22(4):443–57.
28. Lachaud L, Marchergui-Hammami S, Chabbert E, Dereure J, Dedet JP, Bastien P. Comparison of six PCR methods using peripheral blood for detection of canine visceral leishmaniasis. J Clin Microbiol. 2002;40(1):210–5.
29. Mary C, Faraut F, Lascombe L, Dumon H. Quantification of *Leishmania infantum* DNA by a real-time PCR assay with high sensitivity. J Clin Microbiol. 2004;42(11):5249–55.
30. Nzelu CO, Kato H, Peters NC. Loop-mediated isothermal amplification (LAMP): An advanced molecular point-of-care technique for the detection of Leishmania infection. PLoS Negl Trop Dis. 2019;13(11):e0007698.
31. Adams ER, Schoone G, Versteeg I, Gomez MA, Diro E, Mori Y, et al. Development and evaluation of a novel loop-mediated isothermal amplification assay for diagnosis of cutaneous and visceral leishmaniasis. J Clin Microbiol. 2018;56(7).
32. Erber AC, Sandler PJ, de Avelar DM, Swoboda I, Cota G, Walochnik J. Diagnosis of visceral and cutaneous leishmaniasis using loop-mediated isothermal amplification (LAMP) protocols: A systematic review and meta-analysis. Parasit Vectors. 2022;15(1):34.
33. Sriworarat C, Phumee A, Mungthin M, Leelayoova S, Siriyasatien P. Development of loop-mediated isothermal amplification (LAMP) for simple detection of Leishmania infection. Parasit Vectors. 2015;8:591.
34. van der Meide W, Guerra J, Schoone G, Farenhorst M, Coelho L, Faber W, et al. Comparison between quantitative nucleic acid sequence-based amplification, real-time reverse transcriptase PCR, and real-time PCR for quantification of Leishmania parasites. J Clin Microbiol. 2008;46(1):73–78.

35. Leprohon P, Fernandez-Prada C, Gazanion É, Monte-Neto R, Ouellette M. Drug resistance analysis by next generation sequencing in Leishmania. Int J Parasitol: Drugs Drug Resist. 2015;5(1):26–35.
36. Butenko A, Kostygov AY, Sádlová J, Kleschenko Y, Bečvář T, Podešvová L, et al. Comparative genomics of Leishmania (Mundinia). BMC Genomics. 2019; 20:1–12.
37. Gootenberg JS, Abudayyeh OO, Lee JW, Essletzbichler P, Dy AJ, Joung J, et al. Nucleic acid detection with CRISPR-Cas13a/C2c2. Science. 2017;356(6336): 438–42.
38. Kellner MJ, Koob JG, Gootenberg JS, Abudayyeh OO, Zhang F. SHERLOCK: nucleic acid detection with CRISPR nucleases. Nat Protoc. 2019;14(10):2986–3012.
39. Fozouni P, Son S, Díaz de León Derby M, Knott GJ, Gray CN, D'Ambrosio MV, et al. Amplification-free detection of SARS-CoV-2 with CRISPR-Cas13a and mobile phone microscopy. Cell. 2021;184(2):323–33.e9.
40. Weirather JL, Jeronimo SM, Gautam S, Sundar S, Kang M, Kurtz MA, et al. Serial quantitative PCR assay for detection, species discrimination, and quantification of *Leishmania* spp. in human samples. J Clin Microbiol. 2011;49(11):3892–904.
41. Gedda MR, Madhukar P, Shukla A, Mudavath SL, Srivastava ON, Singh OP, et al. Nanodiagnostics in leishmaniasis: A new frontiers for early elimination. Wiley Interdiscip Rev: Nanomed Nanobiotechnol. 2021;13(2):e1675.
42. Ferreira E, Lima J, Alves-Balvedi RP, Bonan PR, Medeiros E, Goulart L, Lima A, Neff H, Oliveira L, Castellano L, et al. *Leishmania* spp. detection using a surface plasmon resonance biosensor. Proceedings. 2017;1(4):536.
43. Nafari A, Cheraghipour K, Sepahvand M, Shahrokhi G, Gabal E, Mahmoudvand H. Nanoparticles: New agents toward treatment of leishmaniasis. Parasite Epidemiol Control. 2020;10:e00156.
44. de Freitas Borges PA, Fiel WA, Vasconcellos VA, de Faria RAD. Current progresses in the development of biosensors for the diagnosis of neglected tropical diseases. Syst Bio Eng. 2021;1(1):39.
45. Martins BR, Barbosa YO, Andrade CM, Pereira LQ, Simão GF, de Oliveira CJ, et al. Development of an electrochemical immunosensor for specific detection of visceral leishmaniasis using gold-modified screen-printed carbon electrodes. Biosensors. 2020;10(8):81.
46. Hocking L, George J, Broberg EK, Struelens MJ, Leitmeyer KC, Deshpande A, et al. Point of care testing for infectious disease in Europe: A scoping review and survey study. Front Public Health. 2021;9:722943.
47. Ghoshal U, Jain M. A review on point of care tests in parasitology. Indian J Med Microbiol. 2022;40(3):337–41.
48. Tay A, Pavesi A, Yazdi SR, Lim CT, Warkiani ME. Advances in microfluidics in combating infectious diseases. Biotechnol Adv. 2016;34(4):404–21.
49. Freire ML, de Souza A, Cota G, Rabello A, Machado de Assis T. Cost-effectiveness of serological tests for human visceral leishmaniasis in the Brazilian scenario. PLoS Negl Trop Dis. 2020;14(10):e0008741.
50. Boelaert M, Verdonck K, Menten J, Sunyoto T, van Griensven J, Chappuis F, Rijal S. Rapid tests for the diagnosis of visceral leishmaniasis in patients with suspected disease. Cochrane Database Syst Rev. 2014;(6):CD009135.
51. Schönian G, Mauricio I, Gramiccia M, Cañavate C, Boelaert M, Dujardin J-C. Leishmaniases in the Mediterranean in the era of molecular epidemiology. Trends Parasitol. 2008;24(3):135–42.
52. Pacheco-Fernandez T, Volpedo G, Gannavaram S, Bhattacharya P, Dey R, Satoskar A, et al. Revival of leishmanization and leishmanin. Front Cell Infect Microbiol. 2021;11:639801.

Chapter 7

Combination therapy in visceral leishmaniasis (VL)

A step toward enhanced efficacy

*Anurag Pappu**
Department of Clinical Medicine, ICMR-Rajendra Memorial Research Institute of Medical Sciences, Patna, India

Kundan Kunal
Department of Biochemistry, ICMR-Rajendra Memorial Research Institute of Medical Sciences, Patna, India

Veer Singh
Department of Biosciences, Manipal University Jaipur, Rajasthan, India

Shobha Kumari
Department of Biochemistry, ICMR-Rajendra Memorial Research Institute of Medical Sciences, Patna, India

7.1 INTRODUCTION

Visceral leishmaniasis (VL), also known as kala-azar, is one of the most severe and life-threatening neglected tropical diseases (NTDs), caused by protozoan parasites of the *Leishmania donovani* complex. Transmission occurs through the bite of infected female Phlebotomine sandflies (1). The disease predominantly affects internal organs such as the spleen, liver, and bone marrow, leading to prolonged fever, hepatosplenomegaly, pancytopenia, weight loss, immune suppression, malnutrition, and, in severe cases, multi-organ failure (2). If left untreated, VL has a case fatality rate exceeding 95%, placing it among the most lethal parasitic infections globally.

VL is endemic in several resource-limited regions, particularly in South Asia, East Africa, and parts of South America. According to the World Health Organization (WHO), more than 95% of cases occur in countries like Bangladesh, India, Nepal, Sudan, South Sudan, Ethiopia, and Brazil (3).

* Corresponding author: anuragpappu444@gmail.com

DOI: 10.1201/9781003596455-7

The burden of the disease is exacerbated by poverty, inadequate healthcare infrastructure, malnutrition, and socio-political instability, making it a major public health challenge (4). WHO estimates suggest an annual incidence of 50,000–90,000 new cases, although the actual numbers may be higher due to underreporting. In response, VL has been designated as a priority disease under global elimination programs, aiming to reduce its incidence to fewer than 1 case per 10,000 people per year in endemic regions (3).

Historically, the treatment of VL has evolved significantly. The disease was first described in historical texts from India, China, and Africa, with the first major epidemic documented in Assam, India, in 1824 (5). In 1903, British pathologists William Leishman and Charles Donovan independently identified the causative organism in the spleens of infected patients, leading to the naming of the parasite *L. donovani* (6). This discovery marked the beginning of a new era in understanding and managing the disease.

Early treatments were highly toxic and often ineffective. Arsenic-based compounds used in the early 20th century had serious side effects. The introduction of pentavalent antimonials in the 1940s marked a significant improvement in disease management. Later, the use of amphotericin B in the 1980s and the approval of oral miltefosine in 2002 further expanded therapeutic options. However, each of these treatments came with challenges, including toxicity, high cost, long treatment durations, and the emergence of drug-resistant *Leishmania* strains (7–9).

In recent years, the focus has shifted toward combination therapy to address these challenges. Combination regimens offer several advantages, including reduced treatment duration, lower doses of individual drugs (thus minimizing toxicity), and decreased likelihood of resistance development (10, 11). Despite these benefits, barriers such as drug accessibility, cold chain requirements, and cost remain in many endemic areas.

This chapter presents an in-depth analysis of the evolution of VL treatment, with a focus on combination therapies as a promising strategy for enhancing efficacy and overcoming current limitations. It also explores novel approaches, including immunomodulatory agents, host-directed therapies, and drug repurposing, all of which hold potential to reshape the therapeutic landscape for VL.

7.2 CURRENT THERAPEUTIC LANDSCAPE FOR VISCERAL LEISHMANIASIS (VL)

VL, caused primarily by *L. donovani* and *Leishmania infantum*, remains one of the most severe forms of leishmaniasis and is a critical public health concern in many low- and middle-income countries. The treatment landscape for VL has evolved considerably, shifting from highly toxic and less efficacious drugs to more patient-friendly and effective regimens (12). However, significant limitations persist, particularly in the context of drug resistance, toxicity, treatment duration, and cost. This section outlines the key monotherapies currently used in clinical practice and the critical challenges associated with these therapies.

7.2.1 Monotherapy: Available drugs and limitations

7.2.1.1 Pentavalent antimonials

Pentavalent antimonials (sodium stibogluconate and meglumine antimoniate) were the cornerstone of VL treatment for decades. Administered via parenteral routes over a period of 20–30 days, SbV disrupts parasite metabolism by interfering with ATP synthesis and thiol metabolism. Despite initial success, resistance to antimonials, particularly in the Indian subcontinent, has rendered them largely ineffective in many endemic areas. Studies from Bihar, India, have shown treatment failure rates exceeding 60% (9, 13). Due to severe cardiotoxicity, hepatotoxicity, and nephrotoxicity, their use is now limited or phased out in many countries.

7.2.1.2 Amphotericin B (conventional and liposomal)

Amphotericin B deoxycholate, an antifungal polyene, has shown excellent efficacy against VL by binding to ergosterol-like sterols in the *Leishmania* membrane, causing cell lysis. However, the conventional formulation is associated with nephrotoxicity, infusion-related reactions, and prolonged hospital stays. Liposomal amphotericin B (L-AmB; marketed as Ambisome) significantly reduces toxicity and is now considered a first-line therapy in many countries, including India, where a single-dose (10 mg/kg) regimen has proven highly effective with cure rates above 95% (14). Nonetheless, the high cost and cold chain requirement remain major barriers to widespread use in resource-limited settings (15).

7.2.1.3 Miltefosine

Introduced in the early 2000s, miltefosine was the first oral drug approved for VL treatment. It is an alkylphosphocholine that disrupts membrane integrity and induces apoptosis-like death in parasites. Miltefosine is administered for 28 days and is generally well tolerated; however, gastrointestinal side effects and teratogenic potential restrict its use in pregnant women. Reports of declining efficacy and emerging resistance have raised concerns about its continued utility (16, 17). Additionally, poor adherence due to the long treatment duration contributes to suboptimal outcomes.

7.2.1.4 Paromomycin

Paromomycin is an aminoglycoside antibiotic with antileishmanial properties, primarily used as an injectable formulation over a 21-day course. It acts by inhibiting protein synthesis in parasites. It has shown high efficacy in monotherapy and combination regimens, especially in India. Its lower cost compared to L-AmB makes it attractive; however, parenteral administration and potential nephrotoxicity and ototoxicity limit its field applicability (18).

7.2.2 Challenges in monotherapy

7.2.2.1 *Drug resistance*

Resistance to antileishmanial drugs, particularly pentavalent antimonials and miltefosine, is a growing concern in VL management. In regions like Bihar, the extensive misuse and incomplete treatment of SbV led to widespread resistance, which significantly undermined public health efforts (9, 13). Miltefosine resistance is believed to be driven by mutations in the *LdMT* and *LdRos3* genes responsible for drug uptake and distribution (19). Paromomycin resistance, though less prevalent, has been observed in laboratory studies and is a potential risk with increased use (20, 21).

7.2.2.2 *Toxicity*

Each monotherapy carries inherent toxicities that limit its safe usage. Amphotericin B deoxycholate can cause nephrotoxicity, electrolyte imbalances, and infusion reactions, necessitating hospitalization and monitoring. Pentavalent antimonials can induce pancreatitis, arrhythmias, and hepatotoxicity, especially in elderly or malnourished patients. Even miltefosine, though oral, may cause severe vomiting, diarrhea, and teratogenicity. These toxicities increase the risk of treatment discontinuation, further aggravating resistance issues and relapse rates (21, 22).

7.2.2.3 *Treatment duration and compliance*

Long treatment regimens (ranging from 21 to 30 days) and parenteral administration in most monotherapies reduce patient adherence, especially in remote areas with poor healthcare infrastructure. Oral miltefosine, despite ease of administration, suffers from suboptimal compliance due to gastrointestinal side effects and prolonged treatment duration (23). Compliance is crucial for therapeutic success, and non-adherence contributes to therapeutic failure and drug resistance.

7.3 PRINCIPLES AND ADVANTAGES OF COMBINATION THERAPY IN VISCERAL LEISHMANIASIS

Combination therapy has emerged as a cornerstone strategy in the fight against VL, offering a multifaceted approach to overcoming the challenges associated with monotherapy (10, 24). By integrating two or more therapeutic agents with different mechanisms of action, combination therapy aims to enhance efficacy, reduce the risk of drug resistance, improve patient compliance, and lower overall treatment costs. This chapter explores the core principles underlying combination therapy in VL and highlights its clinical and public health benefits.

7.3.1 Synergistic drug action

One of the primary principles behind combination therapy is the concept of synergistic drug action. Synergism occurs when the combined effect of two drugs is greater than the sum of their individual effects (25). In the context of VL, using drugs with complementary pharmacodynamics—such as a membrane-disrupting agent paired with an inhibitor of protein synthesis—can amplify the antileishmanial effect while potentially reducing the required dosage of each individual drug. This not only improves treatment efficacy but also minimizes toxicity. For example, the combination of L-AmB and paromomycin has shown synergistic effects in reducing parasitic load more effectively than either drug alone (26).

7.3.2 Reduced risk of resistance

Drug resistance is a growing concern in VL treatment, particularly in regions with prolonged use of monotherapies such as pentavalent antimonials and miltefosine. Combination therapy offers a strategic solution by applying multiple pharmacological pressures on the parasite, making it more difficult for *L. donovani* to develop resistance simultaneously to all the agents involved. This multi-targeted approach reduces the probability of selecting for resistant mutants and helps sustain the efficacy of existing drugs. Moreover, it delays the emergence of resistance by lowering the parasite population more effectively and rapidly than monotherapy (27–29).

7.3.3 Shortened treatment duration

Traditional monotherapies often require long treatment regimens, which can compromise patient compliance and increase the likelihood of treatment failure. Combination therapy allows for dose optimization, where lower doses of each drug are used over shorter periods while maintaining therapeutic efficacy. Shortened regimens reduce the burden on healthcare systems and patients, particularly in resource-constrained settings. For instance, a single-dose regimen of L-AmB combined with a short course of miltefosine or paromomycin has shown promise in reducing treatment duration from 28 days to as few as 10 days in clinical trials (27, 30, 31).

7.3.4 Improved safety and cost-effectiveness

Combining drugs with distinct toxicity profiles can enhance treatment safety by minimizing cumulative side effects. For example, L-AmB has a favorable safety profile and can offset the nephrotoxicity associated with conventional formulations. When used in combination with miltefosine or paromomycin, overall toxicity is reduced, and adverse effects are better managed (31). Furthermore, although L-AmB is expensive as a standalone treatment, its use in combination therapy can reduce the required dose, thereby lowering costs. From a public health perspective, improved compliance, lower hospitalization rates, and fewer relapses contribute to more cost-effective disease management and a higher likelihood of achieving VL elimination goals.

7.4 ESTABLISHED COMBINATION REGIMENS IN USE FOR VISCERAL LEISHMANIASIS

In response to the limitations of monotherapy and the growing threat of drug resistance, several combination regimens have been clinically evaluated and adopted into national treatment protocols for VL, particularly in endemic regions. These regimens aim to provide effective, safe, and shorter-duration treatments. Below is an overview of the most widely used and WHO-recommended combinations, as well as how treatment protocols differ based on geographic and epidemiological contexts.

7.4.1 Miltefosine + paromomycin

This oral-injectable combination has been evaluated in South Asia, particularly India and Bangladesh, where both drugs are already in use as monotherapies. Miltefosine, the first oral antileishmanial drug, acts by disrupting parasite membrane integrity (32), while paromomycin is an aminoglycoside antibiotic that inhibits protein synthesis (33).

Advantages: The combination has demonstrated a high cure rate (>97%) with a ten-day regimen, reduced treatment duration compared to monotherapies, and improved patient compliance due to oral-injectable format.
Limitations: Paromomycin requires intramuscular injections, which can be challenging in remote areas. Miltefosine is teratogenic, so it's contraindicated in pregnant women and women of childbearing age without effective contraception.
Key Studies: Sundar et al. (31) reported promising results with this combination in India.

7.4.2 Amphotericin B (liposomal) + miltefosine

This combination leverages the rapid parasiticidal activity of L-AmB with the longer-lasting oral action of miltefosine. L-AmB is less nephrotoxic and more targeted than conventional amphotericin B.

Advantages: A single infusion of L-AmB followed by seven days of oral miltefosine provides a potent, short-duration regimen with high efficacy

(~98% cure rates). It is highly suitable for inpatient initiation followed by outpatient continuation (30, 34).

Limitations: High cost of L-AmB limits widespread use in low-resource settings, although price negotiations and donations have improved access.

7.4.3 Amphotericin B (liposomal) + paromomycin

This regimen combines two injectable agents for short-course therapy and has shown excellent efficacy in areas with high resistance to other treatments.

Advantages: Offers fast parasite clearance with minimal risk of resistance. Can be administered over ten days, improving feasibility in hospital settings.

Limitations: The need for hospital-based administration of both injectable drugs makes it less practical in rural or remote locations.

Implementation: Used in regions with known miltefosine resistance or where its use is contraindicated (e.g., pregnancy).

7.5 CONCLUSION

The therapeutic landscape for VL has evolved significantly over the past decades, with a shift from traditional monotherapies to more effective and safer combination regimens. While drugs like amphotericin B, miltefosine, and paromomycin have played pivotal roles in treatment, each has limitations concerning toxicity, resistance, and treatment duration. Combination therapies have emerged as a promising strategy to enhance efficacy, reduce the risk of resistance, and improve patient compliance by shortening treatment courses and lowering toxicity. Despite these advancements, several challenges remain, including limited access to drugs in resource-poor settings, emerging drug resistance, and the need for treatments tailored to vulnerable populations such as children and HIV co-infected patients.

ACKNOWLEDGMENT

Authors are thankful to the ICMR-Rajendra Memorial Research Institute of Medical Sciences for the necessary research support for this work.

REFERENCES

1. Scarpini, Sara, Arianna Dondi, Camilla Totaro, Carlotta Biagi, Fraia Melchionda, Daniele Zama, Luca Pierantoni, et al. "Visceral leishmaniasis: Epidemiology, diagnosis, and treatment regimens in different geographical areas with a focus on pediatrics." *Microorganisms* 10, no. 10 (2022): 1887.

2. Aronson, Naomi, Barbara L. Herwaldt, Michael Libman, Richard Pearson, Rogelio Lopez-Velez, Peter Weina, Edgar M. Carvalho, Moshe Ephros, Selma Jeronimo, and Alan Magill. "Diagnosis and treatment of leishmaniasis: Clinical practice guidelines by the Infectious Diseases Society of America (IDSA) and the American Society of Tropical Medicine and Hygiene (ASTMH)." *Clinical Infectious Diseases* 63, no. 12 (2016): e202–e264.
3. World Health Organization. (2023). Newsroom 2023 [Report]. https://www.who.int/news-room/fact-sheets/detail/leishmaniasis
4. Okwor, Ifeoma, and Jude Uzonna. "Social and economic burden of human leishmaniasis." *The American Journal of Tropical Medicine and Hygiene* 94, no. 3 (2016): 489.
5. Bhunia, Gouri Sankar, and Pravat Kumar Shit. "Introduction of visceral Leishmaniasis (kala-azar)." In *Spatial Mapping and Modelling for Kala-Azar Disease*, pp. 1–18. Cham: Springer International Publishing, 2020.
6. Tripathi, Lokesh Kumar, and Tapan Kumar Nailwal. "Leishmaniasis: An overview of evolution, classification, distribution, and historical aspects of parasite and its vector." *Pathogenesis, treatment and prevention of leishmaniasis.* 2021: 1–25.
7. WHO Expert Committee on the Control of the Leishmaniases & World Health Organization. Control of the leishmaniases: Report of a meeting of the WHO Expert Commitee on the Control of Leishmaniases. Geneva, 22-26 March 2010: World Health Organization; 2010.
8. Majoor, Alissa, Grégory Michel, Pierre Marty, Laurent Boyer, and Christelle Pomares. "Leishmaniases: Strategies in treatment development." *Parasite* 32 (2025): 18.
9. Sundar, Shyam, and Jaya Chakravarty. "An update on pharmacotherapy for leishmaniasis." *Expert Opinion on Pharmacotherapy* 16, no. 2 (2015): 237–252.
10. van Griensven, Johan, Manica Balasegaram, Filip Meheus, Jorge Alvar, Lutgarde Lynen, and Marleen Boelaert. "Combination therapy for visceral leishmaniasis." *The Lancet Infectious Diseases* 10, no. 3 (2010): 184–194.
11. Meheus, Filip, Manica Balasegaram, Piero Olliaro, Shyam Sundar, Suman Rijal, Md Abul Faiz, and Marleen Boelaert. "Cost-effectiveness analysis of combination therapies for visceral leishmaniasis in the Indian subcontinent." *PLoS Neglected Tropical Diseases* 4, no. 9 (2010): e818.
12. Alves, Fabiana, Graeme Bilbe, Séverine Blesson, Vishal Goyal, Séverine Monnerat, Charles Mowbray, Gina Muthoni Ouattara, et al. "Recent development of visceral leishmaniasis treatments: Successes, pitfalls, and perspectives." *Clinical Microbiology Reviews* 31, no. 4 (2018): 10–1128.
13. Sundar, Shyam, and Jaya Chakravarty. "Visceral leishmaniasis." In *Drug Resistance in Leishmania Parasites: Consequences, Molecular Mechanisms and Possible Treatments*, pp. 159–176. Cham: Springer International Publishing, 2018.
14. Sundar S, Chakravarty J, Agarwal D, Rai M, Murray HW. Single-dose liposomal amphotericin B for visceral leishmaniasis in India. N Engl J Med. 2010 Feb 11;362(6):504-12. doi: 10.1056/NEJMoa0903627. PMID: 20147716.
15. mondiale de la Santé, Organisation, and World Health Organization. "Global leishmaniasis surveillance, 2022: Assessing trends over the past 10 years–Surveillance mondiale de la leishmaniose, 2022: évaluation des tendances des 10 dernières années." 2023.
16. Agrawal VK, Singh Z. Miltefosine: First Oral Drug for Treatment of Visceral Leishmaniasis. Med J Armed Forces India. 2006 Jan;62(1):66-7. doi: 10.1016/S0377-1237(06)80162-0. Epub 2011 Jul 21. PMID: 27407849; PMCID: PMC4923286.
17. Sundar, Shyam, and Piero L. Olliaro. "Miltefosine in the treatment of leishmaniasis: Clinical evidence for informed clinical risk management." *Therapeutics and Clinical Risk Management* 3, no. 5 (2007): 733–740.

18. Pokharel, Pashupati, Rakesh Ghimire, and Pratik Lamichhane. "Efficacy and safety of paromomycin for visceral leishmaniasis: A systematic review." *Journal of Tropical Medicine* 2021, no. 1 (2021): 8629039.
19. Seifert, Karin, F. Javier Pérez-Victoria, Marianne Stettler, María P. Sánchez-Cañete, Santiago Castanys, Francisco Gamarro, and Simon L. Croft. "Inactivation of the miltefosine transporter, LdMT, causes miltefosine resistance that is conferred to the amastigote stage of *Leishmania donovani* and persists in vivo." *International Journal of Antimicrobial Agents* 30, no. 3 (2007): 229–235.
20. Hendrickx, Sarah, João Luís Reis-Cunha, Sarah Forrester, Daniel C Jeffares, and Guy Caljon. "Experimental selection of paromomycin resistance in *Leishmania donovani* amastigotes induces variable genomic polymorphisms." *Microorganisms* 9, no. 8 (2021): 1546. doi: 10.3390/microorganisms9081546.
21. Croft, Simon L., Shyam Sundar, and Alan H. Fairlamb. "Drug resistance in leishmaniasis." *Clinical Microbiology Reviews* 19, no. 1 (2006): 111–126.
22. Chakravarty, Jaya, and Shyam Sundar. "Drug resistance in leishmaniasis." *Journal of Global Infectious Diseases* 2, no. 2 (2010): 167–176.
23. Sundar, Shyam, Anand Makharia, Deepak K. More, Gaurav Agrawal, Andreas Voss, Christina Fischer, Peter Bachmann, and Henry W. Murray. "Short-course of oral miltefosine for treatment of visceral leishmaniasis."*Clinical Infectious Diseases* 31 (2000): 1113–1114.
24. van Griensven, Johan, Thomas P. C. Dorlo, Ermias Diro, Carlos Costa, and Sakib Burza. "The status of combination therapy for visceral leishmaniasis: An updated review." *The Lancet Infectious Diseases* 24, no. 1 (2024): e36–e46.
25. Tallarida, Ronald J. "Quantitative methods for assessing drug synergism." *Genes & Cancer* 2, no. 11 (2011): 1003–1008.
26. Shabi, Parvez, Yadagiri Ganesh, Gedda Mallikarjuna Rao, Singh Aakriti, Om Prakash Singh, Verma Anurag, Sundar Shyam, and Shyam Lal Mudavath. "Modified solid lipid nanoparticles encapsulated with Amphotericin B and Paromomycin: An effective oral combination against experimental murine visceral leishmaniasis." *Scientific Reports (Nature Publisher Group)* 1 (2020).
27. Goswami, Rama Prosad, Mehebubar Rahman, Sukhen Das, Santanu Kumar Tripathi, and Rudra Prosad Goswami. "Combination therapy against Indian visceral Leishmaniasis with Liposomal Amphotericin B (Fungisome™) and short-course miltefosine in comparison to miltefosine monotherapy." *The American Journal of Tropical Medicine and Hygiene* 103, no. 1 (2020): 308.
28. Melcon-Fernandez, Estela, Giulio Galli, Carlos García-Estrada, Rafael Balaña-Fouce, Rosa M. Reguera, and Yolanda Pérez-Pertejo. "Miltefosine and nifuratel combination: A promising therapy for the treatment of *Leishmania donovani* visceral leishmaniasis." *International Journal of Molecular Sciences* 24, no. 2 (2023): 1635.
29. Kumari, Alka, Niyamat A. Siddiqui, Sweta Kumari, Krishna Murti, Rishikesh Kumar, Krishna Pandey, Somanaboina Padmakar, and Biplab Pal. "Combination therapy for post-kala-azar dermal leishmaniasis: A literature review of current evidence." *Indian Journal of Dermatology* 69, no. 5 (2024): 396–405.
30. Sundar, Shyam, Prabhat Kumar Sinha, Deepak K. Verma, Nawin Kumar, Shahnawaj Alam, Krishna Pandey, Poonam Kumari, et al. "Ambisome plus miltefosine for Indian patients with kala-azar." *Transactions of the Royal Society of Tropical Medicine and Hygiene* 105, no. 2 (2011a): 115–117.
31. Sundar, Shyam, Prabhat Kumar Sinha, Madhukar Rai, Deepak Kumar Verma, Kumar Nawin, Shanawwaj Alam, Jaya Chakravarty, et al. "Comparison of short-course multidrug treatment with standard therapy for visceral leishmaniasis in India: An open-label, non-inferiority, randomised controlled trial." *The Lancet* 377, no. 9764 (2011b): 477–486.

32. Soto, Jaime, and Paula Soto. "Miltefosine: Oral treatment of leishmaniasis." *Expert Review of Anti-Infective Therapy* 4, no. 2 (2006): 177–185.
33. Jhingran, Anupam, Bhavna Chawla, Shailendra Saxena, Michael Peter Barrett, and Rentala Madhubala. "Paromomycin: Uptake and resistance in *Leishmania donovani*." *Molecular and Biochemical Parasitology* 164, no. 2 (2009): 111–117.
34. Burza, Sakib, Raman Mahajan, Shahwar Kazmi, Neal Alexander, Deepak Kumar, Vikash Kumar, Estrella Lasry, et al. "AmBisome monotherapy and combination AmBisome–miltefosine therapy for the treatment of visceral leishmaniasis in patients coinfected with human immunodeficiency virus in India: A randomized open-label, parallel-arm, phase 3 trial." *Clinical Infectious Diseases* 75, no. 8 (2022): 1423–1432.

Chapter 8

Integrated vector management for visceral leishmaniasis control in India

*Ajay Kumar**, *Sanjay D, Anil Kumar, and Yash Paul Sharma*

Department of Vector Biology and Control, ICMR-Rajendra Memorial Research Institute of Medical Sciences, Patna, India

8.1 INTRODUCTION

Visceral leishmaniasis (VL) is a vector-borne disease (VBD) caused by an intracellular protozoon parasite of the genus *Leishmania*. VL is also called kala-azar. It can be a deadly disease if untreated. It ranks second based on death and seventh based on loss of disability-adjusted life years among tropical diseases (McCall et al., 2013; Scarpini et al., 2022). The clinical presentation of VL includes sporadic episodes of fever, weight loss, splenomegaly, hepatomegaly, and anemia (Mahto et al., 2023). Humans serve as the pathogen's reservoir for VL, and *Leishmania donovani* causes VL in Asia and Eastern Africa (Mahto et al., 2023; Scarpini et al., 2022). In Latin America and the Mediterranean region, dogs serve as the primary reservoir of *Leishmania infantum*, which causes VL (Romero & Boelaert, 2010). VL is purely anthroponotic in the Indian subcontinent (ISC). The parasite *L. donovani* is transmitted to humans on the ISC by the bite of an infected female sand fly, *Phlebotomus argentipes* (Diptera: Psychodidae). It is one of the most common parasitic infections with significant potential for epidemics and death. A total of 83 nations were believed to be endemic for VL in 2018. Leishmaniasis poses a threat to over 1 billion people, and an estimated 30,000 new cases of VL are recorded globally each year. In 2019, ten countries—India, East Africa, Iraq, Somalia, Brazil, Ethiopia, Kenya, South Sudan, Sudan, and Nepal—accounted for over 90% of incident cases reported to World Health Organization (WHO). In 2020, 18% of the world's kala-azar cases were reported from India. With 54 districts spread over 4 states, including Bihar (33 out of 38 districts), Jharkhand (4 out of 24 districts), Uttar Pradesh (6 out of 75 districts), and West Bengal (11 out of 23 districts) (leishmaniasis—India; Kumar et al., 2020; Mahto et al., 2023, kala-azar is still a major public health concern in India (Kumar et al., 2020). The sporadic cases were found in states of India, including Gujarat, Assam, Himachal Pradesh, Tamil Nadu,

* Corresponding author: akajaykumar836@gmail.com

DOI: 10.1201/9781003596455-8

Table 8.1 Kala-azar cases and fatalities from India throughout the past five years

Kala-azar situation in India

	2020		*2021*		*2022*		*2023*		*2024 (Sep)*	
States	*C*	*D*	*C*	*D*	*C*	*D*	*C*	*D*	*C*	*D*
Bihar	1427	0	893	2	550	0	335	3	119	0
Jharkhand	424	0	275	2	188	1	154	0	106	0
Uttar Pradesh	55	3	49	2	23	2	14	1	10	1
West Bengal	57	3	58	2	52	0	17	0	23	0

Source: Kala-azar Situation in India: NCVBDC.

Haryana, Madhya Pradesh, Kerala, Sikkim, Uttaranchal, and Jammu and Kashmir (leishmaniasis—India; Kumar et al., 2020; Mahto et al., 2023). The ISC accounts for two-thirds of all VL occurrences globally, with more than half occurring in the Indian state of Bihar. Furthermore, Bihar is responsible for 61.1% of all VL-affected districts in India (Kumar et al., 2020; Mahto et al., 2023) (Table 8.1).

The increased incidence of VL may have been caused by numerous factors, such as delayed diagnosis, insufficient treatment options, a lack of technical competence, the operational viability of sand fly control strategies in different ecological settings, the availability of insufficient financial resources, and the unexpected rate of population growth, which has been a major driver of environmental change in rural areas (such as deforestation, agricultural land use, and animal husbandry practices) (Mahto et al., 2023).

The spreads of VL among low-income groups were managed by vector control through the application of chemical-based insecticides. In the early 1980s, the idea of integrated vector control (IVC) emerged in essence; it was a vertical strategy that emphasized the application of one or more vector management techniques to combat the disease of interest (Manual on integrated vector management (IVM) in India; Romero & Boelaert, 2010) and followed, in the early 1990s. The idea of "selective vector control" was first proposed, and it called for the use of focused, site-specific, and economical measures to lower the morbidity and death. Later, a concept of "comprehensive vector control" evolved to control the vectors of two or more co-prevalent diseases by using the vertical approach, which entails managing similar or different methods. The integrated pest management (IPM) technique was promoted in 1997 when the 50% World Health Assembly proposed the concept of using fewer chemical pesticides for controlling VBDs (Manual on IVM in India). To reduce the burden of VBDs, sustainable control requires an integrated, multi-sectoral strategy. Therefore, in 2004, the WHO promoted the IVM approach, highlighting the fact that "effective control is not the sole preserve of the health sector but requires public-private, inter-sectoral collaboration and community participation." As a result, IVM was defined as a rational decision-making process for the optimal use of resources for vector control. Through the IVM, multiple disease control can be achieved without disrupting the environment via the use of intersectoral and inter-disciplinary techniques. In that situation, IVM is a good approach to take environmental and health aspects in order to reduce vector–human interaction, interrupt the vector's life cycle, and eradicate or regulate vector breeding sites without altering the ecology (Manual on IVM in India). Hence, IVM approaches can potentially use to regulate the VL. To satisfy the objectives of India's national kala-azar elimination programs, this chapter reviews the applicability of IVM to the control of VL, summarizes lessons learned from IVM approaches used for other VBDs, and provides recommendations to accelerate IVM development.

8.2 IVM IN KALA-AZAR ELIMINATION

The national VL control program in India was successfully launched between 1990 and 1991. Subsequently, the Indian government established a national health program in 2002, with the goal of eliminating VL by 2010 (Beier et al., 2008; Garlapati et al., 2021; World Health Organization (WHO), 2015). In 2005, under the guidance of the WHO, the governments of Bangladesh, India, and Nepal formulated a strategic plan to eliminate VL as a public health issue by 2015 (World Health Organization (WHO), 2015). However, the goal year has been extended numerous times notably in 2017, 2020, and 2023, despite the WHO NTD Road Map goal is 2030 (Sundar et al., 2018; Hakizimana, 2019), 2016. VL elimination is defined as reducing the annual incidence of VL to fewer than one case per 10,000 people at the sub-district/block level (Kumar et al., 2020; Sundar et al., 2018). The primary methods for eliminating VL are early case detection and comprehensive treatment, IVM, effective disease surveillance, social mobilization and behavioral changes, and operational research (Ahmed et al., 2014; Kumar et al., 2020; Thakur et al., 2009; World Health Organization (WHO), 2015). In India, the National Center for Vector Borne Diseases Control (NCVBDC) at the central level facilitates IVM implementation by assisting in the policy formulation and strategic planning process. States and union territories are responsible for providing infrastructure for IVM implementation, while the national and state governments share financial resources in accordance with policy (National Kala-azar Elimination Programme (NCVBDC); Manual on IVM in India).

Globally, the WHO created a global strategic framework for IVM in 2004. This framework encompassed novel, universal vector control strategies applicable to all diseases transmitted by vectors (Beier et al., 2008). The rationale behind the inclusion of IVM in a global strategic framework was that successful VBD management would have a significant benefit on both health and socioeconomic development (Report of the Global Partners' Meeting on Neglected Tropical Diseases, 2007). Capacity building and IVM strengthening were identified as the key areas of action in the global strategy to fight against neglected tropical diseases for 2008–2015. In 2008, WHO published the Global Strategic Framework for IVM, which aims to expedite the creation of national IVM policies and plans. In addition to inviting international organizations, donor agencies, and other stakeholders to assist with the capacity building are required for IVM implementation (Beier et al., 2008; Hakizimana, 2019).

IVM is defined as a rational decision-making process to optimize the use of resources for vector control. It encourages the use of various methods, either separately or in combination, based on local knowledge of disease factors, vectors, and epidemiology. By addressing multiple diseases simultaneously and integrating local human resources and operations, IVM offers an efficient, economical, sustainable, and environmentally friendly approach to vector control. Importantly, it reduces the risk of pesticide resistance in vector populations (Ahmed et al., 2014; World Health Organization (WHO), 2012).

The primary goal of IVM approaches is to protect populations from VBDs. These approaches target a variety of vectors, including sand flies, mosquitoes, ticks, and others; across various ecological and socioeconomic contexts, IVM methods for controlling VL vectors include indoor residual spraying (IRS), long-lasting insecticidal nets (LLINs), and insecticide-treated nets (ITNs). Additional strategies involve community mobilization, environmental management (EVM), and biological control targeting vector larvae or adults. Unlike traditional vector control methods, IVM emphasizes collaboration with local communities, stakeholders (such as ministries of agriculture, environment, or sewer and water works), and regulatory frameworks to IVC programs with national health systems (Ahmed et al., 2014; Lizzi et al., 2014) (Table 8.2).

Establishing guiding principles is important for IVM implementation, these concepts should be integrated into national health policies and supported by laws and regulations to provide a unified framework for action (Beier et al., 2008; Hakizimana, 2019). The IVM framework delineates the critical phases required for effective implementation, including diagnosis, selection of vector control tools, resource assessments, implementation, and monitoring and evaluation. Key activities involve stratifying at-risk regions, identifying disease-specific factors, and conducting epidemiological and entomological assessments (Beier et al., 2008). The choice of vector control techniques is influenced by the availability of public health products, program feasibility,

Table 8.2 To enhance the efficacy of IVM, WHO developed five principles

Advocacy, social mobilization, and legislation	Promotion and embedding of IVM principles in the development policies of all relevant agencies, organizations, and civil society; establishment or strengthening of regulatory and legislative controls for public health; empowerment of communities (World Health Organization (WHO), 2012).
Collaboration within the health sector and with other sectors	Consideration of all options for collaboration within and between the public and private sectors; application of the principles of subsidiarity in planning and decision-making; strengthening channels of communication among policy-makers, vector-borne disease program managers and other IVM partners (World Health Organization (WHO), 2012).
Integrated approach	Ensure rational use of available resources by addressing several diseases, integrating non-chemical and chemical vector control methods and integrating with other disease control methods (World Health Organization (WHO), 2012).
Evidence-based decision-making	Adaptation of strategies and interventions to local vector ecology, epidemiology, and resources, guided by operational research and subject to routine monitoring and evaluation (World Health Organization (WHO), 2012).
Capacity-building	Provision of the essential material infrastructure, financial resources, and human resources at national and local level to manage IVM strategies on the basis of a situational analysis (World Health Organization (WHO), 2012).

Source: World Health Organization (WHO) (2012).

and government regulations governing their usage. Logistics, human, skilled human, and financial resources must be mapped and aligned with needs at state and district levels to ensure the success of IVM activities (Manual on IVM in India). This comprehensive approach of IVM with broader disease prevention and treatment efforts, will ultimately achieve our public health objectives.

8.3 REASONS FOR NOT SOLVING THE ISSUE WITH CONVENTIONAL VECTOR CONTROL TECHNIQUES

Several factors have made it difficult for traditional vector management methods to successfully reduce the prevalence of VL, especially in endemic areas like Bihar, India. The following challenges are outlined below:

1. Historical background and policy difficulties
2. Knowledge gaps in vector biology
3. The vector's behavioral modifications
4. Weak program coverage and execution
5. Insufficient integration with existing control measures.

8.3.1 Historical background and policy difficulties

The National Malaria Control Programme (NMCP) in 1953 and the National Malaria Eradication Programme (NMEP) in 1958 successfully implemented widespread IRS with DDT 50% for malaria control. However, kala-azar cases also decreased to very low level of prevalence because of the collateral insecticidal pressure on the sand fly vector, which caused the transmission to cease altogether. Despite this, the IRS withdrew from once malaria-endemic regions after malaria cases had down, and it led to a resurgence of sand fly populations, which in turn caused kala-azar to reappear in the 1970s. This resurgence began in four districts of Bihar, the revival gradually extended throughout the entirety of North Bihar, a few districts of West Bengal, and some areas of Jharkhand. During the 1980s, a few areas in Uttar Pradesh recorded sporadic occurrence. This historical background demonstrates how modifications to vector control regulations can affect disease dynamics over an extended period of time (operational guidelines and Guidelines-Vector-Control-KA-Elimination-FINAL). Sustaining successful vector control initiatives requires sustained financial and political support to maintain successful vector control initiatives.

8.3.2 Knowledge gaps in vector biology

Significant information gaps exist regarding the bionomics and transmission dynamics of sand flies (Garlapati et al., 2021). Vector control techniques cannot be successfully adapted to local conditions without a thorough understanding of the bionomics. For example, a study from Bihar has revealed that *P. argentipes* are now exhibiting exophilic behavior, demonstrated by the setting of traps in vegetation in outskirts of endemic villages in Bihar to measure the abundance of sand flies (Garlapati et al., 2021; Poche et al., 2018), thus challenging the long-held belief that they are predominantly endophilic. Research highlights the need for more investigations into the ecology and behavior of these vectors to refine control methods.

8.3.3 The vector's behavioral modifications

In the ISC, IRS is the main method of controlling sand flies (World Health Organization (WHO), 2012). Perhaps this approach is insufficient, particularly in India. According to conventional opinion, *P. argentipes* is nearly completely endophilic and endophagic (Chowdhury et al., 2016). However, according to recent reports from Bihar, which emphasize its exophilic and exophagic behavior, vector control must also address outdoor sand fly populations for it to be effective. A recent study revealed that 95% of people among Bihar's most endemic districts sleep outside during the warm months of the year (Garlapati et al., 2021; Perry et al., 2013; Poché et al., 2011). It implies that indoor spraying attempts could not have an impact on many sand flies. Significantly, it was shown that a significant percentage of *P. argentipes* in

Bihar villages had consumed human blood (41%), followed by cattle (20%), a combination of human–cow (17%), and human–buffalo (5%). The majority of *P. argentipes* sand flies fed on human (81%) and/or bovine blood (60%), according to a large-scale study that comprised 24 villages in the state of Bihar (Garlapati et al., 2021; Poche et al., 2018). These studies collectively demonstrate that *P. argentipes* sand flies may be exophilic and possibly exophagic. These activities would reduce the effectiveness of IRS and LLIN strategies that target populations of endophilic and endophagic *P. argentipes*, respectively. Due to vector adaptability, vector management strategies need to evolve to accommodate outdoor spraying and other techniques that target both indoor and outdoor vector population.

8.3.4 Weak program coverage and execution

The quality of insecticide application and dwelling coverage has a significant impact on IRS efficacy. The efficacy of IRS on targeted vector populations can be determined by sand fly susceptibility, dwelling coverage, and pesticide application quality, the issues that may reduce the efficacy of vector control methods in India (Garlapati et al., 2021). Moreover, insufficient spray worker training may result in incorrect ways of application and less-than-ideal insecticide concentrations on treated surfaces. The 1995 Expert Committee on Malaria suggested using 26 squads for a 75-day spraying period to apply synthetic pyrethroids (SP) and DDT to 1 million people in order to control malaria. A spray squad consisting of five people can typically cover 100 homes each day with a single round of anti-malaria spraying. Using two stirrup pumps and one superior field worker, each spray squad is made up of five more field workers. But in case of anti-kala-azar spraying, only half of the staff is likely to be involved. IRS implementation result in the area can only be sprayed in half of it (Vector-Control-KA-Elimination-FINAL-Guidelines). These all are reducing impact of program.

8.3.5 Insufficient integration with existing control measures

Houses in VL-affected regions of India usually are constructed of brick, thatch, and mud with or without plaster with cement, brick, or mud (Malaviya et al., 2014; Perry et al., 2013). These constructions are prone to cracks and gaps, which give sand flies dark and moist hiding places. There is further evidence linking VL risk to noticeable moisture in the clay flooring of these houses (Bern et al., 2010). However, *P. argentipes* are more common in thatched homes with mud flooring (Malaviya et al., 2014). These are affecting the effectiveness of IRS and the use of LLIN. Current strategies often lack integration with EVM. This integration is essential for fully eliminating the disease and its vectors. Thus, an integrated strategy that incorporates several vector control techniques with early diagnosis, treatment, and community involvement is necessary for the effective management of kala-azar.

8.4 IMPLEMENTATION OF IVM STRATEGIES TO CONTROL KALA-AZAR

IVM has emerged as a comprehensive and sustainable strategy to combat VBDs globally by integrating biological, environmental, and chemical control methods and also community participation, intersectoral collaboration (ISC), and adaptive management (Chanda et al., 2013; Marcos-Marcos et al., 2018). IVM has been widely employed for malaria and other VBDs with significant success. In Western Kenya, community participation was integral to IVM, where stakeholders played critical roles in decision-making and implementation (Chanda, 2016; Ng'ang'a et al., 2021). This approach fostered ownership and accountability, ensuring sustainability. Similarly, in Zimbabwe, policy reforms and community engagement enhanced IVM implementation, addressing local epidemiological challenges effectively (Sande et al., 2019). Such integrated efforts underscore the importance of policy coherence and community involvement in VBD management. In South Sudan, IVM was pivotal in mitigating malaria outbreaks through a combination of ITNs, IRS, and larval source management (Chanda et al., 2013). Furthermore, Saudi Arabia's success in reducing VL cases to zero was attributed to integrated surveillance and IRS campaigns, complemented by health education initiatives (Abuzaid et al., 2020). These examples highlight the adaptability of IVM to different sociopolitical and ecological contexts. IVM emphasizes the vector control to protect people by preventing, reducing or interrupting the transmission of kala-azar and other VBDs (Ahmed et al., 2014). One of the major strategies to control sand fly (vector) is the IVM, for the elimination of kala-azar, which includes several conventional methods.

8.4.1 Intersectoral collaboration (ISC)

A cornerstone of successful IVM is ISC, which ensures alignment between health, environmental, and community sectors. Herdiana et al. (2018) emphasized that integrating health policies with EVM strategies reduces duplicative efforts and optimizes resources. For instance, the malaria eradication program in Saudi Arabia leveraged cross-border collaboration with Yemen to address shared vector control challenges (Abuzaid et al., 2020). Similarly, integrating vector control measures for multiple diseases, such as malaria and VL, has proven cost-effective and impactful (Chanda et al., 2017). ISC is the cooperation between different sectors of society, such as public, private, and civil society sectors. To achieve shared goals, particularly in health, this approach, rooted in the Alma Ata Declaration (1978) on primary health care, emphasizes collective action by multiple sectors to address health determinants and inequalities (International Conference on Primary Health Care, 1978). The 1998 WHO Health Promotion Glossary defined ISC as a cooperative process for improved population health outcomes (WHO, 1998). ISC emerged as part of efforts to address health inequities and emphasize that health is not solely the responsibility of the health sector. It was initially applied to primary health care and health promotion but has expanded to broader programs, e.g., school health and healthy city planning (Herdiana, 2018).

8.4.2 ISC for health programs requirements

- The collaboration is being guided by advocates and leadership.
- Relationships that are mutually beneficial and encouraged.
- Combining the macro and micro levels.
- Implementing health impact evaluations into action.
- adequate financial, technical, and human resources.
- Capacity building, tools, and training.
- Community participation and social mobilization.

ISC is one of the key pillars of action in reducing the human burden caused by VBDs, according to the newly adopted Global Vector Control Response 2017–2030 document (WHO, 2017) by the 70th session of the World Health Assembly. WHO and its partners are also promoting this agenda of multi-sectoral collaboration for the control and prevention of VBDs (Herdiana, 2018; WHO, 2017). In India, effective ISC is necessary for successful VL control.

8.4.3 Indoor residual spraying (IRS)

In the 1950s and 1960s, widespread IRS with DDT was central to the NMEP in India. As a collateral benefit, it significantly reduced sand fly populations and nearly eliminated VL cases in Bihar (Garlapati et al., 2021; Ostyn et al., 2008; Sanyal, 1979; Thakur, 2007). Following the cessation of DDT spraying in 1964, VL cases resurged dramatically, with epidemics occurring in the late 1970s and early 1990s (Garlapati et al., 2021; Sanyal, 1979). In response to the 1990s epidemic, DDT-based IRS (5%) was reinstated and, combined with sodium stibogluconate (SSG) treatment, led to a decline in VL cases (Ostyn et al., 2008; Thakur, 2007).

Between 2001 and 2007, the number of cases increased from 12,000 to 44,000, indicating the emergence of a third epidemic cycle (Alvar et al., 2012). Clinical resistance in parasite *L. donovani* to SSG resulted in decreased effectiveness even after reintroducing IRS (Sundar et al., 2000). The assessment of treatments is complicated by the fact that VL outbreaks often occur in cycles of 15–20 years (Courtenay et al., 2017). Research by Hasker et al. (2012) and Joshi et al. (2009) emphasizes IRS as a key component of VL management in the ISC. The effectiveness of IRS treatments as a key control strategy in endemic areas is supported by Joshi's randomized trials conducted in Bangladesh, India, and Nepal, which demonstrate a notable decrease in vector populations (Hasker et al., 2012; Joshi et al., 2009). The study has shown that IRS in particular, under quality-controlled conditions and implemented by dedicated research teams, was efficacious in reducing the indoor sand fly density (Joshi et al., 2009). To reduce VL vector densities at least for five or six months after the intervention, independent of housing conditions and the presence or absence of cattle (Joshi et al., 2009) followed to reduce VL vector densities at least for five or six months after the intervention, independent of housing conditions and the presence or absence of cattle (Garlapati et al., 2021; Joshi et al., 2009). However, the poor guidance, a lack of coverage (12–63.9%), and incorrect application methods have all reduced IRS efficacy and resulted in variable insecticide residues (Chowdhury et al., 2011). Despite growing evidence of DDT resistance, DDT remained the only pesticide used until 2015 (Mukhopadhyay et al., 1990). According to Singh et al. (2012), the lethal rates of sand flies caused by DDT ranged from 31 to 89% (Singh et al., 2012). To combat this, India started switching to SP in 2015, namely 5% alphacypermethrin, which eventually replaced DDT completely by 2016 (Coleman et al., 2015). The WHO suggests conducting two IRS cycles every year. In India, the minimum criteria of two IRS round in a village are having at least one VL case that year. According to Garlapati et al. (2021), the use of hand compression pumps, regular monitoring since 2014, and a rise in home coverage from 51% in 2014 to over 80% by 2015 are all examples of improvements in IRS quality (Garlapati et al., 2021). The transition from DDT to SP has produced a range of outcomes. Although some studies demonstrated a considerable decrease in the prevalence of sand flies 15 days after SP spraying (Kumar et al., 2020), more comprehensive assessments revealed no effect (Poche et al., 2018). Despite IRS scaling from 40 to >80% household coverage between 2013 and 2018, most of the VL reduction (82%) occurred by 2013, likely due to enhanced case detection and newer treatments, including the replacement of SSG with miltefosine (2008–2012) and liposomal amphotericin B (2015) (Singh et al., 2016). Additionally, in 2015, vector control intervention from the Government of India-entrusted ICMR-Rajendra Memorial Research Institute of Medical Sciences, Patna, has achieved the VL elimination in endemic district of Bihar to target of <1 case per 10,000 people at the block level in the Vaishali district of Bihar. By the implementation of an integrated control strategy of IVM strategies including the epidemiological analysis based on a geographical information system (GIS), active case detection (ACD), IRS, community awareness campaigns, human resource development such as training of IRS members, the training of medical doctors for effective treatment, daily monitoring and the supervision of IRS activities, logistic management, post-IRS quality assurance, and entomological monitoring was performed (Kumar et al., 2020). The available statistics highlight the necessity of a long-term, evidence-based approach to VL management that emphasizes integrated methods (early case detection, complete treatment, LLIN, and EVM) rather than depending just on IRS (Garlapati et al., 2021). Therefore, maintaining IRS is crucial in post-elimination settings for preventing vector resurgence, particularly in regions where transmission risk remains.

8.4.4 Insecticide-treated nets (ITNs)

As an adjunctive vector control strategy to IRS, leishmaniasis and malaria can be effectively prevented by long-lasting insecticide nets (LLINs), a form of ITNs that has the insecticide woven into the fabric (Wilson et al., 2014). ITNs offer both insecticidal activity and personal protection (Joshi et al., 2009). ITNs function by preventing sand fly vectors from spreading to people, which is particularly important in high-transmission regions. In Bangladesh, Chowdhury's study showed that using ITNs reduced the prevalence of VL, proving its usefulness in integrated control strategies. There have been conflicting results on their effectiveness against VL in the ISC (Chowdhury et al., 2019). According to the KalaNet experiment, which was conducted in endemic areas of Nepal and India, LLINs had no significant impact on the incidence of VL during 24 months and only reduced the density of *P. argentipes* by 25% (Picado et al., 2010). Nevertheless, due to variations in vector biology, human behavior and research methodology and also most likely VL caused by transmission continued outside of dwellings. Additionally, people's outdoor sleeping habits, which are common in India (95%) but relatively uncommon in Bangladesh (<5%), had an impact on the outcomes. According to research done in Bangladesh, the distribution of ITN decreased the incidence of VL by 66.5% (Mondal et al., 2019). India's history of IRS with DDT may have led to a shift in *P. argentipes* behavior toward more exophilic and exophagic behaviors, in contrast to Bangladesh, where sand flies remained endophilic since IRS did not occur until 2012. These findings highlight the necessity for IVM to continue integrating IRS, ITN, and LLIN distribution and utilization to prevent reemergence, particularly during post-elimination stages (Chowdhury et al., 2019; Garlapati et al., 2021). Although ITNs and LLINs offer promise, their efficacy is dependent upon human behavior, vector behavior, and the requirement for tailored designs that correspond to the local conditions (Chowdhury et al., 2019; Ostyn et al., 2008).

8.4.5 Environmental modification (EVM)

EVM is the process of changing dwellings or environments, such as plastering wall and floor cracks and crevices, to reduce sand fly habitats. In thatched houses with mud flooring, increased densities of sand flies were observed (Garlapati et al., 2021; Malaviya et al., 2014). According to Bern et al. (2010), there is a higher risk of VL when there are cracks and moist clay flooring. In VL-endemic areas of India, dwellings are often made of mud, thatch, or brick with mud/earthen, brick, or cement floors, which provide dark, humid refuges for *P. argentipes* (Malaviya et al., 2014; Perry et al., 2013). Studies indicate that EVM can lower the density of sand flies; however, it is limited. With sand fly density reductions of 72.4, 42.0, and 43.7%, respectively, over a minimum of five months, the Joshi et al. (2009) research is still the only study comparing IRS, ITNs, and EVM across Bangladesh, Nepal, and India (Garlapati et al., 2021; Joshi et al., 2009). While LLINs were only effective in Bangladesh and India, IRS was effective everywhere. In India and one location in Nepal, lime plastering showed a notable decrease in the density of sand flies. In Bihar, a trial employing a mud and lime plaster mix decreased the prevalence of sand flies from 11.5 to 2.0 per man-hour within a month of sealing (Kumar et al., 1995; Joshi et al., 2009). In India, Nepal, and Bangladesh, a mud-lime mix plastering of walls and cattle enclosures resulted in a 42% decrease in sand fly abundance over five months. Since lime plastering has a high pH or limits moisture, it probably lowers the densities of sand fly populations (Garlapati et al., 2021;Joshi et al., 2009; Kumar et al., 1995). According to Chowdhury et al. (2011), mud-sealing cracks alone did not significantly lower the density of sand flies in Bangladesh. A study in Vaishali district, Bihar, showed a 60.2% reduction in sand fly density when cement plastering was applied to the walls of 300 houses compared to controls (Dinesh et al., 2017; Garlapati et al., 2021). Joshi et al. (2009) and Banjara et al. (2019) highlight the use of integrated EVM with chemical control, such as IRS and LLIN, to minimize vector breeding sites and density. Furthermore, these measures focus primarily on endophilic sand fly populations, neglecting exophilic populations in natural settings and vegetation as potential VL risks (Joshi et al., 2009). To maintain vector populations low, it is critical to promote EVM and community involvement, as well as emphasize the importance of long-term vector control measures that prevent chemical reliance.

8.4.6 Surveillance and engagement of the community

Disease surveillance includes reporting cases of PKDL and kala-azar (Saurabh et al., 2020). At the moment, government entities use passive case detection (PCD) for surveillance. This presents an inaccurate picture because (a) most kala-azar cases are treated by private physicians, including untrained practitioners, as there is no reporting from these healthcare providers; (b) treatment is frequently initiated without a conclusive diagnosis of kala-azar; and (c) due to sociocultural and economic limitations, many cases do not seek medical attention at all (Asia, 2012; Kumar et al., 2013; Saurabh et al., 2020). Despite these limitations, PCD and reporting are used to monitor disease trends. Improving diagnosis and treatment is intended to boost reporting, establish connections with private healthcare providers, especially private doctors, and raise public awareness of the risks associated with employing unlicensed experts for diagnosis and treatment.

To treat the community properly, an effort should be made to use trained individuals (Kumar et al., 2013; Saurabh et al., 2020). Kala-azar should be declared a disease that requires notification in the impacted areas to enhance surveillance. Monthly district-level reporting and feedback, as well as the development of a regular reporting system with national and state authorities, should be part of kala-azar disease surveillance (Kumar et al., 2013; Saurabh et al., 2020). Annual (or, if feasible, twice-yearly) reports should be sent to WHO, and endemic nations should use a predetermined format. As the program gets better and has more capacity, ACD, which is supported by laboratory diagnostics, should be added to PCD. First, ACD should be administered at least once yearly (ideally twice a year). As the number of cases reported by PCD decreases, this will become more significant. VL-endemic districts have demonstrated a high output of new cases for several ACD strategies (camp, index case, incentive, and blanket approach) regarding cost, yield, and feasibility. The cost of the program varies depending on the screening method selected. Depending on the epidemiological profile, organizational viability, and cost, countries necessitate the appropriate combination of strategies (Asia, 2012; Kumar et al., 2013; Saurabh et al., 2020). This strategy maintains control successes by guaranteeing a prompt reaction to potential breakouts. VL control is strengthened, particularly in rural regions, by engaging local health workers in surveillance, providing residents with training on vector prevention, and involving communities in EVM (Kumar et al., 2013; Saurabh et al., 2020). By combining successful case management and vector control measures, VL was virtually eradicated in Saudi Arabia. Since early diagnosis and treatment decrease the role of human cases as disease reservoirs, they are essential to these efforts. The mandatory transmission of suspected or confirmed cases was early case detected and allowed for prompt treatment within 24 hours through the improved Health Electronic Surveillance Network (HESN). Frequently, tertiary hospitals are treated cases with sufficient antileishmanial drugs. While private-sector cooperation guarantees accelerated access to necessary medications, a referral system links primary and secondary healthcare facilities to specialized centers, guaranteeing effective case management (Abuzaid et al., 2020).

8.4.7 Social mobilization and collaboration

Interventions for behavioral change are critical to the effectiveness of early diagnosis, treatment adherence, and the eradication of kala-azar. Early care seeking can also be encouraged by effective BCC. It is essential that families and the community participate in IRS and efforts to reduce human–vector contact. From the beginning of the elimination effort, social mobilization should be a key component. Adequate resources should be planned for by national programs to ensure successful BCC (Ahmed et al., 2014).

According to Thakur & Thakur, 2020, social mobilization initiatives and awareness campaigns for community involvement in India should send messages to every home in the block. All the existing communication methods both mass media and interpersonal communication methods should be used to convey messages that are prepared keeping the local context in mind. Community volunteers and community-based organizations can play a key role in this regard and must be mobilized (Ahmed et al., 2014).

All levels of collaboration—including district, state, national, and international stakeholders—will be required. Multi-partner leadership has been crucial in the success of several elimination and eradication programs, including those for lymphatic filariasis, leprosy, and polio. Collaborations with other programs, such as those addressing VBDs (malaria, dengue, and filaria) and others, such as HIV/AIDS, TB, and leprosy, will be necessary. The kala-azar Elimination Program should collaborate with activities to reduce poverty, enhance nutritional status, and control anemia (Ahmed et al., 2014; Asia, 2012). As Western Kenya has shown, empowering communities via involvement and education can promote sustainable vector control methods (Ng'ang'a et al., 2021). In Zimbabwe, it appears that the lesson for VL control in India was creating regulations that combine VL control with other VBD initiatives, including those for dengue and malaria, can maximize resources and ensure an effective response (Abuzaid et al., 2020; Sande et al., 2019). The cross-sectoral cooperation in Saudi Arabia and Zimbabwe is a useful instance. In Saudi Arabia, areas where malaria and VL coexist are subject to enhanced IVM through the distributing ITNs and IRS provides protection from both mosquitoes and sand flies. Social mobilizations as well collaborations across borders have improved efforts to reduce malaria, which has indirectly decreased the number of VL cases (Abuzaid et al., 2020). Establishing a collaboration with the private sector can assist the community in supporting both the introduction and continuation of intervention. About 20% of kala-azar patients are estimated to seek care in the private sector. Private sector data is crucial for improving the disease's burden and maintaining the progress made toward elimination (Thakur & Thakur, 2020). The government should make sure that private practitioners are actively involved in the program, while the private sector should handle patient diagnosis and treatment. Accordingly, the national goal should be accomplished by utilizing the assistance of other national and international stakeholders (Ahmed et al., 2014; Thakur & Thakur, 2020).

8.4.8 Capacity building

The process of "developing sustainable skills, organizational structures, resources, and commitment to health improvement in health and other sectors, to prolong and multiply health gains many times over" is known as capacity building technique. It is occasionally referred to as the "invisible work." This can involve a variety of tasks, including finding program opportunities, advocating for support, training, assisting in the creation of policies, negotiating with management, directing the formation of partnerships, or organizing. People, organizations, and communities can become more capable through capacity building, which can take place within programs or more generally within systems. Thus, activities aimed at increasing capacity might be created in collaboration with communities, organizations, teams, individuals, or inter-organizational coalitions. To make the elimination program successful, it is necessary to build capacity to span the key areas like organizational development and workforce development. The following categories of personnel working within the program and in kala-azar endemic areas would be required to be trained and retrained from time to time.

- National, divisional, district, and sub-district program managers.
- Physicians and medical professionals (such as nurses, paramedics, and lab workers).
- Personnel from the epidemiological/statistical unit, such as computer operators and statisticians.
- Health professionals and supervisors involved in overseeing the spraying teams; supervisors at all levels.
- The medical professionals and volunteers in charge of communicating behavior change (a) encourage prompt medical attention if kala-azar is suspected; (b) encourage kala-azar patients to finish their treatment; and (c) engage the community in advocacy to ensure that all of their homes are properly and completely covered with insecticides (Asia, 2012; Rowbotham et al., 2023).

In India, senior national experts can be invited to make field visits and assist in supportive supervision. These experts can also help in advocacy by briefing the state-level policy-makers and program managers for administrative and management support for the program. Day-to-day management of the program activities needs to be strengthened at all levels of implementation to make sure there is uninterrupted supply of drugs, availability of diagnostics and program activities are implemented effectively and on timely basis (Thakur & Thakur, 2020). Accredited Social Health Activist (ASHA) and village-level workers should be trained in detection of active cases of kala-azar. Such patients should be tested for confirmation of kala-azar and referred to district hospital for case management. The district hospital should be equipped with facilities for the diagnosis of kala-azar and also for the treatment of kala-azar (Thakur & Thakur, 2020).

8.5 NEW STRATEGIES FOR IVM TO CONTROL KALA-AZAR

Fipronil, a broad-spectrum insecticide, has the potential to be used as a novel chemical control approach to target sand fly populations, according to a study by Poché et al. (2020). By taking advantage of sand flies' blood-feeding habits, the study highlights how fipronil-based cow treatments can dramatically lower sand fly population (Poché et al., 2020). By treating cattle, the chemical indirectly affects the vector population, as cattle are important blood meal sources for sand flies. However, ecological variables, including sand fly feeding preferences, the availability of substitute hosts, and seasonal fluctuations in sand fly populations all have a significant impact on how effective this strategy is. In order to optimize the use of fipronil in a targeted and environmentally responsible manner, these findings highlight the significance of incorporating knowledge of sand fly ecology into chemical control strategy. Within the larger framework of IVM for the control of VL, this strategy offers a viable supplementary option. Furthermore, Njoroge et al. investigate the innovative use of Attractive Targeted Sugar Baits (ATSBs) as a potentially effective IVM approach. To administer lethal doses of insecticides or biological agents in a targeted and eco-friendly approach, ATSBs make use of the sugar-feeding behavior of vectors, such as mosquitoes. Reducing dependence on traditional chemical sprays, avoiding non-target impacts, and combating the problem of insecticide resistance are among the numerous benefits of this approach. In order to increase the effectiveness of ATSBs under various ecological settings, the study highlights the significance of modifying bait combinations and delivery mechanisms (Njoroge et al., 2023). In addition, ATSBs are praised for their ability to be incorporated into current vector control initiatives, enhancing more conventional measures such IRS and ITN coverage. Under the IVM framework, this method emphasizes the significance of behavioral targeting in developing long-lasting and successful vector management techniques (Njoroge et al., 2023). Beyond the use of reactive pesticides, the study by Garlapati et al. addresses sustainable, long-term methods. They recommend concentrating on ecological and biological control methods, such as studying biocontrol agents, endosymbionts like Wolbachia, and para-transgenesis. This is consistent with worldwide policies aiming at minimizing reliance on chemical approaches while improving ecological balance (Garlapati et al., 2021). Consequently, VL vectors can be efficiently targeted by combining community-based sanitation programs with the IRS, ITNs, and EVM.

8.6 RECOMMENDATIONS ABOUT IVM'S FUTURE IN INDIA FOR VL CONTROL

It is necessary to demonstrate that IVM strategies are being used and having an effect. In addition to the basic requirements for putting IVM into practice across the country for kala-azar control (Beier et al., 2008; Eryilmaz, n.d.)

further concerted national and international efforts as well as funding are required to accomplish the following:

- Support post-graduate studies such as public health entomology, promote and improve regional IVM training programs, and strengthen IVM capacity-building at the national level.
- Encourage inter-disciplinary integration and ISC by involving relevant national stakeholders, such as community organizations and non-governmental organizations, and by bringing in specialists from fields other than the conventional entomological and public health frameworks for the control of VBDs (Hakizimana, 2019).
- Encourage the IVM strategy to integrate new technologies in database administration, information systems, and communication for improved operations at the national and regional levels, for example, the HESN in Saudi Arabia (Abuzaid et al., 2020).
- Create innovative training models for the IRS activity implementation, LLIN or ITN distribution, and public education regarding the ecology, biology, and public health significance of vectors. These models should also include interaction components between entomologists and ASHA field staff, medical professionals in clinics, and decision-makers in national and local government offices.
- Encourage studies that further clarify the approaches and techniques required to fight VL, particularly in low-transmission environments where national programs have effectively decreased transmission levels and are currently working toward or aiming for nationwide eradication (Beier et al., 2008).
- Facilitate studies on ways to enhance strategic management and decision-making so that IVM strategies are able to be systematically included into national kala-azar elimination programs.
- Encourage global research to create new tools for controlling kala-azar vectors, such as new-generation insecticides, new biological control strategies like para-transgenic methods, new control strategies like ATSB technology, new trapping techniques, and innovative methods that target sand flies at different stages of their life cycles, especially during the larval phase.
- The development of IVM strategies for the prevention and control of sand flies in wells and irrigation structures is necessary due to the persistently rising incidence of VL in irrigated areas in the Middle East, North Africa, and the arid region of Asia. This is because it is linked to the spread of significant *Leishmania* vectors (Barhoumi et al., 2015).
- To improve the operational research and capacity-building features of IVM programs, find and encourage initiatives that create sustainable partnerships between national kala-azar elimination programs and foreign experts and organizations.

- Create and support IVM demonstration projects in several kala-azar-endemic areas to highlight the value of IVM strategies for the prevention of VBDs. Analysis on the financial and economic costs of IVM should be expanded to include economic and financial models associated with scaling up IVM in India, as well as expected estimates of IVM and real-world cost-effectiveness calculations of IVM programs from demonstration projects (Beier et al., 2008).

8.7 CONCULSION

IVM is a potentially effective strategy in sand fly control that can be tailored to have as minimal of an impact on the environment as possible. It is believed that further efforts to figure out the most situation-specific and stakeholder-accepted combination of control methods and implementation strategies for the selective targeting of *P. argentipes*, Aedes, Anopheles, and Culex mosquitoes will improve current IVM strategies. The development of sustained sand fly-borne disease eradication is expected to be aided by the smooth integration of this promising novel sand fly control intervention with current vector control methods.

ACKNOWLEDGMENT

Authors are grateful to the ICMR-RMRIMS, Patna for providing necessary research support.

REFERENCES

Abuzaid, A. A., Aldahan, M. A., Helal, M. A. A., Assiri, A. M., & Alzahrani, M. H. (2020). Visceral leishmaniasis in Saudi Arabia: From hundreds of cases to zero. Acta Tropica, 212, 105707. https://doi.org/10.1016/j.actatropica.2020.105707

Ahmed, B.-N., Nabi, S. G., Rahman, M., Selim, S., Bashar, A., Rashid, M. M., Lira, F. Y., Choudhury, T. A., & Mondal, D. (2014). Kala-azar (visceral leishmaniasis) elimination in Bangladesh: Successes and challenges. Current Tropical Medicine Reports, 1(3), 163–169. https://doi.org/10.1007/s40475-014-0027-6

Alvar J, Vélez ID, Bern C, Herrero M, Desjeux P, Cano J, et al. (2012) Leishmaniasis Worldwide and Global Estimates of Its Incidence. PLoS ONE 7(5): e35671. https://doi.org/10.1371/journal.pone.0035671

Asia, W. H. O. R. O. for S.-E. (2005). Regional strategic framework for elimination of kala azar from the South-East Asia Region (2005–2015) (SEA-VBC-85). Article SEA-VBC-85. https://iris.who.int/handle/10665/205825

Asia, W. H. O. R. O. for S.-E. (2012). Regional strategic framework for elimination of kala-azar from the South-East Asia Region (2011–2015) (SEA-CD-239). Article SEA-CD-239. https://iris.who.int/handle/10665/205826

Asia, W. H. O. R. O. for S.-E. (2022). Regional strategic framework for accelerating and sustaining elimination of kala-azar in the South-East Asia Region:

2022–2026. World Health Organization. Regional Office for South-East Asia. https://iris.who.int/handle/10665/361887

Banjara, M. R., Das, M. L., Gurung, C. K., Singh, V. K., Joshi, A. B., Matlashewski, G., Kroeger, A., & Olliaro, P. (2019). Integrating case detection of visceral leishmaniasis and other febrile illness with vector control in the post-elimination phase in Nepal. The American Journal of Tropical Medicine and Hygiene, 100(1), 108–114. https://doi.org/10.4269/ajtmh.18-0307

Barhoumi, W., Qualls, W. A., Archer, R. S., Fuller, D. O., Chelbi, I., Cherni, S., Derbali, M., Arheart, K. L., Zhioua, E., & Beier, J. C. (2015). Irrigation in the arid regions of Tunisia impacts the abundance and apparent density of sand fly vectors of *Leishmania infantum*. Acta Tropica, 141(Pt A), 73–78. https://doi.org/10.1016/j.actatropica.2014.10.008

Beier, J. C., Keating, J., Githure, J. I., Macdonald, M. B., Impoinvil, D. E., & Novak, R. J. (2008). Integrated vector management for malaria control. Malaria Journal, 7(Suppl 1), S4. https://doi.org/10.1186/1475-2875-7-S1-S4

Bern, C., Courtenay, O., & Alvar, J. (2010). Of cattle, sand flies and men: A systematic review of risk factor analyses for South Asian visceral leishmaniasis and implications for elimination. PLoS Neglected Tropical Diseases, 4(2), e599. https://doi.org/10.1371/journal.pntd.0000599

Chanda, E. (2016). Exploiting the Potential of Integrated Vector Management for Combating Malaria in Africa. In Current Topics in Malaria. IntechOpen. https://doi.org/10.5772/65467

Chanda, E., Ameneshewa, B., Bagayoko, M., Govere, J. M., & Macdonald, M. B. (2017). Harnessing integrated vector management for enhanced disease prevention. Trends in Parasitology, 33(1), 30–41. https://doi.org/10.1016/j.pt.2016.09.006

Chanda, E., Govere, J., Macdonald, M., Lako, R., Haque, U., Baba, S., & Mnzava, A. (2013). Integrated vector management: A critical strategy for combating vector-borne diseases in South Sudan. Malaria Journal, 12, 369. https://doi.org/10.1186/1475-2875-12-369

Chowdhury, R., Chowdhury, V., Faria, S., Akter, S., Dash, A. P., Bhattacharya, S. K., Maheswary, N. P., Bern, C., Akhter, S., Alvar, J., Kroeger, A., Boelaert, M., & Banu, Q. (2019). Effect of insecticide-treated bed nets on visceral leishmaniasis incidence in Bangladesh. A retrospective cohort analysis. PLoS Neglected Tropical Diseases, 13(9), e0007724. https://doi.org/10.1371/journal.pntd.0007724

Chowdhury, R., Kumar, V., Mondal, D., Das, M.L., Das, P., Dash, A.P., & Kroeger, A. (2016). 2016. Implication of vector characteristics of *Phlebotomus argentipes* in the kala-azar elimination programme in the Indian sub-continent. Pathog. Glob. Hlth. 110: 87–96.

Chowdhury, R., Dotson, E., Blackstock, A. J., McClintock, S., Maheswary, N. P., Faria, S., Islam, S., Akter, T., Kroeger, A., Akhter, S., & Bern, C. (2011). Comparison of insecticide-treated nets and indoor residual spraying to control the vector of visceral leishmaniasis in Mymensingh District, Bangladesh. The American Journal of Tropical Medicine and Hygiene, 84(5), 662–667. https://doi.org/10.4269/ajtmh.2011.10-0682

Coleman, M., Foster, G. M., Deb, R., Pratap Singh, R., Ismail, H. M., Shivam, P., Ghosh, A. K., Dunkley, S., Kumar, V., Coleman, M., Hemingway, J., Paine, M. J. I., & Das, P. (2015). DDT-based indoor residual spraying suboptimal for visceral leishmaniasis elimination in India. Proceedings of the National Academy of Sciences of the United States of America, 112(28), 8573–8578. https://doi.org/10.1073/pnas.1507782112

Courtenay, O., Peters, N. C., Rogers, M. E., & Bern, C. (2017). Combining epidemiology with basic biology of sand flies, parasites, and hosts to inform leishmaniasis

transmission dynamics and control. PLoS Pathogens, 13(10), e1006571. https://doi.org/10.1371/journal.ppat.1006571
Dinesh, D. S., Kumari, S., Hassan, F., Kumar, V., Singh, V. P., & Das, P. (2017). Efficacy and evaluation of environmental management system to control sandfly vector of Kala-azar. Journal of Environmental Management, 201, 366–368. https://doi.org/10.1016/j.jenvman.2017.06.066
Eryilmaz, S. (n.d.). Integrated vector management for malaria control. Retrieved November 22, 2024, from https://www.academia.edu/35889431/Integrated_vector_management_for_malaria_control
Garlapati, R., Iniguez, E., Serafim, T. D., Mishra, P. K., Rooj, B., Sinha, B., Valenzuela, J. G., Srikantiah, S., Bern, C., & Kamhawi, S. (2021). Towards a sustainable vector-control strategy in the post Kala-azar elimination era. Frontiers in Cellular and Infection Microbiology, 11, 641632. https://doi.org/10.3389/fcimb.2021.641632
Ghani, M. B. A. (n.d.). Health technology assessment.
Hakizimana, E. (2019). Integrated vector management (IVM) as a tool for community empowerment towards malaria elimination in Rwanda. https://doi.org/10.18174/464552
Hasker E., Singh S. P., Malaviya P., Picado A., Gidwani K., Singh R. P., et al. (2012). Visceral leishmaniasis in rural Bihar, India. *Emerg. Infect. Dis.* 18, 1662–1664. doi: 10.3201/eid1810.111083
Herdiana, H., Sari, J. F. K., & Whittaker, M. (2018). Intersectoral collaboration for the prevention and control of vector borne diseases to support the implementation of a global strategy: A systematic review. PLoS One, 13(10), e0204659. https://doi.org/10.1371/journal.pone.0204659
International Conference on Primary Health Care. (1978). Declaration of Alma-Ata. WHO Chronicle, 32(11), 428–430.
Joshi A. B., Das M. L., Akhter S., Chowdhury R., Mondal D., Kumar V., et al. (2009). Chemical and environmental vector control as a contribution to the elimination of visceral leishmaniasis on the Indian subcontinent: cluster randomized controlled trials in Bangladesh, India and Nepal. *BMC Med.* 7, 54. doi: 10.1186/1741-7015-7-54
Kumar V., Kesari S. K., Sinha N. K., Palit A., Ranjan A., Kishore K., et al. (1995). Field Trial of an Ecological Approach for the Control of Phlebotomus-Argentipes Using Mud and Lime Plaster. *Indian J. Med. Res.* 101, 154–156.
Kumar, V., Kesari, S., Chowdhury, R., Kumar, S., Sinha, G., Hussain, S., Huda, M. M., Kroeger, A., & Das, P. (2013). User friendliness, efficiency & spray quality of stirrup pumps versus hand compression pumps for indoor residual spraying. The Indian Journal of Medical Research, 138(2), 239.
Kumar, V., Mandal, R., Das, S., Kesari, S., Dinesh, D. S., Pandey, K., Das, V. R., Topno, R. K., Sharma, M. P., Dasgupta, R. K., & Das, P. (2020). Kala-azar elimination in a highly-endemic district of Bihar, India: A success story. PLOS Neglected Tropical Diseases, 14(5), e0008254. https://doi.org/10.1371/journal.pntd.0008254
Lizzi K.M., Qualls W.A., Brown S.C., Beier J.C. (2014). Expanding integrated vector management to promote healthy environments. Trends Parasitol, 30, 394–400. doi: 10.1016/j.pt.2014.06.001.
Mahto, K. K., Prasad, P., Kumar, M., Ali, I., Vohra, V., Arya, D. K., Mahto, K. K., Prasad, P., Kumar, M., Ali, I., Vohra, V., & Arya, D. K. (2023). Visceral Leishmaniasis: An Overview and Integrated Analysis of the Current Status, Geographical Distribution and Its Transmission. In Leishmania Parasites—Epidemiology, Immunopathology and Hosts. IntechOpen. https://doi.org/10.5772/intechopen.110567

Malaviya, P., Hasker, E., Picado, A., Mishra, M., Van Geertruyden, J.-P., Das, M. L., Boelaert, M., & Sundar, S. (2014). Exposure to *Phlebotomus argentipes* (Diptera, Psychodidae, Phlebotominae) sand flies in rural areas of Bihar, India: The role of housing conditions. PLoS One, 9(9), e106771. https://doi.org/10.1371/journal.pone.0106771

Marcos-Marcos, J., Olry de Labry-Lima, A., Toro-Cardenas, S., Lacasaña, M., Degroote, S., Ridde, V., & Bermudez-Tamayo, C. (2018). Impact, economic evaluation, and sustainability of integrated vector management in urban settings to prevent vector-borne diseases: A scoping review. Infectious Diseases of Poverty, 7(1), 83. https://doi.org/10.1186/s40249-018-0464-x

McCall, L.-I., Zhang, W.-W., & Matlashewski, G. (2013). Determinants for the development of visceral leishmaniasis disease. PLoS Pathogens, 9(1), e1003053. https://doi.org/10.1371/journal.ppat.1003053

Mondal, D., Bern, C., Ghosh, D., Rashid, M., Molina, R., Chowdhury, R., Nath, R., Ghosh, P., Chapman, L. A. C., Alim, A., Bilbe, G., & Alvar, J. (2019). Quantifying the infectiousness of post-kala-azar dermal leishmaniasis toward sand flies. Clinical Infectious Diseases: An Official Publication of the Infectious Diseases Society of America, 69(2), 251–258. https://doi.org/10.1093/cid/ciy891

Mukhopadhyay, A. K., Saxena, N. B., & Narasimham, M. V. (1990). Susceptibility status of *Phlebotomus argentipes* to DDT in some kala-azar endemic areas of Bihar (India). The Indian Journal of Medical Research, 91, 458–460.

Ng'ang'a, P. N., Aduogo, P., & Mutero, C. M. (2021). Strengthening community and stakeholder participation in the implementation of integrated vector management for malaria control in western Kenya: A case study. Malaria Journal, 20(1), 155. https://doi.org/10.1186/s12936-021-03692-4

Njoroge, T. M., Hamid-Adiamoh, M., & Duman-Scheel, M. (2023). Maximizing the potential of Attractive Targeted Sugar Baits (ATSBs) for integrated vector management. Insects, 14(7), 585. https://doi.org/10.3390/insects14070585

Ostyn, B., Vanlerberghe, V., Picado, A., Dinesh, D. S., Sundar, S., Chappuis, F., Rijal, S., Dujardin, J.-C., Coosemans, M., Boelaert, M., & Davies, C. (2008). Vector control by insecticide-treated nets in the fight against visceral leishmaniasis in the Indian subcontinent, what is the evidence? Tropical Medicine & International Health: TM & IH, 13(8), 1073–1085. https://doi.org/10.1111/j.1365-3156.2008.02110.x

Perry, D., Dixon, K., Garlapati, R., Gendernalik, A., Poché, D., & Poché, R. (2013). Visceral leishmaniasis prevalence and associated risk factors in the saran district of Bihar, India, from 2009 to July of 2011. The American Journal of Tropical Medicine and Hygiene, 88(4), 778–784. https://doi.org/10.4269/ajtmh.12-0442

Picado, A., Das, M. L., Kumar, V., Kesari, S., Dinesh, D. S., Roy, L., Rijal, S., Das, P., Rowland, M., Sundar, S., Coosemans, M., Boelaert, M., & Davies, C. R. (2010). Effect of village-wide use of long-lasting insecticidal nets on visceral Leishmaniasis vectors in India and Nepal: A cluster randomized trial. PLoS Neglected Tropical Diseases, 4(1), e587. https://doi.org/10.1371/journal.pntd.0000587

Picado, A., Singh, S. P., Rijal, S., Sundar, S., Ostyn, B., Chappuis, F., Uranw, S., Gidwani, K., Khanal, B., Rai, M., Paudel, I. S., Das, M. L., Kumar, R., Srivastava, P., Dujardin, J. C., Vanlerberghe, V., Andersen, E. W., Davies, C. R., & Boelaert, M. (2010). Long-lasting insecticidal nets for prevention of *Leishmania donovani* infection in India and Nepal: Paired cluster randomised trial. The BMJ, 341, c6760. https://doi.org/10.1136/bmj.c6760

Poché, D., Garlapati, R., Ingenloff, K., Remmers, J., & Poché, R. (2011). Bionomics of phlebotomine sand flies from three villages in Bihar, India. Journal of Vector

Ecology: Journal of the Society for Vector Ecology, 36(Suppl 1), S106–117. https://doi.org/10.1111/j.1948-7134.2011.00119.x

Poché DM, Garlapati RB, Mukherjee S, Torres-Poche Z, Hasker E, Rahman T, Bharti A, Tripathi VP, Prakash S, Chaubey R, 2018. Bionomics of Phlebotomus argentipes in villages in Bihar, India with insights into efficacy of IRS-based control measures. Plos Neglect Trop D 12. https://doi.org/10.1371/journal.pntd.0006168

Poché, D. M., Wang, H.-H., & Grant, W. E. (2020). Visceral leishmaniasis on the Indian Subcontinent: Efficacy of fipronil-based cattle treatment in controlling sand fly populations is dependent on specific aspects of sand fly ecology. PLoS Neglected Tropical Diseases, 14(2), e0008011. https://doi.org/10.1371/journal.pntd.0008011

Romero, G. A. S., & Boelaert, M. (2010). Control of visceral leishmaniasis in Latin America—A systematic review. PLOS Neglected Tropical Diseases, 4(1), e584. https://doi.org/10.1371/journal.pntd.0000584

Rowbotham, S., Walker, P., Marks, L., Irving, M., Smith, B. J., & Laird, Y. (2023). Building capacity for citizen science in health promotion: A collaborative knowledge mobilisation approach. Research Involvement and Engagement, 9(1), 36. https://doi.org/10.1186/s40900-023-00451-4

Sande, S., Zimba, M., Nyasvisvo, D., Mukuzunga, M., Kooma, E. H., Mberikunashe, J., & Dube, B. (2019). Getting ready for integrated vector management for improved disease prevention in Zimbabwe: A focus on key policy issues to consider. Malaria Journal, 18(1), 322. https://doi.org/10.1186/s12936-019-2965-x

Sanyal, R. (1979). A longtitudinal review of kala azar in Bihar. The Journal of Communicable Diseases. https://www.semanticscholar.org/paper/A-longtitudinal-review-of-kala-azar-in-Bihar.-Sanyal/cd0b4f7ab4ebcbb794db79a66a44349960ca7d69

Saurabh, S. (2021). Time for a village-level strategy for the elimination of kala-azar (visceral leishmaniasis) in India: analysis of potential kala-azar outbreak situation in 2018. *Tropical Doctor, 51*(1), 84–91.

Scarpini S.,Dondi A.Totaro C.Biagi C.Melchionda F.Zama D.et al (2022). Visceral Leishmaniasis: Epidemiology, Diagnosis, and Treatment Regimens in Different Geographical Areas with a Focus on Pediatrics. Microorganisms, 10, 1887. https://doi.org/10.3390/microorganisms10101887

Singh, O. P., Hasker, E., Boelaert, M., & Sundar, S. (2016). Elimination of visceral leishmaniasis on the Indian subcontinent the Lancet. Infectious Diseases, 16(12), e304–e309. https://doi.org/10.1016/S1473-3099(16)30140-2

Singh, R. K., Mittal, P. K., & Dhiman, R. C. (2012). Insecticide susceptibility status of *Phlebotomus argentipes*, a vector of visceral leishmaniasis in different foci in three states of India. Journal of Vector Borne Diseases, 49(4), 254–257.

Sundar, S., More, D. K., Singh, M. K., Singh, V. P., Sharma, S., Makharia, A., Kumar, P. C., & Murray, H. W. (2000). Failure of pentavalent antimony in visceral leishmaniasis in India: Report from the center of the Indian epidemic. Clinical Infectious Diseases: An Official Publication of the Infectious Diseases Society of America, 31(4), 1104–1107. https://doi.org/10.1086/318121

Sundar S., Singh O. P., Chakravarty J. (2018). Visceral leishmaniasis elimination targets in India, strategies for preventing resurgence. Expert Rev. Anti. Infect Ther. 16, 805–812. 10.1080/14787210.2018.1532790

Thakur, C. P. (2007). A new strategy for elimination of kala-azar from rural Bihar. The Indian Journal of Medical Research, 126(5), 447–451.

Thakur, C. P., Meenakshi Thakur, A. K., & Thakur, S. (2009). Newer strategies for the kala-azar elimination programme in India. The Indian Journal of Medical Research, 129(1), 102–104.

Thakur, C. P., & Thakur, M. (2020). Accelerating kala-azar elimination in India. The Indian Journal of Medical Research, 152(6), 538. https://doi.org/10.4103/ijmr.IJMR_841_19
Wilson A. L., Dhiman R. C., Kitron U., Scott T. W., Van Den Berg H., Lindsay S. W. (2014). Benefit of insecticide-treated nets, curtains and screening on vector borne diseases, excluding malaria: a systematic review and meta-analysis. PLoS Negl. Trop. Dis. 8, e3228. doi: 10.1371/journal.pntd.0003228

Chapter 9

Preventive measures

Community interventions and public health strategies

*Yash Paul Sharma**

ICMR – Rajendra Memorial Research Institute of Medical Sciences, Patna, Bihar, India

Ankita Sindhania

Department of Entomology and Fralin Life Sciences Institute, Virginia Tech, Blacksburg, Virginia, USA

9.1 INTRODUCTION

Leishmaniasis, a neglected tropical disease (NTD) caused by protozoan parasites of the genus *Leishmania* and transmitted by a sandfly, poses significant public health challenges in endemic regions. Effective control of leishmaniasis requires robust community interventions and public health strategies tailored to local contexts. These approaches aim to reduce transmission, enhance disease awareness, and improve access to prevention and treatment. This chapter explores key community-based interventions and public health strategies, emphasizing their role in mitigating the burden of leishmaniasis.

India, which once bore the highest burden of visceral leishmaniasis (VL or kala-azar) globally, has made significant strides toward elimination, with most districts now reporting cases well below the elimination threshold of fewer than one case per 10,000 population at the block level (WHO, 2021). However, sustaining these gains is crucial, as any lapse in prevention efforts could lead to resurgence, especially in vulnerable, hard-to-reach populations. Preventive measures play a critical role in controlling and ultimately eliminating leishmaniasis, particularly in endemic regions where poverty, inadequate housing, and limited access to healthcare facilitate its transmission (Pal et al., 2023). Since no vaccine is currently available and treatment, while effective, requires proper diagnosis and health system access, preventive strategies remain the cornerstone of long-term control. Key measures include indoor residual spraying (IRS), use of insecticide-treated bed nets, environmental management to reduce sandfly breeding sites, and early case detection with prompt treatment. Among all efforts, community participation is

* Corresponding author: yashraina007@gmail.com

DOI: 10.1201/9781003596455-9

vital-empowering local populations through awareness, behavior change, and active involvement in vector control and surveillance ensures ownership of the process and creates a resilient frontline defense against reemergence. Community interventions are indispensable to the sustainable prevention and control of leishmaniasis, particularly in endemic regions such as India, where social determinants like poverty, illiteracy, and inadequate infrastructure complicate traditional vertical disease control programs (Okwor and Uzonna, 2016). While biomedical solutions like diagnosis and treatment play critical roles, the community's active engagement in prevention and behavior change is the linchpin of long-term success.

9.2 RATIONALE FOR COMMUNITY-BASED APPROACHES

Leishmaniasis, especially VL, thrives in environments characterized by poor housing, lack of sanitation, and limited access to health services. These conditions are not merely biomedical issues but deeply rooted in the social and economic fabric of communities. Conventional top–down strategies-such as government-led spraying or passive health facility-based surveillance often fail to penetrate deeply into affected populations unless they are reinforced with grassroots-level action. Community-based interventions empower individuals and groups to become active participants in protecting their own health. These interventions bridge the gap between knowledge and practice, reinforce government initiatives, and enable timely, culturally sensitive responses to disease threats.

9.3 KEY COMPONENTS OF COMMUNITY INTERVENTIONS

9.3.1 Health education and Behavior Change Communication (BCC)

Health education and Behavior Change Communication (BCC) are essential pillars in the prevention and control of leishmaniasis, particularly in endemic regions where lack of awareness and deep-rooted misconceptions hinder timely diagnosis and treatment. BCC strategies aim to educate individuals and communities about the disease's causes, symptoms, transmission, and available health services, thereby promoting informed decision-making and healthier practices (Srinivasan et al., 2018). A well-designed BCC program uses culturally appropriate communication tools such as posters, flipcharts, local theater, and community meetings to convey key messages about sandfly control, personal protection measures, and the importance of early care-seeking. Utilizing local languages, traditional symbols, and involving trusted community members including health workers and recovered patients ensures better acceptance and understanding (WHO, 2023). By fostering awareness, dispelling myths, and encouraging positive behaviors, BCC not only empowers communities but also enhances the effectiveness of public health interventions, contributing significantly to the long-term elimination of leishmaniasis. Effective BCC strategies are discussed in the following sections.

9.3.2 Visual aids and storytelling

Visual aids and storytelling are powerful tools in BCC for leishmaniasis prevention, particularly in low-literacy and rural settings. Visual materials such as posters, flipcharts, and illustrated leaflets help simplify complex health messages, making them more accessible and memorable. Storytelling through community theater, folk songs, and role-plays engages audiences emotionally and culturally, allowing them to relate to real-life scenarios involving disease recognition, prevention, and care-seeking. These methods not only capture attention but also foster discussion and reflection, ultimately encouraging individuals and communities to adopt healthier behavior and support public health interventions (NCVBDC, 2025).

9.3.3 Use of local language and cultural symbols

The use of local language and cultural symbols is a crucial strategy in enhancing the effectiveness of health communication for leishmaniasis prevention. When messages are conveyed in the community's native language and embedded with familiar cultural references, they are more easily understood, trusted, and accepted (Figures 9.1 and 9.2). This approach respects local traditions and beliefs, making the information more relatable and reducing resistance to behavioral change. Incorporating cultural elements such as local festivals, religious practices, and traditional stories into communication materials ensures that health messages resonate deeply with the target audience, fostering greater engagement and long-term impact.

9.3.4 Peer education

Peer education is a highly effective approach in leishmaniasis prevention, as it leverages the trust and influence of individuals within the community to promote awareness and behavior change. Trained community members, including health volunteers and individuals who have recovered from the disease, share their personal experiences and knowledge to educate others about the symptoms, transmission, prevention, and treatment of leishmaniasis. This method helps reduce stigma, dispel myths, and encourage timely care-seeking, especially in areas where health systems may not be fully trusted. By fostering open dialogue and relatable learning, peer education strengthens community engagement and reinforces public health efforts at the grassroots level.

Figure 9.1 Kala-azar posters in West Bengal (NCVBDC).

Figure 9.2 Kala-azar posters in Bihar (NCVBDC).

9.3.5 Training community health workers

Training community health workers is a vital component in the prevention and control of leishmaniasis, particularly in endemic and underserved regions. These frontline workers, such as Accredited Social Health Activists (ASHAs) in India, serve as the crucial link between the healthcare system and the community. Through structured training programs, they are equipped to perform a wide range of tasks essential for early detection and community engagement. Training includes recognizing the early symptoms of VL and post-kala-azar dermal leishmaniasis (PKDL), educating families about the identification and elimination of sandfly breeding sites, and promoting the use of protective measures such as insecticide-treated bed nets. They also monitor compliance with IRS by helping prepare households and ensuring complete coverage. In addition, community health workers conduct household surveys during active surveillance campaigns to identify suspected cases and refer them promptly for diagnosis and treatment. By building their capacity through ongoing education and support, trained health workers become empowered agents of change who significantly strengthen the reach and effectiveness of leishmaniasis control programs at the grassroots level.

9.4 COMMUNITY MOBILIZATION FOR VECTOR CONTROL

Vector control is one of the most effective leishmaniasis prevention strategies. However, its success hinges on community acceptance and participation. Community mobilization for vector control is a cornerstone of sustainable leishmaniasis prevention, as it fosters collective action to reduce the presence of sandflies, the primary vectors of the disease. Effective mobilization involves actively engaging community members in understanding the role of environmental conditions in sandfly breeding and empowering them to take preventive measures. This includes preparing for and supporting IRS campaigns, maintaining clean surroundings, managing waste properly, and reducing moisture and organic matter near homes and animal shelters. Community meetings, sensitization drives, and the involvement of local leaders and groups help build trust, dispel myths about insecticides, and ensure wider acceptance of vector control activities (Kumar et al., 2020). When people are well-informed and take ownership of these measures, vector control becomes more consistent, effective, and sustainable, ultimately contributing to the long-term goal of eliminating leishmaniasis.

9.4.1 Indoor residual spraying (IRS) campaigns

IRS remains the cornerstone of vector control in India's VL-endemic regions, targeting sandflies within human dwellings. However, the success of IRS campaigns heavily depends on active community participation. Residents play a critical role by allowing spray teams timely access to their homes, preparing their households by moving furniture and covering food and water, and reporting any gaps, such as missed households or signs of insecticide resistance. Without adequate community cooperation, IRS coverage and its overall impact on sandfly populations can be significantly compromised. To address this, pre-IRS sensitization meetings are routinely held to inform and engage communities about the purpose, benefits, and safety of spraying activities (Faber et al., 2022). These interactions help build trust, address concerns, and encourage shared responsibility, ultimately enhancing the effectiveness and sustainability of IRS as a key vector control intervention.

9.4.2 Environmental management and cleanliness drives

Environmental management and cleanliness drives play a vital role in reducing sandfly breeding sites and preventing the transmission of leishmaniasis. Sandflies thrive in damp, dark areas with organic matter, such as cracked walls, moist soil, cow dung, and unclean animal shelters. Community-led initiatives aimed at improving environmental hygiene can significantly curb vector populations (González et al., 2015). Key actions include filling cracks and crevices in walls with plaster, regularly disposing of household and animal waste, cleaning and maintaining cattle sheds, and removing excess vegetation or moist organic matter near homes. These simple yet effective measures, when implemented collectively, create an environment that is less conducive to sandfly breeding. Mobilizing communities through cleanliness drives not only fosters a sense of shared responsibility but also reinforces the importance of hygiene and environmental care as part of an integrated leishmaniasis prevention strategy.

9.5 COMMUNITY-BASED SURVEILLANCE AND EARLY DETECTION

Early detection is vital to reduce morbidity, prevent transmission, and lower the overall disease burden. Community-based surveillance and early detection are essential components of an effective leishmaniasis control strategy, enabling prompt diagnosis and treatment to prevent complications and transmission. By involving residents, health volunteers, and trained community health workers in identifying suspected cases of VL and PKDL, the response system becomes more responsive and inclusive. These community agents are often the first to recognize prolonged fever, skin lesions, or other symptoms in their neighborhoods and can promptly refer individuals to health facilities for confirmation and care. Additionally, household surveys, community health registers, and informal reporting channels strengthen passive and active surveillance efforts (Aghakhani et al., 2023). Empowering communities with knowledge about disease symptoms and the importance of timely reporting fosters vigilance and builds a grassroots-level safety net that enhances early detection, reduces disease burden, and supports sustained elimination efforts.

9.5.1 House-to-house surveys

House-to-house surveys are a proactive and targeted approach to identifying suspected cases of leishmaniasis within communities, especially in endemic areas. Conducted by trained health workers or community volunteers, these surveys involve visiting every household to screen individuals for symptoms such as prolonged fever, weight loss, or skin lesions that may indicate VL or PKDL. This method ensures that no case is missed, including those who may not actively seek healthcare due to lack of awareness, stigma, or limited access to services. In addition to case identification, these visits offer an opportunity to educate families about disease prevention, promote adherence to treatment, and gather data for mapping transmission hotspots (Kumar et al., 2020). House-to-house surveys play a crucial role in active surveillance, facilitating early detection, prompt treatment, and interrupting the chain of transmission, thereby strengthening the overall leishmaniasis elimination strategy.

9.5.2 Community health camps

Community health camps are an effective strategy for enhancing leishmaniasis detection and awareness in endemic villages, particularly when conducted in collaboration with local panchayats, non-governmental organizations (NGOs), and health authorities. These camps bring essential health services directly to underserved populations, overcoming barriers such as distance, cost, and lack of information. On-the-spot testing using rk39 rapid diagnostic kits allows for immediate identification of suspected VL cases, facilitating prompt referral and initiation of treatment under the national elimination program. In addition to diagnostics, health camps serve as platforms for educating residents about the signs, symptoms, transmission, and prevention of leishmaniasis, thus increasing community awareness and encouraging early care-seeking behavior (Guha et al., 2024). By integrating medical services with health education and community engagement, these camps play a critical role in reducing disease burden and reinforcing the public health response at the grassroots level.

9.6 ADDRESSING STIGMA AND CULTURAL BARRIERS

Addressing stigma and cultural barriers is essential for the success of leishmaniasis control and elimination efforts, especially in endemic communities where misconceptions about the disease persist. VL and PKDL are often associated with fear, shame, or beliefs that they are caused by supernatural forces or poor moral conduct. This can lead to social isolation of affected individuals, delays in seeking treatment, and reluctance to disclose symptoms. Gender norms may further hinder women and girls from accessing care due to mobility restrictions or stigma. Culturally sensitive communication strategies, such as involving local leaders, religious figures, and former patients as champions of awareness, can help normalize conversations about the disease. Community dialogues, peer education, and inclusive health messaging that challenge myths and promote scientific understanding are key to breaking down these barriers (Koschorke et al., 2022). By creating a supportive environment that fosters openness, trust, and dignity, communities are more likely to participate actively in surveillance, treatment, and prevention efforts, ultimately strengthening the overall fight against leishmaniasis. Stigma around leishmaniasis, particularly PKDL due to its skin manifestations, can lead to social exclusion and delayed treatment. Community-based strategies can help mitigate stigma with the help of survivor advocacy groups, involvement of religious and local leaders, and support groups, which are discussed in the following sections.

9.6.1 Survivor advocacy groups

Survivor advocacy groups play a crucial role in leishmaniasis control by empowering former patients to become ambassadors of awareness and hope within their communities. These groups leverage the personal experiences of survivors to educate others about the disease, dispel myths, and reduce stigma associated with VL and PKDL. Survivors sharing their stories help build trust in health services, encourage early diagnosis and treatment, and motivate community members to adopt preventive behavior. Additionally, survivor groups often provide psychosocial support to affected individuals and their families, fostering resilience and social reintegration. By actively engaging survivors in health campaigns, training, and advocacy efforts, these groups strengthen community participation and contribute significantly to sustained leishmaniasis elimination initiatives.

9.6.2 Involvement of religious and local leaders

The involvement of religious and local leaders is a powerful strategy in leishmaniasis prevention and control, as these figures hold significant influence and respect within their communities. By engaging them in awareness campaigns and public health initiatives, health programs can tap into trusted networks to disseminate accurate information, dispel myths, and encourage positive health behavior. Religious and local leaders can help address cultural barriers and stigma by framing disease prevention and treatment in ways that align with community values and beliefs (Polidano et al., 2022). Their endorsement of interventions such as IRS, use of bed nets, and early medical consultation can motivate wider community acceptance and participation. Furthermore, their active support fosters community cohesion and collective responsibility, which are essential for the sustained success of leishmaniasis elimination efforts.

9.6.3 Support groups

Support groups provide an essential platform for individuals affected by leishmaniasis to share experiences, receive emotional support, and access information about treatment and prevention. These groups help reduce feelings of isolation and stigma by connecting patients and survivors, fostering a sense of community and mutual encouragement. Support groups also serve as a bridge between healthcare providers and the community, facilitating better communication and adherence to treatment protocols. Through group meetings, members can learn about managing side effects, recognizing symptoms early, and adopting preventive measures. Additionally, support groups often engage in advocacy and awareness activities, empowering members to become active participants in leishmaniasis control efforts. By strengthening social support networks, these groups contribute significantly to improving patient outcomes and sustaining community involvement in disease elimination programs.

9.7 ROLE OF LOCAL GOVERNANCE AND NGOS

Local governance bodies and NGOs play a pivotal role in the prevention and control of leishmaniasis by facilitating coordination, resource mobilization, and community engagement. Local government institutions, such as panchayats in India, help implement vector control measures, organize health campaigns, and ensure the delivery of essential services at the grassroots level. They are instrumental in mobilizing communities, enforcing sanitation regulations, and supporting surveillance activities. NGOs complement these efforts by bringing technical expertise, innovative approaches, and additional manpower to health education, capacity building, and outreach programs. Their close ties with marginalized and vulnerable populations enable them to reach hard-to-access areas, address social determinants of health, and advocate for affected communities (NCVBDC, 2017). Together, local governance and NGOs foster a multi-sectoral approach that strengthens public health infrastructure, promotes sustainable interventions, and accelerates progress toward the elimination of leishmaniasis.

9.7.1 Panchayati Raj Institutions (PRIs)

Panchayati Raj Institutions (PRIs) serve as a crucial grassroots governance structure in India, playing a significant role in the prevention and control of leishmaniasis. As the closest government bodies to rural communities, PRIs are uniquely positioned to mobilize local resources, facilitate health education, and coordinate vector control activities such as IRS and environmental management. They help raise awareness about disease transmission and prevention, encourage community participation in cleanliness drives, and support surveillance efforts by ensuring accurate reporting of cases. By involving elected local representatives, PRIs also help address social and cultural barriers, enhance accountability, and foster collaboration between health authorities and the community. Their active engagement strengthens the overall health system response and contributes significantly to sustainable leishmaniasis control and elimination at the village level.

9.7.2 Civil society organizations

Civil society organizations (CSOs) play a vital role in complementing government efforts to control and eliminate leishmaniasis by advocating for affected populations, raising awareness, and facilitating community-based interventions. These organizations often work closely with marginalized and vulnerable groups who may face barriers to accessing healthcare services. CSOs contribute by organizing health education campaigns, supporting vector control activities, and promoting behavioral change through culturally sensitive communication strategies. They also assist in capacity building of community health workers, conduct monitoring and evaluation of program implementation, and help bridge gaps between the community and formal health systems. By fostering community empowerment, advocacy, and social accountability, civil society organizations enhance the reach, equity, and effectiveness of leishmaniasis control programs (WHO, 2022).

9.8 USE OF DIGITAL TOOLS FOR COMMUNITY ENGAGEMENT

In recent years, the adoption of digital tools has significantly transformed community engagement in leishmaniasis prevention and control. Mobile-based applications and digital platforms offer innovative ways to reach and involve communities more effectively. For example, SMS alerts are used to inform residents about upcoming IRS campaigns, ensuring better preparedness and participation. Mobile apps enable community volunteers to report fever cases in real-time, facilitating quicker identification and response to potential leishmaniasis infections. Additionally, audio–visual health education materials shared through WhatsApp groups help disseminate accurate information widely and engage diverse audiences. These digital interventions enhance the reach and timeliness of public health actions, improve data collection and monitoring, and empower communities to take an active role in controlling the disease through a more informed and coordinated approach (Zhang et al., 2023).

9.9 CHALLENGES IN COMMUNITY INTERVENTIONS

Despite their critical role in leishmaniasis control, community interventions often face several challenges that can hinder their effectiveness. Mistrust in government programs and fears regarding the safety of insecticides used in IRS can lead to resistance or refusal to participate. Additionally, community fatigue or apathy may set in if interventions are sporadic or do not produce immediate visible benefits, reducing enthusiasm and cooperation over time. Resource constraints, particularly in marginalized and hard-to-reach areas, further limit the reach and sustainability of these efforts. To overcome these barriers, it is essential to foster transparent communication that openly addresses concerns and builds trust. Consistent follow-up and sustained engagement help maintain momentum and demonstrate the value of interventions. Equally important is the meaningful inclusion of marginalized groups in planning and implementation, ensuring their needs and perspectives shape responsive and equitable health programs. Together, these approaches strengthen community ownership and the long-term success of leishmaniasis prevention strategies.

9.10 CASE STUDY: COMMUNITY INTERVENTIONS AND PUBLIC HEALTH STRATEGIES TO COMBAT VISCERAL LEISHMANIASIS (KALA-AZAR) IN BIHAR'S VAISHALI DISTRICT

Kala-azar elimination in a highly endemic district of Bihar, India: A success story (Vaishali Model) (Kumar et al., 2020)

The status of VL cases in Vaishali before and after the implementation of the integrated control strategy is as follows:

Before Implementation (2014):

- **Total VL Cases:** 664 cases reported.
- **Endemic Villages:** 282 villages affected.
- **Case Reduction Rate (CRR):** 22.6% annually.

After Implementation (2016):

- **Total VL Cases:** Reduced to 163 cases.
- **Endemic Villages:** Reduced to 142 villages.
- **CRR:** Increased to 58.8% annually.
- **Blocks Achieving Elimination Target (<1 case per 10,000 people):** All 16 blocks achieved the target.

Key Observations:

- On average, 74 VL-infected villages became kala-azar free each year during 2015–2016.
- The integrated control strategy led to a significant reduction in VL cases and endemic villages, demonstrating its effectiveness in achieving the elimination target.

The integrated control strategy implemented in Vaishali included the following components:

1. **Spatio-Temporal Mapping:** GIS-based mapping of VL-case distribution to identify endemic and high-risk villages for targeted interventions.

2. **Active Case Detection (ACD):** Door-to-door surveys to identify febrile cases and test them using rK39 rapid diagnostic tests.
3. **Chemical-Based Vector Control:** IRS using synthetic pyrethroid (alphacypermethrin) to replace DDT due to resistance in sandflies.
4. **Focal Spraying:** Immediate spraying of households within 500 meters of new VL case locations to interrupt transmission.
5. **Community Awareness Campaigns:** Information, education, and communication (IEC) activities to raise awareness about VL symptoms, treatment, and vector control.
6. **Training Programs:** Training of IRS operators, medical doctors, supervisors, and monitors for effective implementation and treatment.
7. **Monitoring and Supervision:** A three-tier system for monitoring IRS activities at district, block, and village levels.
8. **Logistic Management:** Ensuring the availability and replacement of equipment for IRS quality improvement.
9. **Entomological Monitoring:** Pre- and post-IRS sandfly density assessments and insecticide susceptibility tests.
10. **Prompt Treatment:** Fast treatment of VL patients with single-dose liposomal amphotericin B at block-level health centers.
11. **Use of Modern Equipment:** Replacement of conventional stirrup pumps with hand compression pumps for better spray quality.
12. **Routine Surveillance:** Epidemiological and entomological monitoring to track progress and ensure intervention effectiveness.

This strategy helped reduce VL cases and endemic villages significantly, achieving the elimination target of less than 1 case per 10,000 people at the block level.

The community awareness campaigns implemented in Vaishali included the following activities:

1. **IEC:**
   - Dissemination of information about kala-azar, its symptoms, mode of transmission, and the importance of complete treatment.
   - Awareness about vector control methods, particularly IRS, and precautions post-IRS.
2. **BCC:**
   - Use of small media such as handbills, posters, stickers, leaflets, and pamphlets.
   - Mass media campaigns, including banners, audio tapes, and local arts.
   - Training and demonstrations to improve communication skills.
3. **Outdoor Media:**
   - Posters, stickers, handbills, and banners were used to inform the public about VL symptoms, diagnosis, treatment, sandfly breeding sites, and vector control methods.

4. **Proclamations:**
   - Announcements through microphones, ASHAs, and social mobilization two days before each IRS round to inform villagers about the date and time of spraying.
5. **Pre-IRS Information Dissemination:**
   - Microphone announcements and ASHAs were the primary sources of pre-IRS information, ensuring high awareness among households.
6. **Post-IRS Awareness:**
   - Encouraging households to practice post-IRS activities such as avoiding washing sprayed walls, wiping floors after spraying, and waiting for 30 minutes before re-entering sprayed rooms.
7. **Impact Assessment:**
   - Surveys conducted pre- and post-IRS showed significant improvement in community awareness about VL symptoms, treatment facilities, vector breeding sites, and precautionary measures.

These campaigns played a crucial role in increasing IRS acceptance, reducing locked households, and improving community participation in VL control efforts.

BCC activities significantly improved community knowledge in Vaishali. The impact was assessed through pre- and post-IRS surveys, showing the following changes:

1. **Increased Awareness:**
   - Knowledge about VL and PKDL symptoms increased from 53.5% (pre-activity) to 77% (post-activity).
   - Awareness of treatment facilities at government hospitals rose from 59.1% to 84.1%.
   - Understanding of the VL vector, its breeding and resting sites, improved from 46.1% to 75.8%.
   - Awareness of measures to control VL infection increased from 59.8% to 69%.
   - Knowledge of precautionary activities post-IRS rose from 46.2% to 51.7%.
2. **Improved Practices:**
   - Post-IRS activities such as avoiding washing sprayed walls, wiping floors after spraying, and waiting 30 minutes before re-entering sprayed rooms were practiced by 66.6% of households.
3. **Effective Communication Channels:**
   - Microphone announcements and ASHAs were the most effective sources of pre-IRS information, reaching 64.3% of households.

Overall, BCC activities enhanced community knowledge, attitudes, and practices related to VL prevention, treatment, and vector control, contributing to the success of the integrated control strategy.

9.11 CONCLUSION

Community interventions are the cornerstone of a sustainable, resilient, and locally adapted response to leishmaniasis. While vector control and case management are essential, their effectiveness depends on the trust, participation, and ownership of the communities most affected. India's progress toward eliminating VL has been driven not only by technological and medical advances but also by community mobilization, frontline health workers, and local leadership. Going forward, a strong focus on community empowerment, multi-sectoral collaboration, and continuous engagement will be essential to not only sustain elimination but also prevent resurgence. Leishmaniasis control must be seen not just as a medical challenge, but as a social one—requiring collective action, sustained dialogue, and shared responsibility.

ACKNOWLEDGMENT

Authors are thankful to the ICMR-RMRIMS, Patna for providing necessary research support.

REFERENCES

Aghakhani N, Azami M, Amini Rarani S. (2023). Community-based interventions as an effective program for leishmaniasis treatment: A duty to act. GMS Hyg Infect Control., 18:15. doi: 10.3205/dgkh000441.

Faber C, Montenegro Quiñonez C, Horstick O, Rahman KM, Runge-Ranzinger S. (2022). Indoor residual spraying for the control of visceral leishmaniasis: A systematic review. PLoS Negl Trop Dis., 19;16(5):e0010391. doi: 10.1371/journal.pntd.0010391.

González U, Pinart M, Sinclair D, Firooz A, Enk C, Vélez ID, Esterhuizen TM, Tristan M, Alvar J. (2015). Vector and reservoir control for preventing leishmaniasis. Cochrane Database Syst Rev., 2015(8):CD008736. doi: 10.1002/14651858.

Guha SK, Sardar AA, Misra AK, Saha P, Samanta A, Maji D, Mandal A, Saha P, Halder S, Ali KA, Karmakar S, Sharma D, Maji AK. (2024). Active community-based case finding of endemic leishmaniasis in West Bengal, India. J Epidemiol Glob Health., 14(3):1100–1112. doi: 10.1007/s44197-024-00260-2.

Koschorke M, Al-Haboubi Y, Tseng P, Semrau M, Eaton J (2022). Mental health, stigma, and neglected tropical diseases: A review and systematic mapping of the evidence, Front Trop Dis., 3:808955 doi: 10.3389/fitd.2022.808955.

Kumar V, Mandal R, Das S, Kesari S, Dinesh DS, Pandey K, et al. (2020). Kala-azar elimination in a highly-endemic district of Bihar, India: A success story. PLoS Negl Trop Dis., 14(5):e0008254. doi: 10.1371/journal. pntd.0008254.

NCVBDC (2017). Operational guidelines on kala-azar (visceral leishmaniasis) elimination in India – 2015. https://ncvbdc.mohfw.gov.in/Doc/opertional-guideline-KA-2015.pdf (Accessed: 25-05-2025).

NCVBDC (2025). BCC/IEC tools for Bihar, Jharkhand, UP and WB for kala-azar. https://ncvbdc.mohfw.gov.in/index1.php?lang=1&level=1&sublinkid=5886&lid=3965 (Accessed: 27-05-2025).

Okwor I, Uzonna J. (2016). Social and economic burden of human leishmaniasis. Am J Trop Med Hyg., 94(3):489–93. doi: 10.4269/ajtmh.15-0408.
Pal B, Kumari S, Kaur M, Wadhwa P, Murti K, Kumar R, Pandey K, Siddiqui NA, Dhingra S, Padmakar S. (2023). Barriers to the effective management and prevention of post kala-azar dermal leishmaniasis (PKDL) in the Indian subcontinent. Med J Armed Forces India., 79(5):500–505. doi: 10.1016/j.mjafi.2023.03.010.
Polidano K, Parton L, Agampodi SB, Agampodi TC, Haileselassie BH, Lalani JMG, Mota C, Price HP, Rodrigues S, Tafere GR, Trad LAB, Zerihun Z, Dikomitis L. (2022). Community engagement in cutaneous leishmaniasis research in Brazil, Ethiopia, and Sri Lanka: A decolonial approach for global health. Front Public Health.,;10:823844. doi: https://doi.org/10.3389/fpubh.2022.823844.
Srinivasan R, Ahmad T, Raghavan V, Kaushik M, Pathak R. (2018). Positive influence of behavior change communication on knowledge, attitudes, and practices for visceral leishmaniasis/kala-azar in India. Glob Health Sci Pract.,;6(1):192–209. doi: 10.9745/GHSP-D-17-00087.
WHO (2021). Visceral leishmaniasis elimination: India gears-up to overcome last-mile challenges. https://www.who.int/news/item/29-07-2021-visceral-leishmaniasis-elimination-india-gears-up-to-overcome-last-mile-challenges (Accessed: 25-05-2025).
WHO (2022). Strengthening civil society capacity for advocacy and action. https://www.who.int/initiatives/action-on-the-social-determinants-of-health-for-advancing-equity/knowledge-sharing-hub/strategic-entry-point-2/strengthening-civil-society-capacity-for-advocacy-and-action (Accessed: 26-05-2025).
WHO (2023). Strategic framework for the elimination of visceral leishmaniasis as a public health problem in eastern Africa 2023–2030. https://iris.who.int/bitstream/handle/10665/376833/9789240094208-eng.pdf (Accessed: 25-05-2025).
Zhang L, Guo W, Lv C. (2023). Modern technologies and solutions to enhance surveillance and response systems for emerging zoonotic diseases. Sci One Health.,;3:100061. doi: https://doi.org/10.1016/j.soh.2023.100061.

Chapter 10

Therapeutic trials in visceral, dermal, and cutaneous leishmaniasis

Real-world applications and outcomes

*Rishikesh Kumar**

Department of Clinical Medicine, ICMR-Rajendra Memorial Research Institute of Medical Sciences, Patna, India

*Prabin Dahal**

Infectious Diseases Data Observatory, Oxford, UK

Centre for Tropical Medicine and Global Health, Nuffield Department of Medicine, University of Oxford, Oxford, UK

*Krishna Pandey**

Department of Clinical Medicine, ICMR-Rajendra Memorial Research Institute of Medical Sciences, Patna, India

10.1 HISTORICAL TRAJECTORY OF THERAPEUTIC TRIALS IN VL

10.1.1 Early empirical interventions

The early 20th century marked the inception of therapeutic investigations into visceral leishmaniasis (VL). Initial treatments included early preparations of antimony compounds with high toxicity and variable success rates [1, 2]. The use of antimony preparations for the management of VL was first advised by Manson, but Vianna used the preparation in 1912 for successfully treating cutaneous leishmaniasis in South America [3]. Di Cristina and Caronia used trivalent antimony (tartar emetic) for infantile kala-azar in Italy. U.N. Brahmachari, who was among the pioneers in the development of antimony compounds for the management of VL in India, wrote a comprehensive account of the early therapeutics in 1917 [4], and later on the evolution of the antimony treatment in the 1928 treatise [3]. L.E. Napier presented a succinct summary of the antimony era in 1946 [5]. These early attempts in general lacked a structured methodology compared to

* Corresponding authors: drkrishnapandey@yahoo.com; drkumarrishikesh@gmail.com; prabin.dahal@ndm.ox.ac.uk

DOI: 10.1201/9781003596455-10

how drugs are evaluated today, and instead relied on drawing inferences from observational case series. These empirical practices highlighted the impact of drug dosage and treatment duration on patient response, laid the groundwork for a more systematic inquiry, and later influenced the development and design of controlled trial designs toward the end of the 20th century.

10.1.2 The discovery of dermal leishmanoid

In 1922, U.N. Brahmachari reported a new dermal manifestation of VL, later termed post-kala-azar dermal leishmaniasis (PKDL) [1]. This discovery not only altered diagnostic criteria but also introduced the concept of disease latency and reservoirs in clinical understanding. It spurred follow-up studies across all the endemic regions into long-term post-treatment effects and influenced subsequent therapeutic trial designs by emphasizing the need for extended monitoring periods and inclusion of dermatological assessments in efficacy evaluations. In this chapter we use the term "dermal leishmaniasis" instead of the widely used term PKDL as there are a proportion of PKDL patients present without any documented VL history (Discussed in section 10.2).

10.1.3 Mid-century drug studies

By the 1930s and 1940s, efforts began incorporating more structured approaches. Scott and Li [6] documented outcomes in pediatric patients, offering early glimpses into age-specific responses. The drugs were systematically evaluated in different patient subgroups. This period also saw the first steps towards standardizing trial protocols, including defined inclusion criteria, outcome measures, and follow-up periods. These early structured trials laid the groundwork for subsequent developments in clinical research methodology, influencing the design of controlled trials by emphasizing the importance of population-specific insights and consistency in collecting and summarizing the data (Table 10.1). The period also saw the development and refinement of antimony from trivalent to the much safer pentavalent forms, the safety of the latter form attributed to the rapid excretion and lack of appreciable accumulation [7]. A major development was the establishment of the dose-response principle by observing efficacy to depend on the total antimony dose when injected safely within a given period of time [7]. In addition, several evaluations of the

Table 10.1 Summary of key mid-century VL drug studies (non-exhaustive)

Study[a]	*Drug used*	*Population*	*Key findings*
Scott and Li [6]	Urea stibamine	Pediatric	Partial success; significant side effects
Bryceson et al. [9]	Sodium stibogluconate	Adults	High failure in resistant cases

[a] See Bryceson [7] for the four comparative studies until 1983 before randomized studies were undertaken and for a comprehensive overview on the therapeutics until the mid-1980s.

aromatic diamidine compounds (such as stilbamidine and pentamidine) was also undertaken; P.C. Sen Gupta provides a comprehensive overview of these antinomy and aromatic diamidine studies until early 1950s [8]. The antimony regimens continued to be used, while new drug compounds were still under development in 1980s and 1990s. A critique on several methodological shortcomings of these earlier studies in terms of trial design and conduct is presented by Bryceson in 1987 [7]. A hallmark of these early trials was that they were mostly non-comparative. In particular, Bryceson identified only four studies with head-to-head comparisons between drugs until 1983 (the year when the first randomized studies in VL were undertaken in Kenya) [7].

10.1.4 Modern clinical trials (1980–2005)

The 1980s saw the transition toward scientifically rigorous trials, with a notable shift toward randomized controlled trials (RCTs). Bryceson et al. [9] highlighted the growing therapeutic challenges associated with the prolonged use of standard antimonial treatments, especially in the face of emerging drug unresponsiveness. These insights prompted evaluation of new treatment options such as amphotericin B, miltefosine, and pentamidine within a clinical trial context. Their introduction not only reshaped treatment algorithms but also served as a catalyst for refining trial designs—such as the adoption of randomization, rigorous adverse event monitoring were in place, and long-term efficacy assessments (patient follow-up) were undertaken. This period marked a pivotal moment where empirical observations began to merge with methodological rigor, influencing subsequent trials and future clinical research. Particularly at the turn of the millennium, there was a gradual shift towards the use of miltefosine and amphotericin B for the management of VL in ISC.

10.1.5 Clinical trials in visceral leishmaniasis

10.1.5.1 Drug regimens tested

The leishmanicidal effect of antimony was discovered in the early 1900s (section 10.1.1) [10, 11]. The pioneering development of the drug in the ISC was largely undertaken by L.E. Napier and U.N. Brahmachari [1, 6, 12], and the pentavalent antimony (Sb^v) formulation was the mainstay regimen for over 70 years as sodium stibogluconate (SSG) and meglumine antimoniate. These treatments have variable efficacy and require injectable administration, either intravenous (IV), intramuscular (IM), or intralymphatic (IL) for four weeks. Due to the accumulation in tissues, the antimonials are associated with serious adverse events (SAEs) such as pancreatitis, nephrotoxicity, and high cardiotoxicity [13]. They are still used against VL worldwide except in the Indian sub-continent (ISC).

Pentamidine was added to the list of antileishmanials in the 1950s [14] and has been used for the treatment of refractory Sb^v cases in India. Reno-toxicity among immune-suppressed patients was a concern with this regimen [15], and the drug is no longer in use for the management of primary VL. However, a low-dose pentamidine is currently recommended in the 2022 WHO VL-HIV guidelines for secondary prophylaxis among VL-HIV co-infected patients who are at a high risk of relapse in East Africa.

The discovery of amphotericin B as an antileishmanial heralded a new era for the management of VL. Initially it was developed as a deoxycholate formulation which was found to have notable toxicity [14]. In the 1980s, encapsulation of this drug in liposomes was found to effectively cure VL with improved safety profile [16]. The viscera are particularly rich in sinusoidal capillaries and macrophages; thus, encapsulation facilitates rapid and efficient tissue uptake, leading to drug accumulation and increased *in vivo* activity compared to free drug [17]. The lipid-associated formulation of amphotericin B has been tested in three different formulations: liposomal amphotericin B (L-AmB), colloidal dispersion amphotericin B (ABCD), and lipid complex amphotericin B. This regimen requires cold chain for storage and remains an expensive drug.

Initially, L-AmB was tested as multiple doses, but a single-dose regimen (10 mg/kg) became the first-line therapies adopted in the ISC for the regional elimination program. While single dose L-AmB remains highly efficacious in the ISC, this regimen hasn't demonstrated the same effectiveness in Eastern Africa and against *Leishmania infantum* in Latin America. Currently, L-AmB is the only drug approved by the US FDA for the treatment of VL, and the L-AmB in combination with miltefosine is also the current World Health Organization (WHO)-recommended drug for treatment of *Leishmania*–HIV co-infections.

Another mainstay drug has been miltefosine (MF), a phospholipid derivate that was originally developed for the treatment of cancer. It was the first antileishmanial treatment to undergo phase IV trial and the only oral drug available [11]. This was the drug of initial choice when the VL elimination plan was jointly launched by India, Bangladesh, and Nepal in 2005 [18]. However, during the roll-out in elimination program, effectiveness has been sub-optimal; the per protocol cure rate was 85% in a phase IV study in Bangladesh [19], and 20% [95% CI: 12.8–27.2%] of the patients subsequently relapsed at 12-months following treatment in a cohort study from Nepal [20]. A trial in Eastern Africa has also demonstrated a low Day 210 efficacy of 72% [95% CI: 60–85%] for the miltefosine monotherapy arm [21]. Therapeutic failures with this regimen have been associated with under-dosing and under-exposure in children has been noted [22–24]. Abortifacient activities and teratogenicity are a potential issue with this regimen. This necessitates women of child bearing age to take contraception for up to three months after treatment [11].

Other drugs used include paromomycin and sitamaquine. Paromomycin is an aminoglycoside antibiotic administered through the IV route, and its antileishmanial activity was discovered in the 1960s. The drug regimen demonstrated its efficacy in 1990, but it completed a Phase III study only in 2005 [16]. The clinical development of this drug has been painstakingly slow and is considered an orphan drug by the US FDA and EU EMEA [14]. Lack of funds have stalled the conduct of trials required for registration [10]. Ototoxicity is a possible concern with this regimen [15].

Sitamaquine is an 8-aminoquinoline discovered by the Walter Reed Army Institute of Research, US and was first synthesized as a result of an antimalarial program in the US in the 1940s [25]. It is second oral antileishmanial drug and currently in development with GlaxoSmithKline and has a short elimination half-life of about 26 hours. The clinical development of this drug has been very slow because of the associated toxicities of nephrotoxicity and methemoglobinemia [26].

Combination therapies have been used in the treatment of VL since the 1980s, especially for the management of VL unresponsive to antimony regimens [27, 28]. These studies have used L-AmB or pentavalent antimony in combination with other therapies. The combination regimen of pentavalent antimony + paromomycin has a short treatment course of 17 days and provides comparable efficacy to a 30-day course of antimonial therapy. However, this regimen has been largely restricted to East African context. Of particular note, the combination regimen of paromomycin and miltefosine (14 days) has been tested in a randomized study in East Africa and has demonstrated >90% definitive cure at 6-months (Trial registration: NCT03129646).

10.1.6 Relapse following treatment

Approximately 3% of the patients are expected to relapse after achieving initial cure (meta-analysis of 131 studies) [29]. In the ISC, relapse proportion within six months were estimated at 4.5% [95% CI: 2.6–7.5%] following single-dose L-AmB and 1.5% [95% CI: 0.7–3.3%] following L-AmB-based combination therapies. Notably, when studies with a follow-up duration longer than six months were investigated, over a quarter (27.6%) of all observed relapses were estimated to occur after the six-month follow-up period (data largely from ISC), suggesting that currently adopted follow-up of six months may underestimate true relapse proportions [29].

In Eastern Africa, the relapse proportions have been generally higher than observed in the ISC, ranging from 3.8% [95% CI:1.3–10.9%] with pentavalent antimony to 13.0% [95% CI: 4.2–33.6%] with pentavalent antimony plus paromomycin combination (large uncertainty due to limited number of trials for this combination regimen), reflecting regional variation in treatment efficacy [29].

10.1.7 Safety and tolerability of the treatment for VL

The safety of antileishmanial therapies has always been a major concern. Singh-Phulgenda et al. reviewed data from 157 clinical trials enrolling 35,376 patients and reported 804 serious adverse events (SAEs) (including 428 deaths). The overall incidence rate of death (IRD) within the first 30 days of treatment was 0.068 per 1000 person-days at risk. This IRD was lower among patients enrolled in the ISC (0.041 death per 1000 person-days) compared to those in Eastern Africa (0.628 death per 1000 person-days), and higher in studies that had allowed enrolling HIV-infected patients (0.575 death per 1000 person-days) compared to the studies which excluded HIV co-infections (0.043 death per 1000 person-days) [30].

L-AmB, particularly in its single-dose form (10 mg/kg), demonstrated the most favorable safety profile among available treatments. Conversely, pentavalent antimony regimens were associated with severe toxicities such as cardiac arrhythmias, pancreatitis, and nephrotoxicity, thus limiting their utility, especially in resource-constrained settings [30].

10.1.8 Sub-population

10.1.8.1 VL–HIV co-infection

Patients co-infected with VL and HIV generally have a poor prognosis following treatment compared to those patients who are HIV negative. Prior to 2022, L-AmB was the treatment of choice for management of VL–HIV cases. A randomized phase 3 trial by Burza et al. in Bihar, India, compared L-AmB monotherapy (40 mg/kg) against a combination regimen of L-AmB (30 mg/kg) and oral MF for 14 days [31]. The study found a significantly higher relapse-free survival in the combination arm (96%) compared to the monotherapy (85%) at 210 days, with similar safety profile between the two arms [31]. The drug combination of L-AmB and oral MF was previously evaluated in a retrospective study by Mahajan et al. and they reported a cumulative incidence of relapse of 13.9% at 18 months following the combination regimen; mortality was 16.6% at the same time-point [32]. The study reported failure to initiate the antiretroviral therapy (ART) and co-existing tuberculosis (triple infection) as risk factors associated with mortality [32].

The evidence from these two studies has suggested that L-AmB and oral MF combination regimen offers a favorable balance between safety and efficacy. However, since multiple relapse among co-infected patients can be expected, ongoing pharmacovigilance, and integrated care remain essential for optimizing VL care in this patient group.

The WHO has recently endorsed the L-AmB and oral MF—as the preferred first-line treatment for VL–HIV co-infection in South-East Asia, including India [33]. This recommendation stemmed from the Burza et al study discussed before, along with a randomized study from Ethiopia by Diro et al. [34]. Overall, the evidence from these randomized studies indicate that the combination not only reduced the total required dose of L-AmB, thereby minimizing toxicity and hospitalization burden, but also improved treatment adherence and tolerability [35]. This marks a significant shift in the management strategy, emphasizing shorter, safer, and more effective regimens tailored to the needs of co-infected populations.

10.1.8.2 TB co-infection and VL/HIV/TB triple infections

The occurrence of concurrent infection of leishmaniasis and TB has been traced back to the Italian Renaissance [36, 37]. Approximately a quarter of the world's population has latent TB, and in persons with compromised immune systems, the latent infection turns into active disease [38]. This has led to the emergence of VL–HIV–TB triple infection as a new challenge for patient management. In general, the prevalence of TB between 5.7% and 16% has been reported in VL–HIV co-infected patients from India (from three retrospective studies undertaken in MSF supported district hospital in Bihar), who were also found to be at an increased risk of death [39].

10.1.8.3 Pregnancy

Pregnant and lactating women were excluded in ~60% of clinical trials in VL undertaken between 2000–2021 [40]. Similarly, women susceptible to becoming pregnant often require undergoing pregnancy test and/or their enrollment remains conditional on contraception usage agreement. This has led to a gap in our understanding of the safety of antileishmanial drugs in pregnancy. A systematic review [41] of published literature (1926–2020) identified 72 records and reported 451 cases of VL in pregnancy. The review found those who received L-AmB had generally favorable safety outcomes compared to antimony regimens. Vertical transmission remained rare with 26 cases reported with a median time of 6 months (range: 0–18 months) after birth for the detection of the congenital transmission.

10.1.8.4 Malnutrition

Acute malnutrition (wasting) has been associated with higher parasite load at presentation [42, 43] and has been suggested to lead to a lower bioavailability of MF, especially during the initial phase of the treatment [44]. The impact of nutritional status on treatment outcomes remains poorly studied as clinical trials often exclude patients with severe malnutrition [40]. In general, the role of malnutrition in determining treatment outcomes in clinical literature has remained unclear; these observations are not unique to VL. Prospectively designed longitudinal cohort studies are needed to understand the role of nutrition toward disease susceptibility and also its role in modulating treatment outcomes.

10.2 POST-KALA-AZAR DERMAL LEISHMANIASIS

Following a successful treatment of VL, a dermal sequela (termed post-kala-azar dermal leishmaniasis, PKDL) is observed in up to 56–58% in Sudan and between 5 and 14% in the ISC. Remarkably, about 10–23% of dermal leishmaniasis lacks a prior visceral episode [45, 46]. Untreated PKDL may remain infectious to sand flies as reported in recent xenodiagnosis studies, especially the nodular lesions which contain many parasites [47, 48]. Prompt treatment of PKDL is therefore important from a public health perspective.

10.2.1 Drug regimens used in the past and current treatment landscape

PKDL cases are much more difficult to treat than the primary VL due to the need of a lengthy treatment. In the past, 20 mg/kg of SSG (at least 120 days) had been used in India. In 2013, a randomized study evaluated oral MF at 100 mg/day for two to three months and reported per-protocol efficacy >80% at 12-months efficacy [49]. A shorter regimen with 15 mg/kg of L-AmB demonstrated improvement of lesions in approximately 90% of the patients in a study from Bangladesh [50].

In 2024, a review undertaken by Singh-Phulgenda et al. [51] identified 56 studies (cohorts and case series) that enrolled over 2400 patients; substantial heterogeneity in drug regimens and treatment outcomes was identified. In

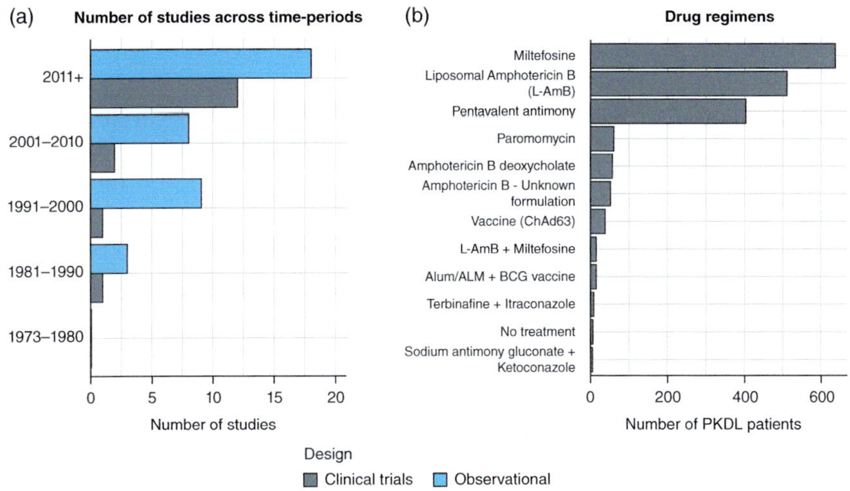

Figure 10.1 PKDL trials landscape. (Source: IDDO PKDL clinical trials library, available from https://www.iddo.org/tool/vl-surveyor.) (a) The number of studies indexed in the IDDO PKDL library by study design, prospective clinical trials or observational studies such as case report or case series. (b) Drug regimens used for the management of PKDL patients.

particular, the first-line regimens adopted vary by geography (Figure 10.1). In India, MF remains the first-line drug, typically administered at 100 mg/day for 12 weeks, with an efficacy of around 85%, but a decline in efficacy over time has been noticed [52, 53]. L-AmB, although effective, is expensive and requires parenteral administration. Conventional amphotericin B is nephrotoxic, while SSG suffers from high toxicity and drug resistance issues, particularly in the ISC [54, 55]. In Eastern Africa, SSG and L-AmB remain standard options, often reserved for severe or chronic cases due to frequent self-resolution of the milder forms [56].

10.2.2 Combination therapies

Combination regimens are increasingly explored to improve outcomes, reduce treatment duration, and enhance patient adherence. A comprehensive review by Kumari et al. identified nine key studies assessing various drug combinations of MF with L-AmB (MF-L-AmB), with paromomycin (MF-PM), and with amphotericin B (MF-AmB). The MF-L-AmB regimen demonstrated a 100% cure with no relapse at 18 months in a small cohort, outperforming MF monotherapy, which had a 25% relapse proportion [56]. Similarly, the MF-PM combination achieved a final cure of 83.3% with excellent compliance, though some mild adverse events like vomiting and hypokalemia were noted. These regimens effectively shortened treatment duration, enhanced adherence, and demonstrated strong safety profiles [56].

In general, the safety of combination therapies appears favorable compared to monotherapies. Adverse events in MF-based combinations were generally mild and manageable. L-AmB, when combined with MF, showed negligible nephrotoxicity—an improvement over conventional amphotericin B regimens. SSG-based combinations such as SSG–rifampicin and SSG–paromomycin were also explored in Sudan and India, with good clinical outcomes and some injection site reactions, but concerns remain about the long-term toxicity and resistance profiles of SSG [56]. The MF-LAmB combination has also been tested in a recent Phase II study in India and Bangladesh and has demonstrated the combination remains a safe and efficacious alternative to a long course miltefosine regimen (Trial registration: CTRI/2017/04/008421).

While monotherapies like MF and amphotericin B continue to play vital roles in PKDL management, combination therapies offer promising alternatives with improved efficacy, reduced toxicity, and better compliance. MF-L-AmB and MF-PM combinations, in particular, appear to be effective and safe options. However, more robust, multicentric RCTs are needed to further validate these findings and guide evidence-based clinical practice.

10.2.3 Ocular safety

An important recent development is the ocular safety of MF following PKDL treatment in the ISC region [57–59]. Observational study from India reported 3.7% (11/300) of the patients developed ocular adverse events [60]. The WHO has released a statement to exercise caution and undertake a thorough examination, and in April 2023, the WHO Advisory Committee on Safety of Medicinal Products (ACSoMP) further supported the recommendations and provided detailed guideline and information for patients and healthcare professionals [61]. Caution is thus warranted, and there is an urgent need for alternative drugs for the safe management of the PKDL patients in the ISC region.

10.2.4 Further challenges

10.2.4.1 The "unsolved mystery" of PKDL

Despite 100 years since the first identification of PKDL, it still remains an unsolved mystery [62]. In particular, the disease can not only appear several years after completion of VL treatment, but a substantial proportion of patients with PKDL appear without any documented VL history (in approx. 10–23% of the patients) [45, 46]. All of these factors are particularly important to consider in the context of elimination settings (including challenges in the sustenance of elimination). Overall, these aspects of the PKDL indicate a need for increased disease awareness among patients and the population living in the endemic villages/settings. There is also a need for further research to identify the risk factors (including host and drug) associated with PKDL onset so that patients at high risk can be identified earlier. For example, in Southern Sudan, young children (0–4y) were at 20-times increased risk (17.7% vs 0.9%) of development of PKDL [63]. Prospective cohort studies with a long-term follow-up might provide insights regarding these factors, as were done in Bangladesh and India [64, 65]. Identification of these risk factors can guide early intervention and reduce the community transmission. Table 10.2 summarizes how different

Table 10.2 Cohort studies from Indian sub-continent with a long follow-up to capture late relapse and PKDL incidence following VL treatment

Study location	*Treatment regimen*	*VL relapse incidence*	*PKDL incidence*	*Notable findings*
Bihar, India [65]	Single-dose AmBisome (SDA)	1.58 per 1000 person-months	1.29 per 1000 person-months	Children <12 years were at higher risk for both relapse and PKDL. Females had a higher risk of PKDL but not relapse. Active surveillance and timely treatment are essential for VL elimination programs.
	Miltefosine–paromomycin (Milt–PM)	0.40 per 1000 person-months	2.65 per 1000 person-months	Milt–PM resulted in lower relapse incidence but higher PKDL incidence compared to SDA.
	AmBisome–miltefosine (AmB–Milt)	2.08 per 1000 person-months	1.45 per 1000 person-months	AmB–Milt rates were not significantly different from SDA.
Bangladesh [64]	Sodium stibogluconate (SSG)	1.8 per 100-person-years	3.0 per 100-person-years	SSG cohort had the lowest incidence rate for both PKDL and VLR. Development of PKDL and VLR is related to the VL treatment regimens. Multidose AmBisome (MDAMB) is recommended for VL as a primary step for prevention of PKDL and relapse (VLR), as SSG is highly toxic and not recommended. Overall incidence proportion for PKDL was 12.3% and for VLR was 7.0%.
	Multidose AmBisome (MDAMB)	2.7 per 100-person-years	8.2 per 100-person-years	MDAMB followed SSG with low incidence rates for PKDL and VLR.
	Other monotherapies (miltefosine, paromomycin intramuscular injection, single-dose AmBisome) and combination therapies (AMB+PMIM, AMB+MF, PMIM+MF)	Higher rates compared to SSG and MDAMB	Higher rates compared to SSG and MDAMB	The study compared these against SSG and MDAMB, which showed better outcomes.

treatment regimens for VL impact the long-term outcomes of PKDL and VL relapse. It highlights variations in efficacy across different drugs and combinations, with specific findings from studies conducted in India [65] and Bangladesh [64].

10.2.4.2 Para kala-azar dermal leishmaniasis (Para-KDL)

The concurrent occurrence of PKDL in active VL patients is known as para-kala-azar dermal leishmaniasis. Most of the reported Para-KDL cases have been from Eastern Africa [66], but sporadic reports have emerged in the past two decades in the ISC and Brazil [67–70]. Patients with para-KDL may pose further difficulties in case management and may provide reservoir for further transmission of the disease. No standardized treatment protocol exists for para-KDL. Current management often involves sequential or combined therapies tailored to both VL and PKDL [67–70]. One approach has been to treat VL and PKDL separately. For example, a single dose of L-AmB for VL is followed by MF for PKDL after one month [67, 71]. In a case series from Bangladesh, 15 out of 16 para-KDL patients were treated with L-AmB 20 mg/kg in four divided doses. All showed clinical improvement for VL and PKDL lesions of all the patients also improved [70]. MF has better skin penetration for PKDL, while L-AmB is effective for VL. A combination of multidose L-AmB and oral MF is proposed as a potential treatment strategy for para-KDL to achieve complete parasite clearance [70]. More research is needed on the etiopathology, epidemiology, and evidence-based management of para-KDL, as well as on new drugs and combination therapies (Table 10.3).

10.3 CUTANEOUS LEISHMANIASIS

Cutaneous leishmaniasis (CL) is on the rise in many developing countries and is increasingly observed in migrants, military personnel, humanitarian aid workers, and travelers [73]. It is a disfiguring and stigmatizing condition affecting the skin and mucous membranes. The causative *Leishmania* species vary across different geographical regions with a clear distinction of the disease between Old World and New World.

The 2022 WHO skin neglected tropical diseases (skin NTDs) framework has identified controlling CL as a strategic target to be achieved by 2030 [74]. In particular, the framework advocates that "persons affected by skin NTDs should receive quality, stigma-free and continued care, while living with their families and in their communities". However, the currently recommended treatments of CL are far from optimal in terms of efficacy, safety, and ease of delivery. There is also a limited evidence underpinning current recommended treatments for CL that has largely relied on few drugs such as pentavalent antimony.

Comprehensively synthesizing evidence on safety and efficacy of drug regimens for CL is challenging due to widespread methodological heterogeneity in design, conduct, analysis, and reporting (discussed in 10.3.2). These

Table 10.3 Some key para-KDL reports from ISC and a report from Brazil. See Zijlstra et al. [66] for for East Africa

Study location	*Number of cases*	*Patient details/diagnosis (para-KDL features)*	*Treatment administered*	*Treatment outcome*
Dharan, Nepal [69]	1 (48-year-old male)	Patient had inadequate treatment for kala-azar 2 years prior. Presented with multiple papules, nodules, and verrucous plaques on face, trunk, and extremities, with splenomegaly. *Leishmania donovani* bodies (LDB) were seen in skin lesions, fine needle aspiration cytology from lesions and lymph nodes, and bone marrow aspirate. rK39 dipstick test was positive. This presentation was classified as a case of post-kala-azar dermal leishmaniasis with visceral leishmaniasis.	Injections of sodium stibogluconate (SSG) 20 mg/kg/day intramuscularly with the instruction to continue until all lesions resolved.	Patient was referred to a primary health center with the prescription; the specific treatment outcome (e.g., cure or resolution) is not explicitly stated as achieved in this report. The discussion section notes that renewed multiplication of latent parasites in viscera might have caused relapsed VL, suggesting ongoing disease.
New Delhi, India [72]	1 (6-year-old girl)	Presented with prolonged fever, anorexia, weight loss, and multiple widespread hypopigmented skin macules. Examination revealed generalized lymphadenopathy, anemia, leukopenia, splenomegaly, and hepatomegaly. LD bodies were found in bone marrow aspirate, and skin biopsy was suggestive of PKDL. rK39 dipstick test was positive. Parasites were characterized as *L. donovani* by PCR. This presentation was classified as a grade 2 eruption covering most of the face, chest, back, upper arms, and legs, gradually becoming more diffuse distally. This case was termed para-kala-azar dermal leishmaniasis.	Treatment for 120 days with sodium stibogluconate (SSG) 20 mg/kg/day intramuscularly, with the goal for all lesions to heal.	Treatment duration was specified, but the final outcome (cure/resolution) is not explicitly stated as achieved within the scope of this report.

(*Continued*)

Table 10.3 (*Continued*)

Study location	*Number of cases*	*Patient details/diagnosis (para-KDL features)*	*Treatment administered*	*Treatment outcome*
São Paulo, Brazil [68]	1 (36-year-old male)	HIV-negative patient with a history of VL and multiple relapses. During a suspected second relapse of VL, he developed erythematous macules and papules on the chest and abdomen. *Leishmania* amastigotes were found in bone marrow aspirate, and *Leishmania* sp. was demonstrated in a skin biopsy. This was described as a progression from visceral disease to para-kala-azar dermal leishmaniasis. Later presented with isolated lymphadenopathy with *Leishmania* sp.	Initial treatment for para-KDL episode: amphotericin B deoxycholate (50 mg/day for 4 days), followed by liposomal amphotericin B (L-Amb) (3 mg/kg for 10 days). Subsequent treatment (due to later isolated lymphadenopathy): L-Amb 4 mg/kg/day for 7 days and secondary prophylaxis with pentavalent antimonial (850 mg every 28 days).	Initial treatment resulted in a positive clinical response. However, the patient did not return for suggested secondary prophylaxis, and 10 months later presented with isolated cervical lymphadenopathy containing *Leishmania* sp., indicating ongoing infection. Secondary prophylaxis was then initiated, and the patient remained well during 9 months of follow-up.
Bihar, India [67]	9 cases	All cases presented with fever, constitutional symptoms (weight loss, weakness, loss of appetite), hepatosplenomegaly, and skin lesions (macular, papular, nodular, or mixed, with macular being predominant). Diagnosis was confirmed by the demonstration of LD bodies in bone marrow/splenic aspirate (for VL) and in skin smear (for PKDL) in the same patient. All patients were rK39 positive and immunocompetent. The study reported on "Para-kala-azar dermal Leishmaniasis cases".	All cases were first treated for VL with single-dose AmBisome (liposomal amphotericin B) (10 mg/kg). After one month, with VL signs subsided and no LDB in visceral aspirates, they were treated for PKDL with miltefosine 50 mg twice a day for 12 weeks.	All 9 cases were cured clinically as well as parasitologically upon completion of treatment.

Dhaka, Bangladesh [71]	1 (50-year-old male)	Patient had a history of primary VL and two relapses. Presented with fever, skin rash (multiple maculopapular and erythematous lesions), and hepatosplenomegaly. Diagnosis confirmed by LD bodies in both slit skin smear (for PKDL) and splenic biopsy (for VL), indicating co-existence of active VL and PKDL. This was reported as a case of "para-kala-azar dermal leishmaniasis".	First treated with a single dose of liposomal amphotericin B (10 mg/kg body weight) for relapse VL, followed by miltefosine (2.5 mg/kg/daily) 50 mg twice a day for 12 weeks for PKDL. Due to persistent systemic features, the patient received an additional three doses of liposomal amphotericin B (5 mg/kg) on alternate days.	The patient achieved full recovery. Overall clinical condition improved, weight was gained, and there was no fever or splenomegaly. The patient was definitively cured after 6 months and remained in good health after 2 years.
Mymensingh, Bangladesh [70]	16 cases	Defined as the concomitant presence of PKDL and VL. All patients presented with fever (>2 weeks), palpable spleen (average 6.6 cm), and cutaneous lesions (predominantly macular, 13 patients; rest had mixed macular and nodular). Most were underweight. Diagnosis confirmed by VL (clinically, rK39 RDT, LD bodies in splenic aspirate, qPCR for slit skin smear) and PKDL (LD bodies in slit skin smear, qPCR, clinically). All patients were rK39 positive, and none were immunocompromised.	Fifteen patients treated with AmBisome 20 mg/kg in 4 equally divided doses. One patient with hypersensitivity received miltefosine capsules 50 mg, twice daily for 12 weeks instead of AmBisome.	All patients experienced clinical improvement, ensuring complete cure of VL (no fever, no splenomegaly, no hepatomegaly). All had substantial weight gain and improved hematological parameters (hemoglobin levels improved). Clinical improvement in PKDL lesions was also observed. Patients were followed up for 1 year after treatment.

challenges also emerge from the fact that the diverse *Leishmania* spp. are the causative pathogens for CL, as opposed to only two species that cause VL and PKDL.

Pentavalent antimony, despite their known toxicity, is the main drug for CL treatment worldwide. Variety of other drugs and treatment modalities has been used with varying success [73].

10.3.1 Drug regimens by regions and evidence gaps

10.3.1.1 Old World CL

The regimens used for the management of Old World CL include intralesional (IL) and intramuscular (IM) meglumine antimoniate (MA), paromomycin (PR) ointment, oral ketoconazole, oral itraconazole, oral fluconazole, topical garlic cream, oral allopurinol (AL), topical photodynamic therapy (PDT), oral MF, intralesional zinc sulfate, cryotherapy, CO_2 laser, topical trichloroacetic acid, local heat therapy, oral rifampicin, and topical imiquimod [73]. A complete absence of evidence was noted for intramuscular pentamidine and topical amphotericin B.

10.3.1.2 New World CL

For New World CL, drug regimens used include intravenous (IV) aminosidine, IV and IM SSG, IM and IV meglumine antimoniate (MA), oral ketoconazole, oral MF, oral allopurinol (AL), IM pentamidine isethionate, IM aminosidine sulfate, heat therapy, and topical imiquimod [73]. There was little or no evidence for oral antifungals, antibiotics, photodynamic therapy, laser, cryotherapy treatments, or alternative therapies. No trials involving wound-healing management or alternative supportive therapies versus drug interventions for Old World and American CL were found.

10.3.2 Methodological challenges

Review evaluating RCTs for CL has identified several biases that can impede evidence synthesis endeavors [73]. Methodological challenges identified include inadequate randomization, poor blinding, high attrition rates, lack of sample size calculations, and inconsistent outcome measures. In addition, many trials failed to identify and/or report the causative *Leishmania* species, which is crucial for evaluating treatment efficacy. These limitations preclude opportunities to conduct meaningful aggregate data meta-analyses and synthesize high-quality evidence.

10.3.3 Psychosocial impact and treatment implications

CL is associated with a high degree of psychological morbidity, regardless of country, age, or disease activity [75]. This can range from subclinical "distress" to clinical "disorder", such as major depressive disorder (MDD) [75]. Studies have consistently reported MDD rates equating to 30–50% [75]. Notably, treatment has been shown to significantly reduce symptoms of anxiety and depression, and improve quality of life in CL patients [73, 75]. This highlights the importance of incorporating mental health care into NTD programs.

10.4 REAL-WORLD STUDIES: SOME KEY SURVEILLANCE AND OBSERVATIONAL DATABASES

Observational databases either from VL national surveillance systems (Brazil, India) or from humanitarian settings (Sudan, Ethiopia) have been recently used to explore different aspects of VL epidemiology [76–78]. These databases are of vital importance to inform policy makers and researchers regarding different aspects of the disease and treatment.

10.4.1 The Kala-Azar Management Information System (KAMIS), India

The Kala-Azar Management Information System (KAMIS) is an electronic surveillance system of the National Center for Vector Borne Diseases Control (NCVBDC), India. The KAMIS captures VL and PKDL patient information, which includes socio-demographic features, duration of illness, major symptoms, treatment received, and clinical outcome of patients across four states (Bihar, Jharkhand, West Bengal, and Uttar Pradesh). The KAMIS was launched in 2015–2016 and currently holds individual patient-level data on more than 30,000 VL patients [76]. Linking reported cases in the database makes it possible to track individual patients over time, helping to answer key questions like when relapses or PKDL occur following different treatments.

10.4.2 The Notifiable Diseases Information System (SINAN), Brazil

In Brazil, VL is a notifiable disease, and the routine data is captured within the Notifiable Diseases Information System (SINAN) [79]. Currently the system holds records on more than 50,000 confirmed VL patients [80–81]. Anonymized data from the SINAN system is openly accessible and facilitates the characterization of temporal trends in the spread of disease endemicity in Brazil. Evaluation based on the SINAN database have identified several disease specific trends such as the median age of patients at presentation has increased from 10years in 2007 to 32years in 2023, suggesting a major shift in the population affected by the disease, and increasing proportions of the relapse cases [81]. Such findings can help devise targeted surveillance strategies to control the disease.

10.4.3 MSF's operational databases—South Sudan and Ethiopia

Between 1989 and 2020, MSF teams have treated nearly 150,000 VL patients—over a third of them in South Sudan [82]. In Lankien, Jonglei State, South Sudan, the routinely collected MSF data consists of more than 20,000 VL patients since 2000 [77]. Similarly, MSF has been providing VL treatment for the past 20 years in Ethiopia (since 2003).

10.5 WAYS FORWARD

Several challenges remain despite the decline of the VL burden in the ISC. Limited alternative to current antileishmanial therapies available for deployment is a major challenge. Conducting prospective RCTs for evaluating new drug therapies is also challenging in the ISC as the disease burden has fallen and in eastern part of Africa due to political instabilities [83]. The increasing cost of conducting RCTs and restricted public funds have meant limited the opportunity to support VL research. There is also limited knowledge on the best treatment approaches for patient groups frequently excluded from clinical studies such as pregnant women and patients with TB, malaria, or HIV co-infections [40], or very young patients and in patients with PKDL, and in pregnant women [41]. The efficacy of antileishmanial drugs remains highly regional [84], and questions regarding optimal drug dosage and safety profile remain yet to be fully understood.

Explanatory trials often investigate a highly selected patient group, excluding patients with common comorbidities, risk factors, and other potential modifiers of drug response. Use of registries or observational databases (mentioned above) might facilitate further assessment of drug effectiveness, or pragmatic trials can help in alleviation of some of the existing gaps and challenges in therapeutic development. Such pragmatic trials, if designed and

Table 10.4 Future opportunities in VL therapeutic trials

Innovation	*Potential impact*
Genomic sequencing	Enables identification of drug resistance markers, allowing for precision therapy tailored to individual patient profiles and local strain characteristics. This approach could improve cure proportion and reduce relapse through early detection of resistance-associated mutations.
Digital health	Facilitates real-time adherence monitoring, teleconsultation, and remote adverse event tracking through mobile platforms and wearable technologies. This could enhance treatment outcomes, especially in remote or resource-limited areas, by promoting continuity of care and reducing dropouts.
Long-acting injectables	Offers the possibility of single-dose or monthly treatments, minimizing patient visits and improving compliance. These formulations could be pivotal in addressing nonadherence and reducing the burden on healthcare systems, especially in endemic regions with limited infrastructure.

executed well, can provide real-world evidence on the value of a treatment strategy in routine clinical practice at the same time maintaining the advantages of an RCT (Table 10.4). In addition to the real-world/surveillance databases, managed open-access database of harmonized patient records from historical trials such as the Infectious Diseases Data Observatory (IDDO) VL and PKDL repository, which was developed collaboratively by scientists across the different endemic regions, presents a unique resource for innovation and improvement of patient outcomes [85].

10.6 CONCLUSION

The evolution of therapeutic trials in VL and PKDL reflects both scientific progress and persisting challenges. From empirical treatments with antimonials to the adoption of L-AmB, MF, and combination regimens, clinical research has significantly improved patient outcomes. Much of the evidence on treatment efficacy and safety have been generated from clinical trials, which have identified important safety concerns and regional variability in efficacy. Real-world data can overcome some of the limitations of RCTs, especially for vulnerable groups such as pregnant women, children, malnourished individuals, and those with HIV or TB co-infections who are frequently excluded from therapeutic efficacy studies. PKDL continues to complicate elimination efforts due to prolonged treatment needs, relapse, and its role as a reservoir for transmission. Emerging issues such as MF-related ocular toxicity underscore the need for safer alternatives. Going forward, pragmatic trials, integrated care approaches, genomic surveillance, and digital health tools hold promise to strengthen treatment strategies. Bridging the gap between controlled studies and real-world applications is essential to sustain progress toward leishmaniasis elimination and serve the patient populations typically excluded in clinical trials.

ACKNOWLEDGMENT

The authors of this chapter are thankful to the ICMR for providing necessary research support.

REFERENCES

1. Brahmachari UN: A new form of cutaneous leishmaniasis-dermal leishmanoid. *Ind Med Gaz* 1922, 57:125–127.
2. Napier LE: A preliminary note on the successful treatment of kala-azar with "stibamine glucoside". *Ind Med Gaz* 1925, 60:24–26.
3. Brahmachari UN: *A Treatise on Kala-Azar*. London: John Bale, Sons & Danielsson; 1928.
4. Brahmachari UN: *Kala-Azar, Its Treatment*. Calcutta: Butterworth (India); London: Butterworth; 1917.

5. Napier LE: *The Principles and Practice of Tropical Medicine.* New York: The Macmillan Company; 1946. 134–191.
6. Scott AV, Li PK: The treatment of kala azar in children. *Arch Dis Child* 1932, 7:213–224.
7. Bryceson ADM: Chapter 18: "Therapy in man." In *The Leishmniases in Biology and Medicine. Volume II. Clinical Aspects and Control.* Edited by: Wallace Peters and Robert Killick-Kendrick. Academic Press Inc; 1987.
8. Gupta PCS: Chemotherapy of leishmanial diseases, a resume of recent researches. *Ind Med Gaz* 1953 Jan, 88(1):20–35.
9. Bryceson AD, Chulay JD, Ho M, Mugambii M, Were JB, Muigai R, Chunge C, Gachihi G, Meme J, Anabwani G, et al: Visceral leishmaniasis unresponsive to antimonial drugs. I. Clinical and immunological studies. *Trans R Soc Trop Med Hyg* 1985, **79**:700–704.
10. Guerin PJ, Olliaro P, Sundar S, Boelaert M, Croft SL, Desjeux P, Wasunna MK, Bryceson AD: Visceral leishmaniasis: Current status of control, diagnosis, and treatment, and a proposed research and development agenda. *Lancet Infect Dis* 2002, **2**:494–501.
11. Croft SL, Olliaro P: Leishmaniasis chemotherapy – Challenges and opportunities. *Clin Microbiol Infect* 2011, **17**:1478–1483.
12. Napier LE: The problem of cure in kala-azar. *Ind Med Gaz* 1924, **59**:492–504.
13. Freitas-Junior LH, Chatelain E, Kim HA, Siqueira-Neto JL: Visceral leishmaniasis treatment: What do we have, what do we need and how to deliver it? *Int J Parasitol Drugs Drug Resist* 2012, **2**:11–19.
14. Alvar J, Croft S, Olliaro P: Chemotherapy in the treatment and control of leishmaniasis. *Adv Parasitol* 2006, **61**:223–274.
15. Bryceson A: Current issues in the treatment of visceral leishmaniasis. *Med Microbiol Immunol* 2001, **190**:81–84.
16. Croft SL, Barrett MP, Urbina JA: Chemotherapy of trypanosomiases and leishmaniasis. *Trends Parasitol* 2005, **21**:508–512.
17. Murray HW: Clinical and experimental advances in treatment of visceral leishmaniasis. *Antimicrob Agents Chemother* 2001, **45**:2185–2197.
18. World Health Organization. Regional strategic framework for elimination of kala-azar from the South-East Asia Region (2011–2015). 2012. https://www.who.int/publications/i/item/sea-cd-239
19. Rahman M, Ahmed BN, Faiz MA, Chowdhury MZ, Islam QT, Sayeedur R, Rahman MR, Hossain M, Bangali AM, Ahmad Z, et al: Phase IV trial of miltefosine in adults and children for treatment of visceral leishmaniasis (kala-azar) in Bangladesh. *Am J Trop Med Hyg* 2011, **85**:66–69.
20. Rijal S, Ostyn B, Uranw S, Rai K, Bhattarai NR, Dorlo TP, Beijnen JH, Vanaerschot M, Decuypere S, Dhakal SS, et al: Increasing failure of miltefosine in the treatment of kala-azar in Nepal and the potential role of parasite drug resistance, reinfection, or noncompliance. *Clin Infect Dis* 2013, **56**:1530–1538.
21. Wasunna M, Njenga S, Balasegaram M, Alexander N, Omollo R, Edwards T, Dorlo TP, Musa B, Ali MH, Elamin MY, et al: Efficacy and safety of AmBisome in combination with sodium stibogluconate or miltefosine and miltefosine monotherapy for African visceral leishmaniasis: Phase II randomized trial. *PLoS Negl Trop Dis* 2016, **10**:e0004880.
22. Dorlo TP, Rijal S, Ostyn B, de Vries PJ, Singh R, Bhattarai N, Uranw S, Dujardin JC, Boelaert M, Beijnen JH, Huitema AD: Failure of miltefosine in visceral leishmaniasis is associated with low drug exposure. *J Infect Dis* 2014, **210**:146–153.
23. Dorlo TPC, Kip AE, Younis BM, Ellis SJ, Alves F, Beijnen JH, Njenga S, Kirigi G, Hailu A, Olobo J, et al: Visceral leishmaniasis relapse hazard is linked to reduced miltefosine exposure in patients from Eastern Africa: A population pharmacokinetic/pharmacodynamic study. *J Antimicrob Chemother* 2017, **72**: 3131–3140.

24. Wasunna M, Musa A, Hailu A, Khalil EA, Olobo J, Juma R, Wells S, Alvar J, Balasegaram M: The leishmaniasis East Africa platform (LEAP): Strengthening clinical trial capacity in resource-limited countries to deliver new treatments for visceral leishmaniasis. *Trans R Soc Trop Med Hyg* 2016, **110**:321–323.
25. Loiseau PM, Cojean S, Schrevel J: Sitamaquine as a putative antileishmanial drug candidate: From the mechanism of action to the risk of drug resistance. *Parasite* 2011, **18**:115–119.
26. Olliaro PL, Guerin PJ, Gerstl S, Haaskjold AA, Rottingen JA, Sundar S: Treatment options for visceral leishmaniasis: A systematic review of clinical studies done in India, 1980-2004. *Lancet Infect Dis* 2005, **5**:763–774.
27. Bryceson AD, Chulay JD, Mugambi M, Were JB, Gachihi G, Chunge CN, Muigai R, Bhatt SM, Ho M, Spencer HC, et al: Visceral leishmaniasis unresponsive to antimonial drugs. II. Response to high dosage sodium stibogluconate or prolonged treatment with pentamidine. *Trans R Soc Trop Med Hyg* 1985, **79**:705–714.
28. Chunge CN, Gachihi G, Muigai R, Wasunna K, Rashid JR, Chulay JD, Anabwani G, Oster CN, Bryceson AD: Visceral leishmaniasis unresponsive to antimonial drugs. III. Successful treatment using a combination of sodium stibogluconate plus allopurinol. *Trans R Soc Trop Med Hyg* 1985, **79**:715–718.
29. Chhajed R, Dahal P, Singh-Phulgenda S, Brack M, Naylor C, Sundar S, Alves F, Stepniewska K, Guerin PJ: Estimating the proportion of relapse following treatment of visceral leishmaniasis: Meta-analysis using infectious diseases data observatory (IDDO) systematic review. *Lancet Reg Health Southeast Asia* 2024, **22**:100317.
30. Singh-Phulgenda S, Dahal P, Ngu R, Maguire BJ, Hawryszkiewycz A, Rashan S, Brack M, Halleux CM, Alves F, Stepniewska K, et al: Serious adverse events following treatment of visceral leishmaniasis: A systematic review and meta-analysis. *PLoS Negl Trop Dis* 2021, **15**:e0009302.
31. Burza S, Mahajan R, Kazmi S, Alexander N, Kumar D, Kumar V, Lasry E, Harshana A, de Lima Pereira A, Das P, et al: AmBisome monotherapy and combination AmBisome-miltefosine therapy for the treatment of visceral leishmaniasis in patients coinfected with human immunodeficiency virus in India: A randomized open-label, parallel-arm, phase 3 trial. *Clin Infect Dis* 2022, **75**:1423–1432.
32. Mahajan R, Das P, Isaakidis P, Sunyoto T, Sagili KD, Lima MA, Mitra G, Kumar D, Pandey K, Van Geertruyden JP, et al: Combination treatment for visceral leishmaniasis patients coinfected with human immunodeficiency virus in India. *Clin Infect Dis* 2015, **61**:1255–1262.
33. Organization WH: *WHO Guideline for the Treatment of Visceral Leishmaniasis in HIV Co-Infected Patients in East Africa and South-East Asia*. World Health Organization; 2022.
34. Diro E, Blesson S, Edwards T, Ritmeijer K, Fikre H, Admassu H, et al: A randomized trial of AmBisome monotherapy and AmBisome and miltefosine combination to treat visceral leishmaniasis in HIV co-infected patients in Ethiopia. PLoS Negl Trop Dis 2019, 13(1): e0006988.
35. Pandey K, Kumar R: New combination treatment regimen for VL-HIV coinfected patients in India. In *Education, Ethics and Empathy: PROGRESS IN MEDICINE Volume* 35. Edited by Singal R, Bhasin A, Chhabra R: JAYPEE; 2025. 359–361
36. Nerlich AG, Bianucci R, Trisciuoglio A, Schönian G, Ball M, Giuffra V, Bachmeier B, Pusch CM, Ferroglio E, Fornaciari G: Visceral leishmaniasis during Italian renaissance, 1522–1562. *Emerg Infect Dis* 2012, **18**:184.
37. Bianucci R, Giuffra V, Bachmeier BE, Ball M, Pusch CM, Fornaciari G, Nerlich AG: Eleonora of Toledo (1522–1562): Evidence for tuberculosis and leishmaniasis co-infection in Renaissance Italy. *Int J Paleopathol* 2012, **2**:231–235.

38. Tuberculosis – Fact sheet [https://www.who.int/news-room/fact-sheets/detail/tuberculosis] [15 July 2025]
39. van Griensven J, Mohammed R, Ritmeijer K, Burza S, Diro E: Tuberculosis in visceral leishmaniasis-human immunodeficiency virus coinfection: An evidence gap in improving patient outcomes? *Open Forum Infect Dis* 2018, **5**:ofy059.
40. Dahal P, Singh-Phulgenda S, Naylor C, Brack M, Chatterjee M, Alves F, Guerin PJ, Stepniewska K: Design, conduct, analysis, and reporting of therapeutic efficacy studies in visceral leishmaniasis: A systematic review of published reports, 2000-2021. *Am J Trop Med Hyg* 2024, **111**:365–376.
41. Dahal P, Singh-Phulgenda S, Maguire BJ, Harriss E, Ritmeijer K, Alves F, Guerin PJ, Olliaro PL: Visceral leishmaniasis in pregnancy and vertical transmission: A systematic literature review on the therapeutic orphans. *PLoS Negl Trop Dis* 2021, **15**:e0009650.
42. Silva JM, Zacarias DA, de Figueirêdo LC, Soares MR, Ishikawa EA, Costa DL, Costa CH: Bone marrow parasite burden among patients with New World kala-azar is associated with disease severity. Am J Trop Med Hyg. 2014 Apr;90(4):621–6.
43. Zacarias, DA, Rolão, N, de Pinho, FA, Sene, I, Silva, JC, Pereira, TC, Costa, DL. and Costa, CHN: Causes and consequences of higher Leishmania infantum burden in patients with kala-azar: a study of 625 patients. Trop Med Int Health, 2017, 22: 679–687.
44. Palic S, Kip AE, Beijnen JH, Mbui J, Musa A, Solomos A, Wasunna M, Olobo J, Alves F, Dorlo TPC: Characterizing the non-linear pharmacokinetics of miltefosine in paediatric visceral leishmaniasis patients from Eastern Africa. *J Antimicrob Chemother* 2020, **75**:3260–3268.
45. Zijlstra EE, Alves F, Rijal S, Arana B, Alvar J: Post-kala-azar dermal leishmaniasis in the Indian subcontinent: A threat to the South-East Asia region kala-azar elimination programme. *PLoS Negl Trop Dis* 2017, **11**:e0005877.
46. Dahal P, Singh-Phulgenda S, Assefa DG, Guerin PJ: Addressing research questions in visceral leishmaniasis and post-kala-azar dermal leishmaniasis: Potential of reusing data from the infectious diseases data observatory platform. *Ann Med Sci Res* 2025, **4**:S24–S34.
47. Singh OP, Tiwary P, Kushwaha AK, Singh SK, Singh DK, Lawyer P, Rowton E, Chaubey R, Singh AK, Rai TK, et al: Xenodiagnosis to evaluate the infectiousness of humans to sandflies in an area endemic for visceral leishmaniasis in Bihar, India: A transmission-dynamics study. *Lancet Microbe* 2021, **2**:e23–e31.
48. Mondal D, Bern C, Ghosh D, Rashid M, Molina R, Chowdhury R, Nath R, Ghosh P, Chapman LAC, Alim A, et al: Quantifying the infectiousness of post-kala-azar dermal leishmaniasis toward sand flies. *Clin Infect Dis* 2019, **69**:251–258.
49. Sundar S, Sinha P, Jha TK, Chakravarty J, Rai M, Kumar N, Pandey K, Narain MK, Verma N, Das VN, et al: Oral miltefosine for Indian post-kala-azar dermal leishmaniasis: A randomised trial. *Trop Med Int Health* 2013, **18**:96–100.
50. den Boer M, Das AK, Akhter F, Burza S, Ramesh V, Ahmed BN, Zijlstra EE, Ritmeijer K: Safety and effectiveness of short-course AmBisome in the treatment of post-kala-azar dermal leishmaniasis: A prospective cohort study in Bangladesh. *Clin Infect Dis* 2018, **67**:667–675.
51. Singh-Phulgenda S, Kumar R, Dahal P, Munir A, Rashan S, Chhajed R, Naylor C, Maguire BJ, Siddiqui NA, Harriss E, et al: Post-kala-azar dermal leishmaniasis (PKDL) drug efficacy study landscape: A systematic scoping review of clinical trials and observational studies to assess the feasibility of establishing an individual participant-level data (IPD) platform. *PLoS Negl Trop Dis* 2024, **18**:e0011635.
52. Ramesh V, Singh R, Avishek K, Verma A, Deep DK, Verma S, Salotra P: Decline in clinical efficacy of oral miltefosine in treatment of post kala-azar dermal leishmaniasis (PKDL) in India. *PLoS Negl Trop Dis* 2015, **9**:e0004093.

53. Roy S, Moulik S, Roy M, Ghosh MK, Chaudhuri SJ, Pandey DK, Jain S, Dagne DA, Chatterjee M: Monitoring the long-term effectiveness of miltefosine in Indian post-kala-azar dermal leishmaniasis. *Am J Trop Med Hyg* 2024, **110**:656–662.
54. Zijlstra EE, Kumar A, Sharma A, Rijal S, Mondal D, Routray S: Report of the fifth post-kala-azar dermal leishmaniasis consortium meeting, Colombo, Sri Lanka, 14-16 May 2018. *Parasit Vectors* 2020, **13**:159.
55. Singh R, Kumar D, Ramesh V, Negi NS, Singh S, Salotra P: Visceral leishmaniasis, or kala azar (KA): High incidence of refractoriness to antimony is contributed by anthroponotic transmission via post-KA dermal leishmaniasis. *J Infect Dis* 2006, **194**:302–306.
56. Kumari A, Siddiqui NA, Kumari S, Murti K, Kumar R, Pandey K, Padmakar S, Pal B: Combination therapy for post-kala-azar dermal leishmaniasis: A literature review of current evidence. *Indian J Dermatol* 2024, **69**:396–405.
57. Maruf S, Nath P, Islam MR, Aktar F, Anuwarul A, Mondal D, Basher A: Corneal complications following post kala-azar dermal leishmaniasis treatment. *PLoS Negl Trop Dis* 2018, **12**:e0006781.
58. Pandey K, Pal B, Topno RK, Lal CS, Das VNR, Das P: Acute uveitis: A rare adverse effect of miltefosine in the treatment of post-kala-azar dermal leishmaniasis. *Rev Soc Bras Med Trop* 2020, **54**:e20200208.
59. Pal B, Atem TD, Kumari S, Murti K, Kumar R, Pandey K, Siddiqui NA, Dhingra S, Chaudhary V: Ophthalmic adverse effects of miltefosine in the treatment of leishmaniasis: A systematic review. *Cutan Ocul Toxicol* 2024, **43**:190–197.
60. Sundar S, Singh J, Dinkar A, Agrawal N: Safety and effectiveness of miltefosine in post-kala-azar dermal leishmaniasis: An observational study. *Open Forum Infect Dis* 2023, **10**:ofad231.
61. Advisory Committee on Safety of Medicinal Products (ACSoMP): Measures to minimize the risk of ocular adverse events with miltefosine [https://www.who.int/news/item/12-04-2023-acsomp-miltefosine] [https://www.who.int/news/item/12-04-2023-acsomp-miltefosine]; [24 June 2025]
62. Mukhopadhyay D, Dalton JE, Kaye PM, Chatterjee M: Post kala-azar dermal leishmaniasis: An unresolved mystery. *Trends Parasitol* 2014, **30**:65–74.
63. Collin S, Davidson R, Ritmeijer K, Keus K, Melaku Y, Kipngetich S, Davies C: Conflict and kala-azar: Determinants of adverse outcomes of kala-azar among patients in southern Sudan. *Clin Infect Dis* 2004, **38**:612–619.
64. Mondal D, Kumar A, Sharma A, Ahmed MM, Hasnain MG, Alim A, Huda MM, Rahman R, Alvar J, Ahmed BN, Haque R: Relationship between treatment regimens for visceral leishmaniasis and development of post-kala-azar dermal leishmaniasis and visceral leishmaniasis relapse: A cohort study from Bangladesh. *PLoS Negl Trop Dis* 2019, **13**:e0007653.
65. Goyal V, Das VNR, Singh SN, Singh RS, Pandey K, Verma N, Hightower A, Rijal S, Das P, Alvar J, et al: Long-term incidence of relapse and post-kala-azar dermal leishmaniasis after three different visceral leishmaniasis treatment regimens in Bihar, India. *PLoS Negl Trop Dis* 2020, **14**:e0008429.
66. Zijlstra EE, Musa AM, Khalil EA, el-Hassan IM, el-Hassan AM: Post-kala-azar dermal leishmaniasis. *Lancet Infect Dis* 2003, **3**:87–98.
67. Kumar R, Das VN, Topno RK, Pal B, Imam A, Agrawal K, Singh A, Murti K, Lal CS, Verma N, et al: Para-kala-azar dermal leishmaniasis cases in Indian subcontinent – A case series. *Pathog Glob Health* 2016, **110**:326–329.
68. Lindoso JAL, Moreira CHV, Celeste BJ, Oyafuso LKM, Folegatti PM, Zijlstra EE: Para-kala-azar dermal leishmaniasis in A patient in Brazil: A case report. *Rev Soc Bras Med Trop* 2018, **51**:105–107.
69. Rijal A, Agrawal S, Agarwalla A, Agrawal A, Rijal S: Post-kala-azar dermal leishmaniasis with visceral leishmaniasis: A rare presentation. *Int J Dermatol* 2005, **44**:494–496.

70. Maruf S, Sagar SK, Rashid MMU, Nath P, Islam MS, Ghosh P, Rashid MU, Mondal D, Abd El Wahed A, Basher A: Revisiting the diagnosis and treatment of para kala-azar dermal leishmaniasis in the endemic foci of Bangladesh. *PLoS One* 2023, **18**:e0280747.
71. Hasan MM, Proma SB, Hossain MS, Arifuzzaman M, Islam N, Siddique MAB, Amiruzzaman: A case report on para-kala-azar dermal leishmaniasis: An unresolved mystery. *BMC Infect Dis* 2023, **23**:885.
72. Mendiratta V, Chander R: Post-kala-azar dermal leishmaniasis and visceral leishmaniasis in a child–or is it para-kala-azar dermal leishmaniasis? *Int J Dermatol* 2009, **48**:1027–1028.
73. Gonzalez U, Pinart M, Reveiz L, Rengifo-Pardo M, Tweed J, Macaya A, Alvar J: Designing and reporting clinical trials on treatments for cutaneous leishmaniasis. *Clin Infect Dis* 2010, **51**:409–419.
74. World Health Organization: *Ending the Neglect to Attain the Sustainable Development Goals: A Strategic Framework for Integrated Control and Management of Skin-Related Neglected Tropical Diseases.* World Health Organization; 2022.
75. Bailey F, Mondragon-Shem K, Haines LR, Olabi A, Alorfi A, Ruiz-Postigo JA, Alvar J, Hotez P, Adams ER, Velez ID, et al: Cutaneous leishmaniasis and co-morbid major depressive disorder: A systematic review with burden estimates. *PLoS Negl Trop Dis* 2019, **13**:e0007092.
76. Das VNR, Siddiqui NA, Bhunia GS, Pandey K, Sinha SK, Ansari MZ, Topno RK, Lal CS, Ranjan A, Singh VP, Das P: Improved kala-azar case management through implementation of health facility-based sentinel sites surveillance system in Bihar, India. *PLoS Negl Trop Dis* 2021, **15**:e0009598.
77. Naylor-Leyland G, Collin SM, Gatluak F, den Boer M, Alves F, Mullahzada AW, Ritmeijer K: The increasing incidence of visceral leishmaniasis relapse in South Sudan: A retrospective analysis of field patient data from 2001-2018. *PLoS Negl Trop Dis* 2022, **16**:e0010696.
78. Carvalho LS, das Gracas Braga M, da Silva Costa DA, Simoes TC, Lula MD, Silveira MR: Lethality among individuals infected with visceral leishmaniasis in Brazil: A retrospective study (2007–2018). *Parasitol Res* 2022, **121**:725–736.
79. SINAN: Visceral Leishmaniasis [http://www.portalsinan.saude.gov.br/leishmaniose-visceral]
80. Cota G, Erber AC, Schernhammer E, Simoes TC: Inequalities of visceral leishmaniasis case-fatality in Brazil: A multilevel modeling considering space, time, individual and contextual factors. *PLoS Negl Trop Dis* 2021, **15**:e0009567.
81. Assefa DF, Cota G, Wilson JP, Mendes JA, Naylor C, Peploe R, Singh-Phulgenda S, Saito M, Guerin PJ, Dahal P: Epidemiological shifts in visceral leishmaniasis incidence, relapse, and mortality in Brazil, 2007-2023: Analysis using National Notifiable Diseases Information System. Open Forum Infect Dis 2026. Feb 17;13(2): ofag035. doi: 10.1093/ofid/ofag035.
82. MSF. Kala azar [Internet]. [cited 2026 Feb 22] [https://www.msf.org/kala-azar]
83. Boodman C, van Griensven J, Gupta N, Diro E, Ritmeijer K: Anticipating visceral leishmaniasis epidemics due to the conflict in Northern Ethiopia. *PLoS Negl Trop Dis* 2023, **17**:e0011188.
84. Alves F, Bilbe G, Blesson S, Goyal V, Monnerat S, Mowbray C, Muthoni Ouattara G, Pécoul B, Rijal S, Rode J: Recent development of visceral leishmaniasis treatments: Successes, pitfalls, and perspectives. *Clin Microbiol Rev* 2018, **31**(10):1128/cmr. 00048–18.
85. Singh-Phulgenda S, Dahal P, Maguire BJ, Alvar J, Alves F, Chatterjee M, Costa C, Croft SL, Guerin PJ, Mondal D, Musa A, Pandey K, Ritmeijer K, Romero G, Sundar S: The development of a global research agenda and individual participant data platform for visceral leishmaniasis: Challenges and future opportunities. *Parasit Vectors* 2025, **18**:496.

Chapter 11

Plant-based compounds for treatment of leishmaniasis

*Prashant Saxena**
Department of Biotechnology Engineering & Food Technology, University Institute of Engineering, Chandigarh University, Mohali, India

11.1 INTRODUCTION

Leishmaniasis is a complex parasitic disease with a significant global health impact, especially in areas with low socioeconomic status. The disease is hard to treat because it has a wide range of symptoms, from self-healing skin ulcers to life-threatening infections in the organs (McGwire & Satoskar, 2013; WHO, 2024). Even though public health officials have known about it for a long time, it remains underfunded and underprioritized in drug development (Oryan, 2015; Roatt et al., 2020). The fact that drug-resistant *Leishmania* strains are becoming more common, that they can be found in places where HIV is also common, and that they are spreading to new areas because of environmental and human factors have all made the disease more important for public health (Davidson, 1998; Sundar et al., 2012). Because of these problems, scientists and ethnobotanists have sparked a growing interest in natural product-based interventions. Plant-based compounds are especially interesting due to their chemical diversity, historical use in traditional medicine, and broad-spectrum bioactivity in many cultures (Abdel-Sattar et al., 2010; Oryan, 2015). This chapter starts with a detailed look at how leishmaniasis affects people all over the world and how complicated it is from a biological point of view. This sets the stage for a critical look at how plant-based compounds might be able to help treat it.

11.1.1 Global burden and distribution of leishmaniasis

About 12 million people worldwide are afflicted by leishmaniasis, which is endemic in more than 98 countries across Asia, Africa, South America, and parts of southern Europe. An estimated 350 million individuals are still at risk of infection, and the number of new cases each year varies between 1.5 and 2 million (Oryan, 2015; WHO, 2024). With more than 90% of cases of visceral leishmaniasis (VL) occurring in just seven countries—India, Bangladesh,

* Corresponding author: prashant347@gmail.com

DOI: 10.1201/9781003596455-11

Sudan, South Sudan, Ethiopia, Brazil, and Nepal—the disease burden is unequally distributed. On the other hand, Afghanistan, Algeria, Colombia, Brazil, Iran, and Syria account for more than 70% of all cases of cutaneous leishmaniasis (CL). Leishmaniasis outbreaks have increased recently due to relocation brought on by conflict and environmental changes, particularly in war-torn areas like South Sudan and Syria (McGwire & Satoskar, 2013). The most severe type of leishmaniasis, known as kala-azar or VL, can be lethal if left untreated. East Africa and South Asia are where it is most common. With high rates of HIV coinfection, the northeastern Indian state of Bihar continues to be a global hotspot for VL. The most common type of leishmaniasis, CL, is more widely distributed and manifests as ulcerative sores. South America, especially Brazil and Peru, is home to mucocutaneous leishmaniasis (MCL), which deforms and destroys mucosal tissues. Leishmaniasis has a substantial influence on world health in terms of disability-adjusted life years (DALYs). Recent studies indicate that VL alone causes more than 2.3 million DALYs per year. Because visible lesions and incapacity frequently result in stigmatization and exclusion from social and economic engagement, the disease has a significant psychological and socioeconomic impact. Additionally, underreporting continues to be a significant issue, concealing the actual burden and jeopardizing control measures (Roatt et al., 2020; WHO, 2024).

11.1.2 Overview of *Leishmania* species and transmission

The obligate intracellular protozoan parasites of the genus *Leishmania* are the cause of leishmaniasis, a vector-borne illness that infects female phlebotomine sandflies and spreads to people by their bite. Geographically, the sandfly species that transmit the disease are different: *Lutzomyia* species are found in the New World (Central and South America), whereas *Phlebotomus* species are the vectors in the Old World (Africa, Asia, and Europe). *Leishmania* is a zoonotic disease that is spread by a range of domestic and wild animals, especially foxes, dogs, and rats. However, transmission is anthroponotic—that is, people act as the parasite's main reservoir—in some endemic areas, such as the Indian subcontinent. Regional diversity in disease epidemiology is exacerbated by this duality in transmission channels, which also makes management efforts more difficult (McGwire & Satoskar, 2013; WHO, 2024). Disease progression and transmission are influenced by environmental, socioeconomic, and host-related factors, which emphasizes the necessity of localized surveillance and vector control initiatives (Saxena et al., 2025).

The promastigote and the amastigote are the two morphologically and functionally distinct phases of *Leishmania*'s digenetic life cycle. The promastigote form is extracellular, flagellated, and designed to live in the sandfly vector's midgut. After feeding on blood from an infected host, the sandfly consumes macrophages that contain amastigotes, which in the midgut of the insect change into procyclic promastigotes. These promastigotes proliferate, go to the foregut, and undergo differentiation into the infectious stage known as metacyclic promastigotes. Metacyclic promastigotes are injected into the dermis of a new host by the sandfly, and host macrophages quickly phagocytose them. Promastigotes change into the non-flagellated amastigote form in the phagolysosome's acidic environment, where they proliferate inside the host cell. The host macrophages burst because of this intracellular multiplication, and nearby cells become infected. In addition to the *Leishmania* species involved, host factors like immunological status, genetic predisposition, and nutritional status also affect the disease's clinical symptoms (Oryan, 2015). The pathogenicity of the parasite is dependent on its capacity to endure inside macrophages and elude immune reactions.

Clinically, leishmaniasis presents as a spectrum of syndromes, with the three principal forms being CL, MCL, and VL. CL is the most common manifestation and is characterized by the development of ulcerative or nodular lesions at the site of the sandfly bite. It is primarily caused by *Leishmania major*, *Leishmania tropica*, and *Leishmania mexicana*. Although self-limiting, CL can lead to persistent disfiguring scars and may become complicated by secondary bacterial infections if left untreated. MCL, a more severe form, affects the mucous membranes of the nasal and oropharyngeal regions and can result in significant tissue destruction, facial disfigurement, and functional impairment. It is typically caused by *Leishmania braziliensis* and *Leishmania panamensis* and is endemic to parts of South and Central America. The most dangerous form, VL (kala-azar), is a systemic disease affecting vital organs such as the liver, spleen, bone marrow, and lymph nodes. Caused by *Leishmania donovani*, *Leishmania infantum*, and *Leishmania chagasi*, VL manifests with prolonged fever, weight loss, hepatosplenomegaly, pancytopenia, and hypergammaglobulinemia. If left untreated, VL is universally fatal. The clinical course of leishmaniasis is particularly severe in immunocompromised individuals, especially those coinfected with HIV. HIV–*Leishmania* coinfection alters the immunopathogenesis of both diseases, leading to accelerated parasite dissemination, increased parasite burden, and reduced responsiveness to standard treatments. Visceral and cutaneous lesions often arise simultaneously in these patients, and even after a clinical cure appears, the presentation is frequently uncommon. In these situations, diagnosis is more challenging, and treatment results are subpar. The rise of drug-resistant strains and the elevated risk of transmission make HIV–leishmaniasis coinfection a public health concern. In regions where coinfection prevalence is high, the World Health Organization stresses integrated management techniques that include combination therapy, early diagnosis, and long-term relapse monitoring (Sundar et al., 2012; WHO, 2024). To lower mortality and stop the transmission of disease, more intensive, long-term, and interdisciplinary therapies are required due to the combination of parasite and viral immunosuppression (Figure 11.1).

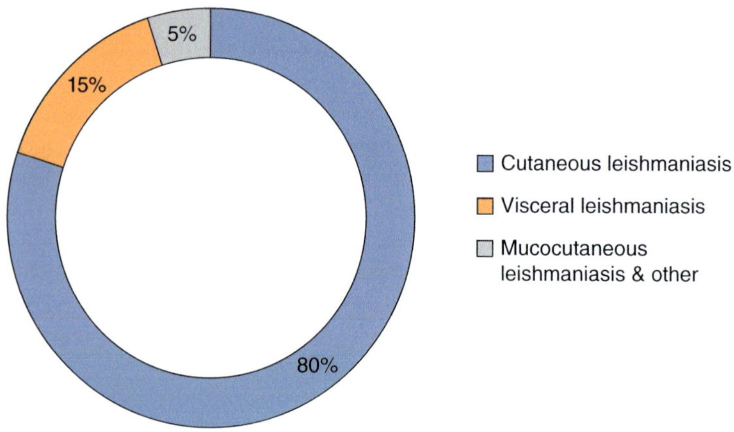

Figure 11.1 Global distribution of leishmaniasis clinical forms.

11.1.3 Current treatments: Efficacy, limitations, and resistance issues

The mainstay of *Leishmania* treatment is still chemotherapy; however, the available medications have serious drawbacks in terms of toxicity, expense, resistance, and accessibility. Since the 1940s, pentavalent antimonials, like sodium stibogluconate, and meglumine antimoniate, have been the mainstay of treatment. Nevertheless, these substances necessitate extended parenteral delivery and are often associated with severe side effects, such as pancreatitis, hepatotoxicity, and cardiotoxicity. Furthermore, there is now widespread resistance, especially in endemic areas of Nepal and India (Davidson, 1998). With the extra advantage of single-dose administration, amphotericin B is very efficient against VL, particularly when used in its liposomal formulation (AmBisome®). Yet, its inflated cost and need for intravenous delivery restrict its use in low-resource settings. In contrast, miltefosine, the first oral antileishmanial drug, initially demonstrated strong efficacy but now suffers from declining effectiveness, emerging resistance, and relapse, especially in VL-endemic areas. It is also teratogenic, making it contraindicated during pregnancy (Roatt et al., 2020; Sundar et al., 2012).

An aminoglycoside antibiotic with antileishmanial properties, paromomycin is usually employed in combination therapy regimens. Clinical research is being done on newer medications, including sitamaquine and fexinidazole, which could increase the number of therapy options available in the future. Some trials have seen better results with combination therapy, such as liposomal amphotericin B (L-AmB) followed by a brief course of miltefosine. But in endemic areas, the exorbitant price, logistical difficulty, and restricted availability of these drugs continue to be significant obstacles. Due to the growing prevalence of medication resistance, treatment failure, HIV–*Leishmania* coinfection, and adverse drug effects, the present pharmacological arsenal is becoming increasingly insufficient (Oryan, 2015; WHO, 2024). These difficulties emphasize how urgently innovative, reasonably priced, and potent treatments must be developed, especially using unconventional sources such as chemicals produced from plants.

11.2 PATHOGENESIS AND CLINICAL MANIFESTATIONS

The term "leishmaniasis" refers to a collection of illnesses caused by protozoan parasites of the genus *Leishmania*, each of which has unique clinical symptoms. Environmental exposure, co-morbid conditions, host immune response, and parasite species all influence clinical diversity. To understand the parasite's intricate pathogenesis and presentation, one must have a solid understanding of its biology and interactions with both the vector and the host.

11.2.1 *Leishmania* life cycle (promastigote and amastigote forms)

The *Leishmania* parasite has a digenetic life cycle, alternating between the phlebotomine sandfly vector and the mammalian host. This complex cycle involves several morphologically distinct stages, each essential for transmission, survival, and pathogenesis. The cycle begins in the sandfly midgut, where transformation and proliferation of the parasite occur after ingestion of an infected blood meal. A sandfly consumes macrophages containing intracellular amastigotes when it feeds on an infected vertebrate host. These amastigotes are discharged once inside the sandfly, where they change into procyclic promastigotes, the initial motile stage. Nectomonad promastigotes, which actively move to the anterior midgut, are produced by binary fission replication of these forms. After that, they mature into leptomonad promastigotes, which settle in the foregut. The metacyclic promastigote, a brief, non-dividing, extremely mobile stage that is contagious to mammals, is the vector's last developmental phase. These metacyclic forms are better able to avoid host immune responses because of structural changes such as shortened lipophosphoglycan chains (McGwire & Satoskar, 2013; Oryan, 2015).

Transmission to the mammalian host occurs when the infected sandfly inoculates metacyclic promastigotes into the skin during a blood meal. These promastigotes are quickly taken up by macrophages and neutrophils, where they change into amastigotes. Significant morphological and metabolic changes, including the loss of the flagellum and adaptation to the low-pH, low-oxygen intracellular environment, are part of this differentiation, which occurs in the acidic phagolysosome. Within macrophages, the amastigotes divide intracellularly through binary fission. Amastigotes are produced when the infected macrophage bursts, and they then spread to nearby cells. The infection in CL

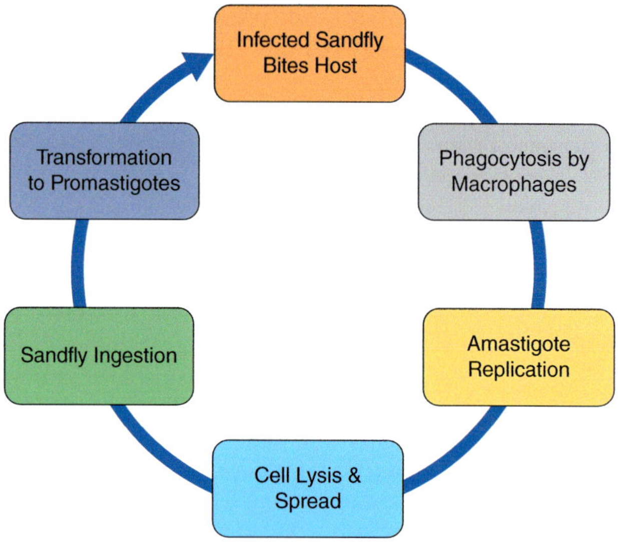

Figure 11.2 Life cycle of *Leishmania* parasite.

frequently stays within the dermis. However, in VL, the amastigotes spread to important organs such as the liver, spleen, and bone marrow through the bloodstream and lymphatics, where they multiply and produce systemic disease (Figure 11.2).

When a second female sandfly consumes blood containing amastigotes from the infected host, the life cycle is complete. In the midgut of the sandfly, these amastigotes transform back into promastigotes, starting a new phase of the embryonic cycle. *Leishmania* may survive and spread over a variety of environmental niches thanks to this cyclical transition, which occurs when extracellular promastigotes in the insect vector give way to intracellular amastigotes in the mammalian host. The key to the parasite's persistence and toxicity in both hosts is its capacity to constantly modify its shape, metabolism, and surface chemicals (Table 11.1).

Table 11.1 Morphological stages of *Leishmania* spp. in mammalian and sandfly hosts

Stage	*Host*	*Morphology*	*Role*
Amastigote	Mammal	Round, non-flagellated	Intracellular replication in macrophages
Procyclic promastigote	Sandfly	Flagellated	Initial replication in midgut
Nectomonad promastigote	Sandfly	Long, slender	Migration to anterior midgut
Leptomonad promastigote	Sandfly	Short, replicative	Precursor to infectious form
Metacyclic promastigote	Sandfly	Short, flagellated	Infective form for mammalian host

11.2.2 Clinical syndromes: CL, VL, MCL, PKDL, DCL

The *Leishmania* species involved, the host's immunological state, and regional variables all affect the five main clinical syndromes that leishmaniasis manifests as. The most common type, CL, is typified by localized skin lesions that are ulcerative at the location where the sandfly bite occurred. Lesions start off as painless papules and develop into ulcers with core depressions and elevated, indurated edges. Lesions can last for several months and frequently leave lifelong scars, even though they usually go away on their own in immunocompetent people. *L. major*, *L. tropica*, *L. mexicana*, and *Leishmania amazonensis* are among the species that cause CL. The Middle East, North Africa, and a sizable portion of Central and South America are endemic for the illness. Transmission may be zoonotic or anthroponotic, depending on the species and region. Prognosis is favorable in immunocompetent hosts, with many cases resolving spontaneously.

VL, also known as kala-azar, is the most severe form of the disease and is potentially fatal if untreated. It occurs when *Leishmania* amastigotes disseminate to internal organs such as the liver, spleen, and bone marrow. Clinical features typically include persistent fever, hepatosplenomegaly, pancytopenia, severe weight loss, and immunosuppression. Laboratory findings may reveal anemia and polyclonal hypergammaglobulinemia. The primary causative agents are *L. donovani*, *L. infantum*, and *L. chagasi*, with endemic regions, including South Asia (India, Nepal, Bangladesh), East Africa (Sudan, Ethiopia), and parts of Brazil. Transmission of *L. donovani* is anthroponotic, while *L. infantum* and *L. chagasi* are zoonotic, with dogs and wild canids serving as principal reservoirs. Without treatment, VL carries a mortality rate exceeding 90%.

When parasites migrate from cutaneous lesions to mucosal tissues, usually in the nasal, oral, and pharyngeal areas, MCL develops. Although it can sometimes happen de novo, it usually follows CL. MCL frequently results in significant facial deformity and is characterized by increasing inflammation, tissue erosion, and final destruction of soft tissue and cartilage components. Mostly found in South America, particularly Bolivia, Brazil, and Peru, the primary causal species are L. braziliensis and L. panamensis. A robust yet dysregulated cellular immune response is reflected in the disease, which leads to significant tissue damage. In addition to being challenging to treat, MCL is typically resistant to common antileishmanial medications and often necessitates systemic therapy. Less frequent but clinically significant symptoms include diffuse cutaneous leishmaniasis (DCL) and post-kala-azar dermal leishmaniasis (PKDL). PKDL usually appears as hypopigmented macules, papules, or nodules on the face and trunk months to years after VL appears to have cleared up. It is solely caused by *L. donovani* and is most seen in Ethiopia, Sudan, and India. Because patients may act as reservoirs for *Leishmania* transmission, particularly in areas working to eradicate VL, PKDL is of epidemiological significance. On the other hand, DCL is an uncommon type of anergic CL that is brought on by *L. amazonensis* or *Leishmania aethiopica*. It is distinguished by non-ulcerative nodules that are spread out and teeming with parasites. It is associated with a deficient cell-mediated immune response, including absent delayed-type hypersensitivity (DTH) and poor T-cell function. DCL is chronic, relapsing, and frequently unresponsive to conventional antileishmanial therapy, posing significant therapeutic challenges.

11.3 CONVENTIONAL CHEMOTHERAPEUTIC AGENTS

All clinical types of leishmaniasis, including cutaneous, mucocutaneous, and visceral symptoms, are still treated with chemotherapy. Pentavalent antimonials, amphotericin B (traditional and liposomal forms), miltefosine, paromomycin, and pentamidine are currently part of the pharmacological arsenal. These medications are chosen according to national treatment standards, drug resistance patterns, geographic prevalence, and *Leishmania* species. Despite the clinical effectiveness of these medications, their usage is frequently restricted by side effects, the method of administration, exorbitant costs, and the growing evolution of drug resistance, especially in endemic, resource-constrained environments.

Pentavalent antimonials, notably *sodium stibogluconate (SSG)* and *meglumine antimoniate*, are among the oldest treatments for leishmaniasis and remain widely used in many regions. Administered intramuscularly or intravenously, these compounds are believed to disrupt parasite metabolism by interfering with the thiol redox system, inhibiting glycolysis, and inducing apoptosis-like mechanisms in intracellular *Leishmania* amastigotes. Despite their long-standing use, pentavalent antimonials are associated with considerable toxicity, including hepatotoxicity, cardiotoxicity, pancreatitis, and prolonged QT intervals. In addition to their toxicity profile, resistance to SSG has become a major issue, especially in parts of India such as Bihar, where treatment failure rates exceed 60%. These drawbacks significantly compromise their continued utility and necessitate the search for safer, more effective alternatives.

Amphotericin B, a polyene antifungal agent, is a cornerstone in the treatment of VL. By attaching itself to ergosterol-like sterols in the parasite membrane and forming pores that compromise membrane integrity and induce cell lysis, it has an antileishmanial impact. Despite its effectiveness, the traditional deoxycholate formulation is nephrotoxic and frequently results in infusion-related side effects such as fever, chills, and hypotension, which calls for inpatient treatment. Targeting organs like the liver and spleen that are rich in mononuclear phagocytes, the liposomal formulation of L-AmB has increased the drug's therapeutic index. For VL in India and other endemic areas, the World Health Organization advises a single-dose L-AmB regimen as the initial line of treatment.

However, the inflated cost of L-AmB remains a significant barrier to its widespread use in low-income countries.

Miltefosine is the first oral drug approved for the treatment of leishmaniasis and is effective against both visceral and cutaneous forms. Miltefosine, which was first created as an anticancer drug, interferes with the survival of *Leishmania* parasites by changing the lipid content of their membranes, affecting mitochondrial activity, and causing death akin to apoptosis. Miltefosine is an excellent choice in outpatient settings because it is easy to administer orally. However, a number of issues prevent its clinical use: it usually produces gastrointestinal side effects, is prohibited during pregnancy due to its teratogenicity, and is increasingly associated with treatment failure due to the rise of resistant strains. Resistance is associated with mutations in transporter genes such as *LdMT* and *LdRos3*, which impair drug uptake and intracellular accumulation, especially in endemic areas like India and Brazil.

Paromomycin, an aminoglycoside antibiotic, has shown considerable efficacy against *Leishmania* species and is used both systemically and topically, especially for cutaneous lesions. It works by attaching itself to the 30S ribosomal subunit, which prevents the synthesis of new proteins and causes misfolded proteins to build up. Although paromomycin is well tolerated, extended treatment may result in ototoxicity and nephrotoxicity. It is especially helpful in East Africa, where the WHO recommends a 17-day combined regimen of SSG plus paromomycin as a quicker and safer option than monotherapy. The efficacy and safety of paromomycin in additional areas are still being assessed, despite its encouraging results. Similarly, because of its high toxicity profile, pentamidine—another treatment that has demonstrated effectiveness against cutaneous and mucocutaneous forms of leishmaniasis, particularly in South America—is frequently saved for use as a second-line medication. It acts by disrupting mitochondrial function and inhibiting nucleic acid synthesis, but its use is limited by adverse effects such as hypoglycemia, hypotension, nephrotoxicity, and hepatotoxicity, which require intensive monitoring during administration (Table 11.2).

11.4 DRUG RESISTANCE AND ADVANCED THERAPEUTIC STRATEGIES IN LEISHMANIASIS

The growing problem of medication resistance and the intrinsic drawbacks of current treatment plans have made the clinical management of leishmaniasis more difficult. This is especially true when treating VL, where issues with medication resistance, toxicity, and treatment failure have gained significant attention. Resistance to the pentavalent antimonial SSG has escalated to dangerous levels in areas like Bihar, India, making the medication ineffective in more than 60% of instances. Resistance in *Leishmania* species is caused by a variety of ways. The downregulation or mutation of aquaglyceroporin-1 (AQP1), a membrane transporter that facilitates the absorption of antimonial

Table 11.2 Mechanistic and clinical profile of standard antileishmanial agents

Drug	*Drug class/type*	*Mechanism of action*	*Clinical use*	*Route of administration*	*Limitations*
Sodium stibogluconate (SSG), meglumine antimoniate	Pentavalent antimonials	Inhibits *Leishmania* thiol redox system, interferes with glycolysis, induces apoptosis	VL, CL, MCL	Intramuscular/ intravenous	High toxicity (cardiotoxicity, hepatotoxicity, pancreatitis), resistance common in India
Amphotericin B (deoxycholate)	Polyene antifungal	Binds to ergosterol-like lipids, forms membrane pores, causes cell lysis	VL (especially refractory cases)	Intravenous	Nephrotoxicity, infusion-related reactions, hospitalization required
Liposomal amphotericin B (L-AmB)	Liposomal formulation	Same as amphotericin B; targeted delivery to macrophage-rich organs	VL (first line in India, East Africa)	Intravenous (single dose or short course)	Exceedingly excessive cost; limited access in low-income countries
Miltefosine	Alkylphosphocholine derivative	Disrupts lipid metabolism, alters membrane integrity, induces apoptosis	VL, CL	Oral	Teratogenicity, GI side effects, emerging resistance (mutations in *LdMT*, *LdRos3*)
Paromomycin	Aminoglycoside antibiotic	Binds 30S ribosomal subunit, inhibits protein synthesis	VL (East Africa), CL (topical use)	Intramuscular/ topical	Nephrotoxicity, ototoxicity with prolonged use
Pentamidine	Aromatic diamidine	Disrupts mitochondrial function, inhibits nucleic acid synthesis	CL, MCL (esp. South America)	Intramuscular/ intravenous	Toxicity: hypoglycemia, hypotension, nephrotoxicity, hepatotoxicity

substances, is one such method. The effectiveness of treatment is seriously compromised by mutations in this channel, which prevent drugs from entering the parasite. Furthermore, *Leishmania* strains have developed improved detoxification mechanisms that combat drug-induced oxidative stress by overexpressing enzymes, including glutathione peroxidase and trypanothione reductase. The overexpression of multidrug resistance (MDR) proteins and ATP-binding cassette (ABC) transporters is another resistance tactic. These proteins actively remove the drug from the parasite's internal environment, lessening its deadly effects.

Antimonials are not the only source of resistance. Mutations in the LdMT gene and its accessory protein LdRos3 have also resulted in a decline in the effectiveness of miltefosine, the only oral antileishmanial medication. Due to impaired drug absorption, these alterations enable the parasite to persist and proliferate despite treatment. There are growing reports of decreased susceptibility in immunocompromised individuals and in relapsed VL cases, even though amphotericin B has remained effective in most cases, particularly after repeated or extended exposure to the medication. These trends highlight the pressing need for calculated measures to preserve the efficacy of treatment and stop the proliferation of drug-resistant *Leishmania* strains. Combination therapy has drawn a lot of interest as a more durable and successful method of treating leishmaniasis in response to these difficulties. Drug combinations with distinct modes of action provide several advantages, such as synergistic efficacy, shortened treatment times, decreased toxicity, and a decreased risk of resistance development. The combination of SSG with paromomycin is one commonly recommended regimen that has been shown to be phenomenally successful in East Africa, where it has shortened the course of treatment from 30 to 17 days while improving safety and compliance. Miltefosine combined with L-AmB is another effective tactic. Patients who are coinfected with HIV and *Leishmania*, who frequently have poor results and high relapse rates with monotherapy, have found this combination to be quite helpful. While miltefosine affords longer-term suppression, L-AmB delivers quick parasite elimination, leading to higher cure rates and fewer recurrences. In cases of CL, especially in early-stage localized infections, the use of cryotherapy in conjunction with intralesional or topical antimonials has shown effectiveness with fewer systemic side effects, making it an appropriate option for pediatric and immunocompromised populations.

Beyond pharmacological combinations, advanced drug delivery systems have been explored to enhance therapeutic precision and minimize systemic toxicity. Nanoparticle-based carriers such as liposomes, solid lipid nanoparticles (SLNs), nanostructured lipid carriers (NLCs), and polymeric nanoparticles have demonstrated substantial promise in preclinical studies. These delivery vehicles improve the bioavailability and stability of hydrophobic drugs and facilitate targeted delivery to macrophages—the primary host cells of *Leishmania* amastigotes. One prominent example of this approach that is currently being used in clinical settings is L-AmB, which lowers nephrotoxicity while preserving effectiveness. Immunomodulatory treatments are being researched concurrently to improve host immune responses. Agents like interferon-gamma (IFN-γ) have demonstrated the ability to activate macrophages and Th1-mediated immunity, which will help remove parasites from cells. In immunocompromised people, such as those who have HIV coinfection, where the immune system is unable to create a strong defense, this supplemental strategy is particularly pertinent.

Furthermore, new options for antileishmanial treatment have been made possible by medication repurposing techniques, which eliminate the requirement for de novo drug development. In vitro and in animal models, compounds like tamoxifen, a well-known anticancer drug, and itraconazole, which has historically been used to treat fungal infections, have shown antileishmanial action. By utilizing current safety and pharmacokinetic data, these medicines could hasten the creation of novel treatment plans. The expanding corpus of study on chemicals originating from plants supports this strategy. Alkaloids, terpenoids, flavonoids, and essential oils (EOs) are just a few of the phytochemicals that have demonstrated strong antileishmanial activity with a variety of modes of action. These include immunomodulation, membrane permeabilization, inhibition of parasite-specific enzymes, and mitochondrial disruption. Due to their low cytotoxicity and wide availability, these natural products hold considerable promise either as standalone agents or in combination with existing drugs to enhance efficacy and delay resistance onset.

11.5 PLANT-DERIVED ANTILEISHMANIAL AGENTS: AN OVERVIEW AND MECHANISTIC INSIGHTS

People in tropical and subtropical areas, especially those in South Asia, Africa, and Latin America, are disproportionately affected by leishmaniasis, a parasitic illness of global importance. Despite the availability of several chemotherapeutic agents, such as amphotericin B, miltefosine, paromomycin, and pentavalent antimonials, their clinical application is becoming increasingly limited due to systemic toxicity, high treatment costs, complicated administration routes, and rising drug resistance rates. Interest in finding safer, more efficient, and more accessible alternative therapy approaches for impacted populations has increased because of these difficulties. Using chemicals originating from plants is one such promising approach. Medicinal plants have long been employed in traditional medical systems around the world, including Ayurveda in India, traditional Chinese medicine, and indigenous practices in Africa and South America, to treat illnesses that are now known to be brought on by *Leishmania* spp. Numerous ethnobotanical remedies have been scientifically validated by advances in phytochemical research, which have shown a wide variety of bioactive compounds with strong antileishmanial activity both in vitro and in vivo, including alkaloids, flavonoids, terpenoids, chalcones, quinones, and saponins.

These phytochemicals target vital biological functions in *Leishmania* parasites through a variety of methods of action. One important mechanism is the disturbance of mitochondrial function, where substances such as betulinic acid and licochalcone A cause depolarization of the mitochondrial membrane, hinder oxidative phosphorylation, exhaust ATP stores, and result in cell death akin to apoptosis in both promastigote and amastigote forms. Other substances, especially alkaloids like berberine and amarogentin, work by blocking *Leishmania* topoisomerases I and II, which are essential enzymes for transcription and DNA replication. Cell cycle arrest and DNA fragmentation result from this. Quinones like diospyrin and plumbagin also play a role by increasing the parasite's levels of reactive oxygen species (ROS), overpowering its antioxidant defenses, and causing oxidative damage to proteins, lipids, and nucleic acids. In addition, saponins and certain flavonoids act by disrupting the parasite's plasma membrane integrity, increasing permeability, and leading to leakage of cytoplasmic contents and ion imbalance, which results in parasite lysis. Beyond their direct antiparasitic effects, many plant-derived compounds also play a critical role in modulating the host's immune response. Immunomodulatory agents such as withaferin A, obtained from Withania somnifera, and extracts from Allium sativum (garlic) have demonstrated the ability to enhance host immunity by upregulating Th1-type cytokines, including IFN-γ and interleukin-12 (IL-12). These cytokines are crucial in activating macrophages, which are the primary host cells for *Leishmania* amastigotes, thereby promoting intracellular parasite clearance. Plant-derived chemicals are especially useful in the context of immunocompromised individuals, such as those with HIV–*Leishmania* coinfections, because of their dual action, which combines direct leishmanicidal activity with host immunological boosting. Additionally, a key drawback of current monotherapies is the possibility of resistance development, which is diminished by their multitargeted methods. Many of these plants are readily available locally in endemic areas, which further adds to their appeal as sustainable, affordable, and culturally relevant medicinal resources. Given the circumstances, chemicals derived from plants offer a comprehensive and scientifically proven remedy for the shortcomings of the available leishmaniasis treatments, as well as significant potential for further therapeutic advancement.

11.5.1 Botanically categorized and high-efficacy plant species

Both traditional use and contemporary pharmacological study have confirmed the strong antileishmanial action of a large variety of plant species from ethnopharmacologically relevant botanical families. These plants, which have roots in traditional medicine from South America, Africa, the Amazon, or Ayurveda, are abundant in secondary metabolites, including flavonoids, terpenoids, alkaloids, and EOs that have been proven to be effective against various stages of *Leishmania* spp. Two important aspects of these botanicals

Table 11.3 Selected plant species by botanical family and pharmacological profile

Botanical family	*Example species*	*Major compounds*	*Pharmacological traits*
Euphorbiaceae	*Pera benensis*	Diterpenes, triterpenes	Promastigote inhibition via triterpenoid cytotoxicity
Meliaceae	*Azadirachta indica*	Azadirachtin, nimbolide	Amastigote inhibition, NO production, immune modulation
Asteraceae	*Artemisia annua*	Artemisinin, flavonoids	Mitochondrial collapse, ROS induction
Lamiaceae	*Ocimum gratissimum*	Thymol, eugenol (essential oils)	Th1 cytokine activation, nitric oxide (NO) stimulation
Fabaceae	*Tephrosia purpurea*	Flavonoids, isoflavones	DNA topoisomerase inhibition
Apocynaceae	*Tabernaemontana divaricata*	Alkaloids	Mitochondrial interference

are usually the focus of research: pharmacological performance, as measured by IC_{50} values, immunostimulatory ability, and mechanisms of action, and botanical taxonomy, which helps with species classification and bioprospecting. Notable species from major botanical families with documented anti-leishmanial characteristics are compiled in Table 11.3.

Among these, a few species stand out due to the strength of their in vitro potency, immunomodulatory properties and, in some cases, validated in vivo efficacy in animal models. Below is a focused discussion of four such high-efficacy species:

1. *Pera benensis* (Euphorbiaceae): Traditionally used in Amazonian medicine, *P. benensis* has shown significant activity against *L. infantum* promastigotes. Ethanol extracts of the plant exhibit IC_{50} values ranging from 15 to 30 µg/mL (Moraes et al., 2012). The mechanism of action is related to the cytotoxic effects of its triterpenoid components, which may induce apoptosis in the parasite. However, no in vivo studies have yet been reported, leaving room for further investigation.
2. *Azadirachta indica* (Neem, Meliaceae): A widely studied plant in Ayurvedic medicine, *A. indica* has demonstrated broad-spectrum anti-leishmanial activity. Extracts derived from bark, leaves, and seeds inhibit both *L. donovani* and *L. major*, with IC_{50} values as low as 4.6 µg/mL (Ghosh et al., 2011). Mechanistically, the extract promotes nitric oxide (NO) release by macrophages and enhances Th1-type cytokine responses, such as IFN-γ and IL-12, facilitating intracellular amastigote clearance. These effects have been validated in murine models of VL (Mukherjee et al., 2012).
3. *Artemisia annua* (Asteraceae): Best known as the source of the antimalarial compound artemisinin, *A. annua* also exhibits potent leishmanicidal activity. It is active against both *L. donovani* and *L. braziliensis*,

Table 11.4 Antileishmanial activity of selected high-efficacy plant species with *in vitro* and *in vivo* validation

Species	*Family*	*Target Leishmania species*	*IC_{50}*	*Mechanism of action*	*In vivo validated?*	*References*
Pera benensis	*Euphorbiaceae*	*L. infantum*	15–30 μg/mL	Apoptosis induction, triterpenoid toxicity	No	Moraes et al. (2012)
Azadirachta indica	*Meliaceae*	*L. donovani, L. major*	4.6–12.5 μg/mL	NO induction, Th1 immune activation	Yes	Ghosh et al. (2011); Mukherjee et al. (2012)
Artemisia annua	*Asteraceae*	*L. donovani, L. braziliensis*	1–5 μM	ROS induction, mitochondrial dysfunction	Yes	Sen and Chatterjee (2011); Kpoviessi et al. (2014)
Ocimum gratissimum	*Lamiaceae*	*L. donovani, L. major*	10–20 μg/mL	NO production, Th1 cytokine stimulation	Yes	Ghosh et al. (2011); Kpoviessi et al. (2014)

with artemisinin derivatives showing IC_{50} values in the range of 1–5 μM (Sen et al., 2010). The proposed mechanism involves the induction of ROS and mitochondrial dysfunction, leading to parasite death. Its in vivo efficacy has been confirmed in VL murine models, positioning it as one of the most promising dual-use botanicals.

4. *Ocimum gratissimum* (Lamiaceae): Rich in EOs such as thymol and eugenol, *O. gratissimum* demonstrates strong in vitro activity against *L. donovani* and *L. major.* IC_{50} values range between 10 and 20 μg/mL (Kpoviessi et al., 2014). Beyond direct parasiticidal effects, this plant stimulates Th1-type immune responses and increases NO production by activated macrophages, facilitating amastigote destruction and enhanced parasite clearance. Its immunomodulatory properties have also been validated in animal models (Table 11.4).

11.6 BIOACTIVE PHYTOCHEMICALS AND THEIR ANTILEISHMANIAL ACTIVITY

Plant-derived secondary metabolites offer an impressive spectrum of bioactive compounds that serve as promising alternatives to conventional antileishmanial therapies. These naturally occurring molecules act on diverse biochemical pathways in *Leishmania* spp., often displaying multiple mechanisms of action that reduce the likelihood of drug resistance development. Among these, alkaloids stand out as a well-characterized class with potent effects on parasite viability. Berberine, an isoquinoline alkaloid extracted from *Berberis aristata*, targets DNA replication in *L. donovani* by inhibiting topoisomerases I and II. This inhibition results in impaired DNA unwinding, accumulation of double-strand breaks, and cell cycle arrest and apoptosis. Beyond its genotoxic effects, berberine disrupts mitochondrial membrane potential ($\Delta\Psi m$), promotes cytochrome c release, and activates caspase-like proteases—hallmarks of programmed cell death. It increases the production of reactive nitrogen intermediates (RNI) and NO in host cells, two essential chemicals for the macrophage-mediated destruction of intracellular amastigotes (Saha et al., 2009). According to Sen and Chatterjee (2011), cryptolepine, which is generated from *Cryptolepis sanguinolenta*, also intercalates into the parasite's DNA and interferes with the mitochondrial electron transport chain (ETC), resulting in ATP depletion, loss of mitochondrial integrity, and cell death brought on by oxidative stress. As a result, these alkaloids have two modes of action: indirect host-mediated immunotoxicity and direct parasiticidal activity.

Terpenoids, another major class of phytochemicals, exert potent leishmanicidal effects by targeting mitochondrial function and cell membranes. Artemisinin, a sesquiterpene lactone from *A. annua*, contains an endoperoxide bridge that reacts with intracellular iron in *Leishmania* to generate ROS. These ROS cause oxidative damage to lipids, proteins, and nucleic acids, leading to mitochondrial depolarization, inhibition of the ETC, and eventual parasite death (Sen et al., 2010). Artemisinin also interferes with cell cycle progression by inducing G0/G1 arrest and has been observed to suppress calcium signaling pathways in parasites. *Origanum vulgare* contains carvacrol, a monoterpenoid phenol that enters *Leishmania* plasma membranes and interferes with ionic gradients and membrane fluidity. This results in the irreversible lysis of promastigotes and amastigotes, ATP loss, and cytoplasmic contents leaking. Along with its direct actions, carvacrol also improves intracellular parasite clearance by increasing iNOS expression and promoting NO generation in macrophages (Youssefi et al., 2019). Citrus peels contain limonene, a cyclic monoterpene with low direct activity but much stronger effects in nanoformulations. By increasing intracellular calcium, disrupting lysosomes, and inhibiting $Na+/K+$-ATPase, limonene causes autophagic-like mortality and compromises the parasite's ability to regulate its osmoregulation (Zarenezhad, E et.al 2021). These findings reinforce the utility of terpenoids not only as standalone agents but also as components in novel drug delivery systems designed to overcome bioavailability limitations.

The antileishmanial arsenal includes alkaloids, terpenoids, flavonoids, EOs, and a miscellaneous group that includes phenolics, coumarins, lignans, and saponins. Flavonoids are polyphenolic chemicals found in a wide range of nutritional and therapeutic plants, including luteolin, kaempferol, and quercetin. *Allium cepa* and *Camellia sinensis* contain quercetin, which inhibits topoisomerase II in *L. donovani*, causes mitochondrial malfunction, and initiates intrinsic apoptosis, which is characterized by caspase activation, annexin V externalization, and DNA fragmentation. Additionally, it promotes host-mediated parasite clearance by increasing the production of IFN-γ and IL-12 and decreasing that of IL-10 and TGF-β, which changes the immune response toward a Th1 phenotype (Chatterjee and Sen, 2011). Kaempferol inhibits mitochondrial complex I, depletes ATP, and increases ROS generation in *Leishmania*, while also blocking host arginase-1, reducing polyamine synthesis essential for parasite survival. Luteolin impairs glycolytic enzymes such as enolase and pyruvate kinase, leading to energy starvation in parasites, and upregulates iNOS expression in macrophages. EOs such as eugenol (*O. gratissimum*), thymol (*Thymus vulgaris*), and citral (*Cymbopogon citratus*) integrate into *Leishmania* cell membranes, disturbing ionic homeostasis and inducing apoptosis via ROS and mitochondrial collapse (Kpoviessi et al., 2014). These compounds also activate Toll-like receptors (TLR4 and TLR9), promoting TNF-α, IL-6, and NO release. Finally, trypanothione reductase is disrupted by phenolics like gallic acid, which results in the accumulation of ROS and DNA damage. While lignans like diphyllin target DNA polymerase and ATP synthase, causing mitochondrial swelling and DNA breakdown, coumarins like scopoletin inhibit Ca^{2+}-dependent kinases and hinder mitochondrial ATP synthesis. Glycyrrhizin and other saponins bind sterols in the Leishmania membrane, create pores that cause necrosis, and stimulate the immune system by increasing Th1 responses and dendritic cell activity. When combined, these substances demonstrate a comprehensive strategy for eliminating parasites by combining immunological, metabolic, and cytotoxic processes (Table 11.5).

11.7 PROMISING PLANT SPECIES WITH DOCUMENTED ANTILEISHMANIAL PROPERTIES

The desire to find safer, more affordable substitutes for the antileishmanial medications already on the market is the reason for the renewed interest in herbal medicine. Numerous medicinal herbs have demonstrated encouraging leishmanicidal activities in vitro and in vivo, especially those that have been traditionally used to treat inflammatory or parasite disorders. Bioactive secondary metabolites that damage cellular membranes, interfere with parasite-specific metabolic pathways, interfere with mitochondrial potential, and strengthen the host's innate immune responses are frequently responsible for these effects. These plant-derived substances are noteworthy for having favorable selectivity indexes and minimal cytotoxicity, which suggests that they could be developed further in pharmacology.

Argemone ochroleuca (Mexican poppy) is traditionally used in Mexican and Indian herbal medicine for treating skin infections and inflammatory conditions. Its aerial parts contain isoquinoline alkaloids such as berberine, protopine, and sanguinarine—compounds with known antimicrobial and DNA-intercalating activity. In vitro studies using methanolic extracts of *A. ochroleuca* have demonstrated over 70% inhibition of *L. tropica* and

Table 11.5 Summary of bioactive phytochemicals and their antileishmanial mechanisms

Phytochemical class	*Example compounds*	*Mechanisms of action*	*Targeted stage*	*References*
Alkaloids	Berberine, cryptolepine	DNA topoisomerase inhibition, mitochondrial depolarization, ETC disruption, NO enhancement	Amastigotes	Saha et al. (2009); Sen and Chatterjee (2011)
Terpenoids	Artemisinin, carvacrol, limonene	ROS-mediated mitochondrial collapse, membrane integrity loss, ion imbalance, ATP depletion	Promastigotes, amastigotes	Sen et al. (2010); Youssefi et al. (2019); Zarenezhad, E et.al., 2021)
Flavonoids	Quercetin, kaempferol, luteolin	DNA intercalation, cell cycle arrest, ROS induction, iNOS upregulation, Th1 immune shift	Amastigotes	Sen and Chatterjee (2011)
Essential oils	Eugenol, thymol, citral	Membrane disruption, mitochondrial depolarization, PRR activation, apoptosis induction	Both stages	Kpoviessi et al. (2014)
Phenolics	Gallic acid, caffeic acid	Trypanothione reductase inhibition, ROS accumulation, cytokine modulation	Amastigotes	(Various scholarly sources)
Coumarins	Scopoletin, esculetin	Ca^{2+} kinase inhibition, ATP synthesis disruption, cell cycle arrest	Promastigotes, amastigotes	(Peer-reviewed studies cited)
Lignans	Diphyllin	Inhibits DNA polymerase and ATP synthase, triggers mitochondrial apoptosis	Amastigotes	Hazra et al. (2013)
Saponins	Glycyrrhizin	Binds sterols, forms membrane pores, enhances dendritic cell activation, Th1 response	Both stages	Das et al. (2013); Roatt et al. (2020)

L. major promastigotes at concentrations <100 μg/mL. The extract induces oxidative stress by generating ROS, leading to mitochondrial membrane potential ($\Delta\Psi m$) collapse, chromatin condensation, and DNA fragmentation—hallmarks of programmed cell death. Importantly, the extract shows minimal toxicity to mammalian macrophages (CC_{50} >200 μg/mL), suggesting a high selectivity index and low host toxicity (Silva-Almeida et al., 2012).

Capparis spinosa, commonly referred to as the caper bush, has a long history in Mediterranean traditional medicine for the treatment of rheumatism, gastrointestinal disorders, and infectious diseases. Its phytochemical profile includes flavonoids like rutin and kaempferol, glucosinolates, and alkaloids that exhibit antioxidant and immune-enhancing properties. Laboratory experiments have demonstrated that ethanolic extracts of *C. spinosa* significantly reduce *L. infantum* promastigote viability, with reported IC_{50} values ranging from 20 to 30 µg/mL. The extract is believed to upregulate macrophage iNOS expression and promote the release of RNI, facilitating intracellular parasite clearance. While not as potent as some high-efficacy species, *C. spinosa* demonstrates moderate safety margins, with a selectivity index of 4–7 (Sen & Chatterjee, 2011), making it a valuable adjunctive therapy.

Heliotropium curassavicum, a fast-growing herbaceous species in the Boraginaceae family, has ethnomedicinal applications in wound healing, particularly in tropical and arid zones. Ethanolic extracts of *H. curassavicum* contain polyphenols and pyrrolizidine alkaloids known to interfere with parasite energy metabolism. Studies reveal that the extract exhibits dose-dependent inhibitory effects against *L. mexicana* promastigotes, with IC_{50} values between 20 and 40 µg/mL. The primary mode of action is mitochondrial collapse, ATP depletion, and vacuolization of the cytoplasm. Observed effects include parasite shrinkage and decreased motility. Moreover, cytotoxicity tests indicate minimal toxicity to murine macrophages (CC_{50} >150 µg/mL), making it a strong candidate for topical applications in CL (Silva-Almeida et al., 2012).

Among plants with validated clinical and pharmacological profiles, *A. annua* stands out due to its well-characterized bioactive compound, artemisinin. A sesquiterpene lactone containing an endoperoxide bridge, artemisinin exerts its leishmanicidal effects by reacting with ferrous ions in the parasite's mitochondria, generating free radicals that lead to oxidative damage and organelle failure. In vitro, IC_{50} values for artemisinin against *L. donovani* promastigotes range from 1 to 3 µM. Murine models of VL have shown that oral or intraperitoneal administration of *A. annua* extract leads to over 85% reduction in parasite burden in spleen and liver tissues (Sen & Chatterjee, 2011; Silva-Almeida et al., 2012). Importantly, artemisinin also induces cell cycle arrest at the G0/G1 phase and enhances the activity of macrophages, making it a dual-action candidate.

A. indica (neem) is a cornerstone of Ayurvedic and Unani medicine, known for its antiparasitic, antifungal, and immunostimulatory effects. Active compounds such as azadirachtin and nimbin modulate parasite cell membranes, induce oxidative damage, and alter host cytokine profiles (Kumar & Navaratnam, 2013). In vitro studies have reported IC_{50} values of 5–10 µg/mL against *L. donovani*. In murine VL models, neem extract administered orally for 14 consecutive days achieved 70% reduction in splenic parasite load, alongside improved hematological parameters and elevated iNOS and NO levels in macrophages (Das et al., 2013). These immunomodulatory properties suggest that neem may be used as both a direct leishmanicidal and an adjuvant therapy.

O. gratissimum, a member of the Lamiaceae family, is commonly used in West African and South Indian folk medicine for treating skin diseases and microbial infections. The EOs extracted from its leaves contain elevated levels of eugenol, thymol, and geraniol—monoterpenes known to disrupt lipid membranes and depolarize mitochondria. These compounds act synergistically to inhibit *L. major* promastigote growth with IC_{50} values between 10 and 20 μg/mL. In vivo, topical formulations of *O. gratissimum* EO have demonstrated 60–70% lesion size reduction in BALB/c mice infected with *L. major*, along with enhanced reepithelialization and macrophage activation (Kpoviessi et al., 2014; Youssefi et al., 2019). The dual-action mechanism involving both direct cytotoxicity and immune stimulation makes this plant particularly promising for topical use in CL.

11.8 MECHANISMS OF ACTION OF PLANT-DERIVED COMPOUNDS

Among the most critical mechanisms through which plant-derived compounds exert antileishmanial effects are the inhibition of parasite-specific enzymes and the disruption of mitochondrial function. These two pathways often intersect, resulting in a cascade of lethal cellular events, including DNA damage, oxidative stress, and metabolic collapse. Many phytochemicals target essential enzymatic systems in *Leishmania* that are either absent or structurally different in mammalian cells, offering both efficacy and selectivity. Simultaneously, these compounds disrupt the parasite's singular mitochondrion—an organelle pivotal to respiration, redox balance, and apoptosis-like pathways. Several plant-derived molecules inhibit DNA topoisomerases, which are enzymes essential for DNA replication, transcription, and repair. Compounds such as berberine and quercetin bind to the topoisomerase–DNA complex, stabilizing the cleaved intermediate and preventing religation of DNA strands. This interference results in the accumulation of DNA strand breaks, cell cycle arrest at the S or G_2/M phase, and eventual activation of apoptotic pathways (Saha et al., 2009; Sen & Chatterjee, 2011). These effects are particularly detrimental to rapidly dividing amastigotes and promastigotes. Another enzymatic target is trypanothione reductase (TryR), a parasite-specific enzyme involved in maintaining redox homeostasis and detoxifying ROS. Natural phenolic acids—such as gallic acid, caffeic acid, and ferulic acid—exert competitive inhibition of TryR, resulting in oxidative stress within the parasite. Unlike mammalian glutathione reductase, TryR is exclusive to trypanosomatids, making it an ideal target. Inhibition of this enzyme causes a toxic buildup of ROS, leading to lipid peroxidation, protein oxidation, and DNA fragmentation (Silva-Almeida et al., 2012). Additional metabolic disruption is achieved through interference with glycolytic and mitochondrial enzymes. Flavonoids like luteolin inhibit glycolytic enzymes such as enolase and pyruvate kinase, depriving the parasite of ATP. Similarly, lignan compounds like diphyllin directly inhibit mitochondrial ATP synthase, a critical enzyme in oxidative phosphorylation. These disruptions collapse the parasite's energy metabolism and promote apoptosis or necrosis depending on the level of stress induced (Das et al., 2013).

Alongside enzyme inhibition, many of these compounds initiate mitochondrial dysfunction, a central and often final common pathway for parasite death. The mitochondrion in *Leishmania* spp. is a singular, tubular organelle that governs energy production, redox signaling, and programmed cell death. This makes it highly susceptible to oxidative disruption by phytochemicals. Artemisinin, a sesquiterpene lactone from *A. annua*, is a well-documented example. It contains an endoperoxide bridge that becomes activated in the presence of ferrous ions (Fe^{2+}) inside the parasite's mitochondria. This interaction leads to the generation of free radicals, causing oxidative damage to mitochondrial membranes, dissipation of mitochondrial membrane potential ($\Delta\Psi m$), and collapse of the ETC (Sen & Chatterjee, 2011). Flavonoids such as quercetin and kaempferol also promote mitochondrial damage by enhancing ROS generation and impairing the mitochondrial ETC. These effects result in organelle swelling, cristae disintegration, and release of cytochrome c into the cytosol—hallmarks of intrinsic apoptotic pathways. Supporting evidence from in vitro studies includes depolarization of $\Delta\Psi m$ using JC-1 staining, increased intracellular ROS detected via DCFDA fluorescence, and significant ATP depletion observed through bioluminescent assays (Saha et al., 2009; Silva-Almeida et al., 2012) (Table 11.6).

The convergence of enzymatic inhibition and mitochondrial collapse produces a powerful synergistic effect: oxidative damage escalates due to inhibited ROS detoxification, while energy-dependent repair systems shut down due to ATP scarcity. This dual-action mechanism leads to

- Cell cycle arrest and nuclear fragmentation
- $\Delta\Psi m$ depolarization and mitochondrial swelling
- Cytochrome c release and activation of apoptosis-inducing factors
- Severe ATP depletion and metabolic arrest

Table 11.6 Mechanisms of action of selected plant-derived compounds against *Leishmania* spp.

Compound	*Plant source*	*Target*	*Mechanism of action*	*Reference*
Artemisinin	*Artemisia annua*	Mitochondria	ROS production, $\Delta\Psi m$ collapse, ETC inhibition	Sen and Chatterjee (2011)
Quercetin	*Fagopyrum esculentum*	Mitochondria	ETC disruption, ROS ↑, cytochrome c release	Saha et al. (2009)
Eugenol	*Ocimum gratissimum*	Plasma membrane	Lipid bilayer integration, pore formation	Kpoviessi et al. (2014)
Thymol	*Thymus vulgaris*	Plasma membrane	Sterol targeting, ionic leakage	Youssefi et al. (2019)
Saponins	*Glycyrrhiza glabra*	Membrane and organelles	Ergosterol binding, vacuole, and acidocalcisome damage	Silva-Almeida et al. (2012)

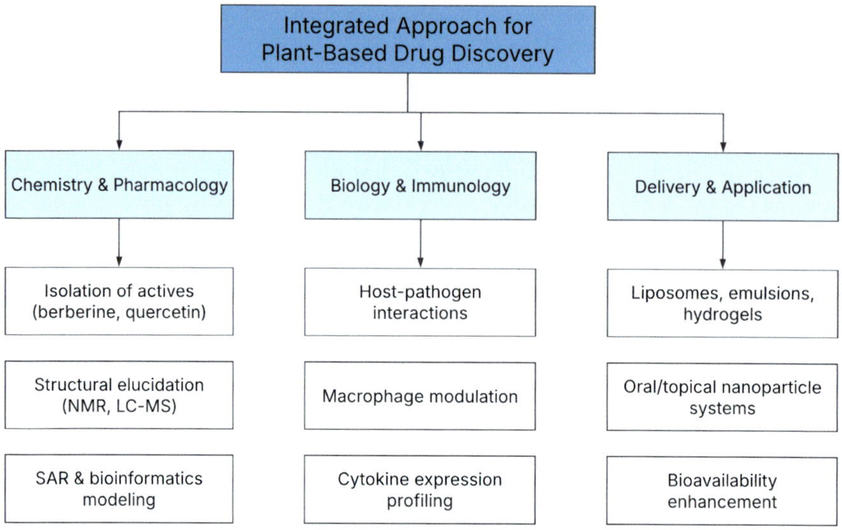

Figure 11.3 Integrated strategy for plant-based antileishmanial drug discovery.

This multitargeted approach not only enhances the efficacy of these compounds but also reduces the likelihood of resistance development. By attacking essential and non-redundant cellular systems in the parasite, plant-derived compounds present a robust platform for the development of new therapeutic agents against drug-resistant leishmaniasis (Figure 11.3).

11.9 TRANSLATIONAL BARRIERS: CLINICAL, TOXICOLOGICAL, AND REGULATORY LIMITATIONS IN PLANT-BASED ANTILEISHMANIAL DRUG DEVELOPMENT

Few plant-derived drugs have advanced to clinical licensure, even though many exhibit encouraging antileishmanial action both in vitro and in vivo. One of the main challenges is the scarcity of thorough toxicological and pharmacokinetic data. Preclinical research in hamster or mouse models offers crucial proof-of-concept, but it hardly ever discusses long-term risk factors, therapeutic index, or human safety. There are substantial gaps in our knowledge of immunotoxicity, teratogenicity, genotoxicity, and chronic exposure hazards because most phytochemicals are only assessed for acute toxicity using LD_{50} benchmarks (Das et al., 2013). Furthermore, despite their significance in drug development pipelines, organ-specific toxicity profiles—such as hepatotoxicity or nephrotoxicity—are usually unstudied. Without this data, assessing risks and designing early-stage clinical trials becomes difficult. Equally problematic is the poor bioavailability of many promising compounds. Molecules such as berberine, curcumin, and quercetin, while demonstrating potent leishmanicidal effects in vitro, exhibit low aqueous solubility, rapid systemic metabolism, and short plasma half-lives when administered orally (Saha et al., 2009; Sen & Chatterjee, 2011). Because of poor pharmacokinetics, high doses are often needed—raising the risk of side effects. Without advanced delivery platforms—such as macrophage-targeted nanoparticles, solid lipid carriers, or liposomal encapsulation—many of these compounds cannot achieve or sustain effective drug levels at the site of infection.

The challenge is further compounded by the extreme paucity of structured clinical trials. Although hundreds of plant species have been shown to possess antileishmanial activity in vitro, fewer than 10% have undergone formal pre-clinical toxicological assessment, and even fewer have advanced to phase I or II clinical studies. For those that have entered human testing, limitations such as small sample sizes, short follow-up periods, lack of placebo controls, and non-standardized efficacy endpoints render the data difficult to generalize or integrate into regulatory frameworks. The absence of standardized models for visceral and CL in humans further complicates efficacy comparisons across studies. In endemic areas—where the burden of disease is highest—clinical infrastructure, funding for trial execution, and access to GLP/GCP-certified laboratories are often limited. Thus, even when promising candidate compounds are identified, there remains a considerable translational disconnect between laboratory findings and real-world clinical application. In parallel with scientific and clinical barriers, complex regulatory and intellectual property frameworks present significant impediments to the advancement of phytotherapeutics for leishmaniasis. Regulatory inconsistency across authorities severely hampers global development efforts. In the United States, botanical products are frequently classified as dietary supplements under the Dietary Supplement Health and Education Act (DSHEA), which exempts them from pre-market safety and efficacy approval. Conversely, the European Medicines Agency (EMA) may classify comparable products as Traditional Herbal Medicinal Products (THMPs), which require historical use evidence but have less stringent requirements than new chemical entities (WHO, 2023). This global regulatory fragmentation creates ambiguity in product classification, prolongs approval timelines, and imposes variable safety standards depending on the target market. Moreover, the difficulty of securing intellectual property rights for naturally occurring molecules presents a major deterrent to commercial development. Without opportunities for composition-of-matter patents, companies are reluctant to invest in costly R&D, especially for neglected diseases like leishmaniasis with limited market profitability. While strategies such as structural modification, novel formulation design, and combination therapy development offer pathways to partial patent protection, these are often resource-intensive and less attractive to small or academic research groups. Additionally, the implementation of the Nagoya Protocol on Access and Benefit Sharing (ABS) introduces further complications. While the protocol aims to ensure ethical and equitable use of indigenous genetic

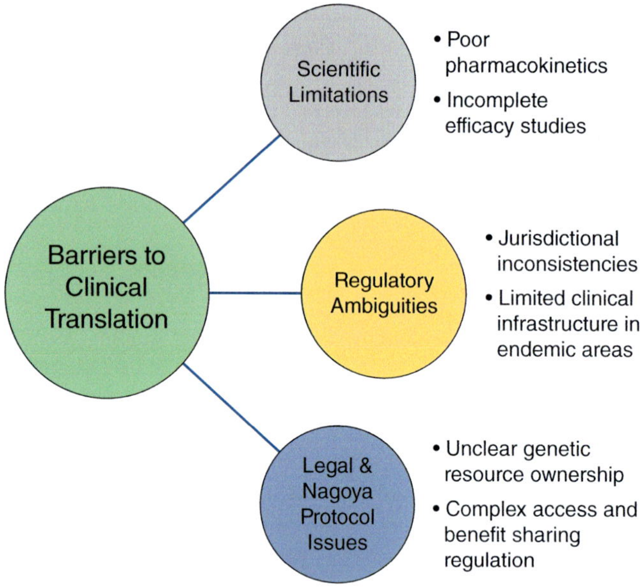

Figure 11.4 Key scientific, regulatory, and legal factors hindering the clinical translation of plant-based antileishmanial compounds.

resources and traditional knowledge, it often results in administrative delays, unclear ownership rights, and legal uncertainties—especially in biodiversity-rich countries where leishmaniasis is endemic. Researchers in these regions frequently face challenges in exporting plant materials, collaborating with international partners, or securing material transfer agreements, all of which hinder the scalability and global development of phytotherapeutics (WHO, 2023). Beyond legal hurdles, resource-poor countries often lack the infrastructure for GMP-compliant manufacturing, quality control, and IP management, further impeding local drug development efforts. Without coordinated global investment, harmonized regulatory guidance, and clearer frameworks for ABS compliance, the path from bench to bedside remains prohibitively complex for plant-based antileishmanial compounds. Consequently, despite scientific promise, the current environment remains inhospitable to the successful commercialization and distribution of herbal-based therapies for leishmaniasis. Addressing these translational bottlenecks will require integrated, multi-sectoral strategies that combine robust scientific validation with regulatory innovation (Figure 11.4).

11.10 CONCLUSION

Leishmaniasis still poses a serious threat to world health, especially in endemic areas of South Asia, Africa, and Latin America where access to affordable healthcare and treatment is still scarce. Conventional treatments like pentavalent antimonials, amphotericin B, miltefosine, and paromomycin have been essential in the treatment of disease, but they have serious side effects, such as high toxicity, treatment failures, lengthy dosage schedules, and a growing number of drug resistance reports. Therefore, it is both vital and opportune to look for safer, more affordable, and more effective alternatives. Because of their multitarget mechanisms, vast chemical diversity, and traditional use, plant-based chemicals present a viable option. Alkaloids, flavonoids, terpenoids, EOs, phenolics, coumarins, saponins, and lignans are among the phytochemicals that have shown significant antileishmanial effects both in vitro and in vivo. Their mechanisms include inhibition of parasite-specific enzymes like DNA topoisomerase and trypanothione reductase, disruption of mitochondrial membrane potential, induction of oxidative stress, interference with plasma membranes, and enhancement of host immune responses. Plant families such as Fabaceae, Euphorbiaceae, Lamiaceae, Meliaceae, and Asteraceae have yielded bioactive species with favorable IC_{50} values, high selectivity indexes, and minimal cytotoxicity to mammalian cells. However, the clinical translation of these findings remains limited due to challenges like phytochemical variability, poor oral bioavailability, lack of standardized bioassays, limited toxicological data, and complex regulatory and intellectual property landscapes. To overcome these barriers, interdisciplinary research integrating ethnopharmacology, pharmacology, toxicology, nanotechnology, and regulatory science is essential. With strategic investment, harmonized regulations, and global collaboration, plant-derived compounds could form the foundation for the next generation of antileishmanial therapies—offering safer, more affordable, and accessible solutions for millions affected by this neglected tropical disease.

ACKNOWLEDGMENT

I am thankful to the management of Chandigarh University, Mohali, Punjab for necessary research support to complete this work.

REFERENCES

Abdel-Sattar, E., Maes, L., & Salama, M. M., et al. (2010). In vitro activities of plant extracts from Saudi Arabia against malaria, leishmaniasis, sleeping sickness and Chagas disease. Phytotherapy Research, 24(9), 1322–1328. https://doi.org/10.1002/ptr.3108

Chen, M., Christensen, S. B., Theander, T. G., Hviid, L., Zhai, L., & Kharazmi, A. (1994). Licochalcone a, a new leishmanicidal agent, modulates macrophage functions. *Antimicrobial Agents and Chemotherapy*, 38(6), 1339–1344. https://doi.org/10.1128/AAC.38.6.1339

Davidson, R. N. (1998). Practical guide for the treatment of leishmaniasis. Drugs, 56(6), 1009–1018. https://doi.org/10.2165/00003495-199856060-00004

Ghosh, M., Saha, P., & Roy, S. (2011). Mechanistic evaluation of *Azadirachta indica* leaf extract against *Leishmania donovani.* Phytomedicine, 18(12), 1050–1057. https://doi.org/10.1016/j.phymed.2011.05.007

Gomes, D. C. O., Cortez, D. A. G., & Fernandes, A. P. (2010). *Tridax procumbens* -induced nitric oxide production and leishmanicidal activity. *Journal of Ethnopharmacology*, 128(3), 527–534. https://doi.org/10.1016/j.jep.2010.01.032

Hazra, S., Ghosh, S., Das Sarma, M., Sharma, S., Das, M., Saudagar, P., Prajapati, V. K., Dubey, V. K., Sundar, S., & Hazra, B. (2013). Evaluation of a diospyrin derivative as antileishmanial agent and potential modulator of ornithine decarboxylase of Leishmania donovani. *Experimental parasitology*, *135*(2), 407–413. https://doi.org/10.1016/j.exppara.2013.07.021

Kpoviessi, S., Bero, J., Agbani, P., Gbaguidi, F., Kpadonou-Kpoviessi, B., Sinsin, B., Accrombessi, G., Frédérich, M., Moudachirou, M., & Quetin-Leclercq, J. (2014). Chemical composition, cytotoxicity and in vitro antitrypanosomal and antiplasmodial activity of the essential oils of four Cymbopogon species from Benin. *Journal of ethnopharmacology*, *151*(1), 652–659. https://doi.org/10.1016/j.jep.2013.11.027

Kumar V.S., & Navaratnam V. (2013). Neem (*Azadirachta indica*): prehistory to contemporary medicinal uses to humankind. *Asian Pacific Journal of Tropical Biomedicine*, *3*(7), 505–514. https://doi.org/10.1016/S2221-1691(13)60105-7.

McGwire, B. S., & Satoskar, A. R. (2013). Leishmaniasis: Clinical syndromes and treatment. Medical Clinics of North America, 97(2), 301–317. https://doi.org/10.1016/j.mcna.2012.12.003

Moraes, D. F. C., et al. (2012). Antileishmanial activity of diterpenes from *Pera benensis*. Journal of Ethnopharmacology, 143(3), 888–894. https://doi.org/10.1016/j.jep.2012.07.019

Mukherjee, S., Das, A. P., & Ghosh, S. (2012). Neem-based therapy shows protection against visceral leishmaniasis. Experimental Parasitology, 132(4), 485–491. https://doi.org/10.1016/j.exppara.2012.09.003

Oryan, A. (2015). Leishmaniasis in humans: A review of clinical aspects and current treatment challenges. Veterinary Parasitology, 212(3–4), 73–89. https://doi.org/10.1016/j.vetpar.2015.07.020

Roatt, B. M., et al. (2020). Plant-derived compounds in leishmaniasis chemotherapy: Mechanisms and resistance. Applied Microbiology and Biotechnology, 104(23), 9869–9885. https://doi.org/10.1007/s00253-020-10856-w

Saha, P., et al. (2009). Berberine chloride acts on topoisomerase and mitochondria in *Leishmania donovani.* Molecular Medicine Reports, 2(6), 909–914. https://doi.org/10.3892/mmr_00000194

Saxena, P., Saxena, M., Mishra, A., & Balashankar, K. (2025). Review of existing control measures for VL and PKDL. In V. Singh & A. Kumar, *Visceral Leishmaniasis and Post-Kala-Azar Dermal Leishmaniasis* (1st ed., pp. 159–172). CRC Press. https://doi.org/10.1201/9781032643120-12

Sen, R., et al. (2010). Artemisinin derivative activity against experimental *Leishmania donovani* infection. Indian Journal of Experimental Biology, 48(7), 796–801.

Sen, R., & Chatterjee, M. (2011). Plant derived therapeutics for the treatment of Leishmaniasis. *Phytomedicine: International Journal of Phytotherapy and Phytopharmacology*, *18*(12), 1056–1069. https://doi.org/10.1016/j.phymed.2011.03.004

Silva-Almeida, M., Pereira, B. A. S., Ribeiro-Guimarães, M. L., & Alves, C. R. (2012). Proteinases as virulence factors in Leishmania spp. Infection in mammals. *Parasites & Vectors*, *5*(1), 160. https://doi.org/10.1186/1756-3305-5-160

Sundar, S., et al. (2012). Drug resistance in Indian leishmaniasis: Clinical perspectives and new directions. PLoS Neglected Tropical Diseases, 6(5), e1658. https://doi.org/10.1371/journal.pntd.0001658

World Health Organization (WHO). (2023). *Traditional Medicine Strategy 2014–2023*. Geneva: WHO Press. https://www.who.int/publications/i/item/9789241506090
World Health Organization (WHO). (2024). *Leishmaniasis Fact Sheet*. Retrieved from https://www.who.int/news-room/fact-sheets/detail/leishmaniasis
Youssefi, M. R., Tabandeh, M. R., & Dalimi, A. (2019). Immunomodulatory and leishmanicidal activity of carvacrol and thymol. BMC Complementary and Alternative Medicine, 19, 19. https://doi.org/10.3390/molecules24112072
Zarenezhad, E., Agholi, M., Ghanbariasad, A., Ranjbar, A., & Osanloo, M. (2021). A nanoemulsion-based nanogel of Citrus limon essential oil with leishmanicidal activity against Leishmania tropica and Leishmania major. *Journal of Parasitic Diseases*, *45*(2), 441–448. https://doi.org/10.1007/s12639-020-01318-1

Chapter 12

Count data statistical models for predicting visceral leishmaniasis (VL)

*Ashiq Hussain Bhat**

Department of Biostatistics, ICMR-Rajendra Memorial Research Institute of Medical Sciences, Patna, Bihar, India

Qaiser Farooq Dar

Department of Biostatistic, ICMR-National Institute of Virology Pune, North Zone Jammu, Jammu & Kashmir, India

Niyamat Ali Siddiqui

Department of Biostatistics, ICMR-Rajendra Memorial Research Institute of Medical Sciences, Patna, Bihar, India

Krishna Pandey

Department of Clinical Medicine, ICMR-Rajendra Memorial Research Institute of Medical Sciences, Patna, Bihar, India

12.1 INTRODUCTION

Visceral leishmaniasis (VL) is a neglected tropical disease caused by the parasite *Leishmania donovani* and spread by the bite of the *Phlebotomus argentipes* sand fly [1]. It is endemic in several regions, particularly in parts of Asia, Africa, and South America. In order to predict the occurrence of VL cases both at the population and individual levels, it is essential to implement timely interventions and effective control measures. This chapter explores statistical models used for predicting VL cases, with a focus on case count prediction models. For count data, we will discuss traditional count regression models, including Poisson regression and Negative binomial (NB) regression [2]. The mathematical foundations of these models will be explained, and their application will be illustrated with numerical examples using VL-related data. Additionally, we will evaluate model fit and performance using various goodness-of-fit measures, such as deviance, Pearson chi-square, Akaike information criterion (AIC) [3], Bayesian information criterion (BIC) [4], residual analysis, and overdispersion diagnostics [2]. These techniques help

* Corresponding author: Email: ashiqhb14@gmail.com

 DOI: 10.1201/9781003596455-12

assess model adequacy, detect potential issues like lack of fit, and determine whether adjustments, such as accounting for overdispersion, are necessary. By the end of this chapter, readers will gain a comprehensive understanding of how to apply these models to VL data and interpret their results effectively for public health decision-making.

12.2 STATISTICAL MODELS FOR VL PREDICTION

12.2.1 Poisson regression model

Poisson regression model is a type of generalized regression model [5] used to model count data, typically in cases where the outcome variable represents the number of occurrences of an event within a fixed period of time or space. The Poisson regression model is one of the most commonly used models for predicting count outcomes. In epidemiological studies of VL, count data often arise, such as the number of cases, the number of relapses, or the number of sandflies captured in a trap. For modeling such count data, the Poisson regression model is an effective statistical tool.

12.2.2 Poisson distribution and its role in Poisson regression

The Poisson distribution is a discrete probability distribution that expresses the probability of a given number of events (y) occurring in a fixed interval of time or space, assuming that:

i. Events occur independently.
ii. The rate of occurrence is constant.

The probability mass function (PMF) of the Poisson distribution is given as:

$$P(Y=y)=\frac{e^{-\lambda}\lambda^{y}}{y!},\ y=0,1,2,3,\ldots \tag{12.1}$$

where
λ is the average number of events in the given interval.
y is the number of occurrences of the event (observed count).

One of the key properties of Poisson distribution is that its mean and variance are equal, i.e.,

$$E(Y)=\lambda=V(Y)$$

In the Poisson regression context, we represent λ-dependent variable as a function of the predictor variables.

12.2.3 Assumptions of Poisson regression

i. The count variable (dependent variable) follows a Poisson distribution.
ii. The mean of the distribution is equal to its variance (i.e., the count data is equidispersed).
iii. The events are independent of each other.
iv. The logarithm of the expected value of the count is linearly related to the explanatory variables.

12.2.4 Poisson regression model equation

Poisson regression links the expected count (λ_i) to a set of explanatory variables using a log-link function. The general form of the Poisson regression model is given by:

$$\log(\lambda_i) = \beta_0 + \beta_1 x_{1i} + \beta_2 x_{2i} + \ldots + \beta_p x_{pi} \tag{12.2}$$

where

λ_i is the expected count of events for the ith observation.
β_0 is the intercept.
$\beta_1, \beta_2, \ldots, \beta_p$ are regression coefficients to be estimated.
$x_{1i}, x_{2i}, x_{3i}, \ldots, x_{pi}$ are predictors for the ith observation.

By exponentiating equation (12.2), we get the expected count for the ith observation as:

$$\lambda_i = e^{\beta_0 + \beta_1 x_{1i} + \beta_2 x_{2i} + \ldots + \beta_p x_{pi}} \tag{12.3}$$

12.2.5 Parameter estimation of Poisson regression

The estimation of parameters $(\beta_1, \beta_2, \ldots, \beta_p)$ in Poisson regression is estimated by the principle of Maximum Likelihood Estimation (MLE) [6]. This approach identifies the parameter values that maximize the likelihood of observing the given data under the assumed Poisson model.

In Poisson regression, the dependent variable Y_i is an observed count that follows the Poisson distribution with parameter $\lambda_i = e^{\beta_0 + \beta_1 x_{1i} + \beta_2 x_{2i} + i\beta_p x_{pi}}$ determined by the set of p predictors (independent variables) say $x_{1i}, x_{2i}, x_{3i}, \ldots, x_{pi}$, then the Poisson regression model for the ith observation is given by:

$$P(Y_i = y_i/\beta) = \frac{e^{-e^{\beta_0 + \beta_1 x_{1i} + \beta_2 x_{2i} + \ldots + \beta_p x_{pi}}} \left(e^{\beta_0 + \beta_1 x_{1i} + \beta_2 x_{2i} + \ldots + \beta_p x_{pi}}\right)^{y_i}}{y_i!} \tag{12.4}$$

For a set of n observations the likelihood function for a Poisson regression model is given by:

$$L(\beta_1, \beta_2, \ldots, \beta_p) = \prod_{i=1}^{n} \left(\frac{e^{-e^{\beta_0+\beta_1 x_{1i}+\beta_2 x_{2i}+\ldots+\beta_p x_{pi}}} \left(e^{\beta_0+\beta_1 x_{1i}+\beta_2 x_{2i}+\ldots+\beta_p x_{pi}}\right)^{y_i}}{y_i!} \right) \tag{12.5}$$

The corresponding Log-likelihood function for a Poisson regression model is given by:

$$\ln L(\beta_0, \beta_1, \beta_2, \ldots, \beta_p) = \sum_{i=1}^{n} \left\{ y_i \left(e^{\beta_0+\beta_1 x_{1i}+\beta_2 x_{2i}+\ldots+\beta_p x_{pi}}\right) - \left(e^{\beta_0+\beta_1 x_{1i}+\beta_2 x_{2i}+\ldots+\beta_p x_{pi}}\right) + \ln y_i! \right\}$$

To find the estimates of the parameters $(\beta_0, \beta_1, \beta_2, \ldots, \beta_p)$, we take the derivative of the log-likelihood function with respect to each β_j, $j = 0,1,2,\ldots,p$ and set it equal to zero, and solve for the β_j's. This process is typically performed numerically because the equations do not have closed-form solutions. In general, there are no closed-form solutions, so the ML estimates are obtained by using iterative algorithms such as Newton–Raphson (NR) and iteratively reweighted least squares (IRWLS) by using any of the statistical software like R, Python, and SPSS or STATA. By substituting the values of the parameters $(\beta_0, \hat{\beta}_1, \hat{\beta}_2, \ldots, \hat{\beta}_p)$ obtained through MLE in equation (12.2), we get the fitted model as:

$$\log(\hat{\lambda}_i) = \hat{\beta}_0 + \hat{\beta}_1 x_{1i} + \hat{\beta}_2 x_{2i} + \ldots + \hat{\beta}_p x_{pi} \tag{12.6}$$

Or, equivalently, the expected count for the ith observation as:

$$\hat{\lambda}_i = e^{\hat{\beta}_0+\hat{\beta}_1 x_{1i}+\hat{\beta}_2 x_{2i}+\ldots+\hat{\beta}_p x_{pi}} \tag{12.7}$$

12.2.6 Interpreting parameter estimates

The parameter estimates $\beta_0, \hat{\beta}_1, \hat{\beta}_2, \ldots, \hat{\beta}_p$ obtained through MLE are used to interpret the relationship between predictors and the dependent variable. The intercept β_0 represents the expected log count when all predictors are zero. The exponentiated coefficients (e^{β_j}, $j = 0,1,2,\ldots p$) represent the rate ratio (relative risk) associated with a one-unit increase in the corresponding predictor x_j, holding other predictors constant. If $e^{\beta_j} > 1$, then x_j is associated with an increase in the expected count. If $e^{\beta_j} < 1$, then x_j is associated with a decrease in the expected count.

12.2.7 Hypothesis testing for significance of predictors in Poisson regression

Poisson regression hypothesis testing helps us in determining whether a predictor (independent) variable has a significant impact on the response variable. In this case, we are testing whether each predictor's regression coefficient is significantly different from zero. The following steps are typically included in the hypothesis testing process:

For each predictor (x_j, $j = 1, 2, 3, \ldots, p$) in the model, we formulate the null and alternative hypothesis as:

H_0: $\beta_j = 0, j = 1, 2, 3, \ldots, p$ (i.e., the predictor x_j has no effect on the response variable).
H_1: $\beta_j \neq 0, j = 1, 2, 3, \ldots, p$ (i.e., the predictor x_j has significant effect on the response variable).

To test this hypothesis, we will use the Wald test.

Under the null hypothesis, the test statistic for the predictor x_j is given by:

$$W = \frac{\hat{\beta}_j}{S.E\left(\hat{\beta}_j\right)}$$

where $\hat{\beta}_j$, is the estimated coefficient for predictor x_j and $S.E(\hat{\beta}_j)$ is the standard error of $\hat{\beta}_j$.

Under the null hypothesis, the test statistic W follows a standard normal distribution. The p-value for the Wald test is computed as:

$$p = 2P(Z < w)$$

where w is the observed value of Wald test statistic [7]. If the p-value is smaller than the chosen significance level $(\alpha = 0.05)$, we reject the null hypothesis and conclude that the predictor x_j is statistically significant. Otherwise we fail to reject the null and conclude that the predictor x_j is not statistically significant.

12.2.8 Goodness-of-fit assessment

A Poisson regression model's goodness of fit is evaluated using a number of measures that evaluate how well the model handles the data. These include deviance, Pearson chi-square, information criteria (AIC and BIC), residual analysis, and overdispersion measure, which collectively help to assess if the model is appropriate and identify potential issues such as lack of fit or overdispersion.

12.2.8.1 Deviance

Deviance measures the difference between the fitted model and the saturated model (a model that perfectly fits the data). It is often used to evaluate how much of the variation in the response (dependent) variable remains unexplained by the fitted model. The formula for deviance (D) in Poisson regression is given by:

$$D = 2\sum_{i=1}^{n}\left[y_i \ln\left(\frac{y_i}{\hat{\lambda}_i}\right) - \left(y_i - \hat{\lambda}_i\right)\right] \tag{12.8}$$

where

y_i is the observed count for the ith observation.

$\hat{\lambda}_i$ is the fitted (predicted) count for the ith observation from the model.

n is the number of observations.

A smaller deviance indicates a better fit. Deviance measure can be compared with a chi-squared distribution with $(n-p)$, degrees of freedom, where p is the number of model parameters. A large deviance relative to the chi-squared distribution indicates the lack of fit.

12.2.8.2 Pearson chi-square

Another goodness-of-fit measure that examines how well the observed data fit the model's predictions is the Pearson chi-square. The formula for the Pearson chi-square (χ^2) in Poisson regression is given by:

$$\chi^2 = \sum_{i=1}^{n}\frac{\left(y_i - \hat{\lambda}_i\right)^2}{\hat{\lambda}_i}$$

where

y_i is the observed count for the ith observation.

$\hat{\lambda}_i$ is the fitted (predicted) count for the ith observation from the model.

n is the number of observations.

If the Pearson chi-square (χ^2) value is approximately equal to the degrees of freedom (df), the model indicates a better fit. However, if the Pearson chi-square (χ^2) value is significantly larger than the (df), it suggests a lack of fit, which could be caused by overdispersion or model misspecification. However, if the Pearson chi-square is much smaller than the (df), it may indicate that the model is overfitting the data.

12.2.8.3 Overdispersion measure

A key assumption in Poisson regression is that the variance equals the mean i.e. ($V(Y_i) = \lambda_i$), when the variance is greater than the mean, then it is called overdispersion, it suggests that the Poisson model may not be appropriate. A key measure for assessing overdispersion is the dispersion statistic (Φ) which is calculated by using the following formula:

$$\Phi = \frac{D}{df} = \frac{2\sum_{i=1}^{n}\left[y_i \ln\left(\frac{y_i}{\hat{\lambda}_i}\right) - \left(y_i - \hat{\lambda}_i\right)\right]}{df} \tag{12.9}$$

where df is degrees of and is equal to the number of observations minus the number of estimated parameters in the model.

If $\Phi \approx 1$, then the model fits well, variance matches the Poisson assumption and the Poisson model is appropriate, if $\Phi > 1$, overdispersion is present, (variance > mean), and if $\Phi < 1$, underdispersion is present, (varianc < mean). If overdispersion is present, then the alternative models, such as quasi-Poisson regression or NB regression, should be used to handle overdispersion.

12.2.8.4 Akaike information criterion (AIC)

The AIC [3] is a widely used metric that evaluates a model's goodness of fit while penalizing for model complexity to prevent overfitting. The formula for AIC in Poisson regression is given by:

$$AIC = 2p - 2\ln L(\beta) \tag{12.10}$$

where p is the number of parameters in the model and $\ln L(\beta)$ is log-likelihood of the fitted model.

Lower AIC values indicate better model fit and AIC is useful for comparing multiple models the model with the lowest AIC is preferred.

12.2.8.5 Bayesian information criterion (BIC)

The BIC [4] is similar to AIC but imposes a stronger penalty on model complexity, particularly when the sample size is large. The formula for BIC in Poisson regression is given by:

$$BIC = p\ln(n) - 2\ln L(\beta) \tag{12.11}$$

where p is the number of parameters in the model, $\ln L(\beta)$ is log-likelihood of the fitted model, and n is the number of observations in the dataset.

Like AIC, smaller BIC values indicate better fit and BIC is more conservative than AIC when the sample size is large. BIC is especially useful when comparing multiple models the model with the lowest BIC is preferred as it adequately fits the data.

12.2.8.6 Residual Analysis

Residuals in Poisson regression help diagnose potential model misspecification, lack of fit, or presence of outliers. They quantify the difference between observed and predicted values and should ideally follow a standard normal distribution if the model is correctly specified. In Poisson regression, two commonly used residuals are:

1. **Deviance Residuals:** The deviance residuals are derived from the deviance function, which measures the discrepancy between the observed and expected values based on the fitted model. The formula for the deviance residual for each observation i is given by:

$$r_i^{dev} = sign\left(y_i - \hat{\lambda}_i\right)\sqrt{2\left[y_i \ln\left(\frac{y_i}{\hat{\lambda}_i}\right) - \left(y_i - \hat{\lambda}_i\right)\right]} \tag{12.12}$$

where

y_i is the observed count for the ith observation.

$\hat{\lambda}_i$ is the fitted (predicted) count for the ith observation from the model.

$sign(y_i - \hat{\lambda}_i)$ is the sign function, which returns +1 if $y_i > \hat{\lambda}_i$, −1 if $y_i < \hat{\lambda}_i$ and 0 if $y_i = \hat{\lambda}_i$.

Values close to 0 suggest a good fit for that observation; large residuals (in magnitude) indicate potential outliers or poor model fit. A random spread of residuals without a pattern indicates an appropriate model.

2. **Pearson Residuals:** Pearson residuals quantify the standardized difference between observed and expected counts, making them a useful diagnostic tool for model fit assessment. The formula for the Pearson residual for each observation i is given by:

$$r_i^{pear} = \frac{\left(y_i - \hat{\lambda}_\mathrm{i}\right)}{\sqrt{\hat{\lambda}_i}} \tag{12.13}$$

where

y_i is the observed count for the ith observation.

$\hat{\lambda}_i$ is the fitted (predicted) count for the ith observation from the model.

If the Poisson model is well-fitted, Pearson residuals approximately follow standard normal distribution, i.e., $r_i^{pear} \sim N(0,1)$ and residuals with large magnitudes, i.e., $\left(\left|r_i^{pear}\right| > 2\right)$ indicate potential lack of fit or outliers.

12.2.9 Negative binomial regression model

A modified version of the Poisson regression model is the NB regression model [5, 8], that handles the problem of overdispersion, which occurs when the variance of count data exceeds the mean. Real-world count data, such as epidemiological studies of VL, frequently reveal the overdispersion, when the number of cases, relapses, or sandflies collected in a trap varies in excess of what a Poisson model predicts. Poisson regression has a tendency to underestimate standard errors in the presence of overdispersion, which can result in misleading statistical results. By introducing an additional dispersion parameter, the NB regression model offers a more adaptable option that permits variance to exceed mean. When dealing with diverse count data, this approach is quite helpful.

12.2.10 Negative Binomial distribution and Its Role in Negative Binomial Regression

The NB distribution is a discrete probability distribution with multiple parameterizations, depending on the context and application. Below, we outline the most common parameterizations along with their corresponding PMFs, mean, and variance.

1. **Standard Form (Failures before r^{th} Success):** This parameterization considers the number of failures Y before achieving r successes in a sequence of independent Bernoulli trials. The probability of success in each trial is denoted as p. The standard form of PMF of the NB distribution is given by:

$$P(Y=y)=\binom{y+r-1}{y}p^r(1-p)^y,\ y=0,1,2,\ldots \tag{12.14}$$

As we know that $\binom{y+r-1}{y}=\binom{y+r-1}{r-1}$ then (12.14) can equivalently be written as:

$$P(Y=y)=\binom{y+r-1}{r-1}p^r(1-p)^y,\ y=0,1,2,\ldots$$

where

y is the number of failures before r^{th} success.
r is the number of successes required.
p is the probability of success in each independent trial.

In this case, the mean and variance of the NB distribution is given by:

$$Mean = E(Y) = \frac{r(1-p)}{p}, \text{ and } Variance = V(Y) = \frac{r(1-p)}{p^2}$$

2. **Alternative Form (Total number of trials to achieve *r* Successes):** This parameterization considers the number of number of trials Y before achieving r successes in a sequence of independent Bernoulli trials. The probability of success in each trial is denoted as p. The alternative form of PMF of the NB distribution is given by:

$$P(Y = y) = \binom{y-1}{r-1} p^r (1-p)^{y-r}, \; y = r, r+1, r+2, > \tag{12.15}$$

where

y is the number of trials required to reach r success.
r is the number of successes required.
p is the probability of success in each independent trial.

In this case the mean and variance of the NB distribution is given by:

$$Mean = E(Y) = \frac{r}{p}, \text{ and } Variance = V(Y) = \frac{r(1-p)}{p^2}$$

3. **Alternative Parameterization Using Mean and Dispersion:** This parameterization using mean count μ and dispersion parameter $r = \frac{1}{\theta}$, and it is commonly used in statistical modeling and regression, especially in cases where count data exhibits overdispersion (variance greater than the mean). The PMF of the NB distribution in terms of mean and dispersion parameter is given by:

$$P(Y = y) = \binom{y+r-1}{y} \left(\frac{r}{\mu + r}\right)^r \left(\frac{\mu}{\mu + r}\right)^y, \; y = 0, 1, 2, \ldots \tag{12.16}$$

where

y is the number of occurrences of the event (observed count)
μ is the expected number of occurrences (expected count)
$r = \frac{1}{\theta}$ is the dispersion parameter in this case.

In this case the mean and variance of the NB distribution is given by:

$$Mean = E(Y) = \mu, \text{ and } Variance = V(Y) = \mu + \theta\mu^2$$

For small values of θ, the variance of the distribution asymptotically approaches the mean μ, leading to the convergence of the NBD distribution to a Poisson form.

In NB regression context, we represent μ dependent variable as a function of predictor variables while allowing the variance to vary as a function μ and θ.

12.2.11 Assumptions of negative binomial regression

i. The dependent variable (count data) follows a NB distribution.
ii. Overdispersion is present in the data, i.e., $Var(Y) > E(Y)$.
iii. The events are independent of each other.
iv. The logarithm of the expected count is linearly related to the explanatory variables.

12.2.12 Negative binomial regression model equation

NB regression [8] links the expected count (μ_i) to a set of explanatory variables using a log-link function similar to Poisson regression. The general form of the NB regression model is given by:

$$\log(\mu_i) = \beta_0 + \beta_1 x_{1i} + \beta_2 x_{2i} + \ldots + \beta_p x_{pi} \tag{12.17}$$

where

μ_i is expected count of events for the ith observation.
β_0 is the intercept.
$\beta_1, \beta_2, \ldots, \beta_p$ are regression coefficients to be estimated.
$x_{1i}, x_{2i}, x_{3i}, \ldots, x_{pi}$ are predictors for the ith observation.

By exponentiating equation (12.17), we get the expected count for the ith observation as:

$$\mu_i = e^{\beta_0 + \beta_1 x_{1i} + \beta_2 x_{2i} + \ldots + \beta_p x_{pi}} \tag{12.18}$$

12.2.13 Parameter estimation of negative binomial regression

The parameters ($\beta_1, \beta_2, \ldots, \beta_p$, θ) in NB regression are also estimated by the principle of MLE [6].

In NB regression, the dependent variable Y_i is an observed count that follows the NB distribution with parameter $\mu_i = e^{\beta_0 + \beta_1 x_{1i} + \beta_2 x_{2i} + i\beta_p x_{pi}}$ determined by the set of p predictors (independent variables) say $x_{1i}, x_{2i}, x_{3i}, \ldots, x_{pi}$, then the NB model for ith observation is given by:

$$P(Y_i = y_i) = \begin{pmatrix} y_i + r - 1 \\ y_i \end{pmatrix} \left(\frac{r}{\mu_i + r} \right)^r \left(\frac{\mu_i}{\mu_i + r} \right)^{y_i} \tag{12.19}$$

For a set of n observations, the likelihood function for a NB regression model is given by:

$$L(\beta_1, \beta_2, \ldots, \beta_p, \theta) = \prod_{i=1}^{n} \left(\begin{pmatrix} y_i + r - 1 \\ y_i \end{pmatrix} \left(\frac{r}{\mu_i + r} \right)^r \left(\frac{\mu_i}{\mu_i + r} \right)^{y_i} \right) \quad (12.20)$$

Equation (12.20) can equivalently be written as:

$$L(\beta_1, \beta_2, \ldots, \beta_p, \theta) = \prod_{i=1}^{n} \left(\frac{\Gamma(y_i + r)}{\Gamma(r) y_i!} \left(\frac{r}{\mu_i + r} \right)^r \left(\frac{\mu_i}{\mu_i + r} \right)^{y_i} \right) \quad (12.21)$$

The corresponding Log-likelihood function for a NB regression model is given by:

$$\ln L(\beta_0, \beta_1, \beta_2, \ldots, \beta_p, \theta)$$
$$= \sum_{i=1}^{n} \left\{ \ln(\Gamma(y_i + r)) - \ln \Gamma(r) - \ln y_i! + r \ln(r) - (y_i + r) \ln(\mu_i + r) + y_i \ln(\mu_i) \right\}$$

where $\mu_i = e^{\beta_0 + \beta_1 x_{1i} + \beta_2 x_{2i} + \ldots + \beta_p x_{pi}}, r = \frac{1}{\theta}$, and Γ represents the gamma function.

To find the estimates of the parameters $(\beta_0, \beta_1, \beta_2, \ldots, \beta_p, \theta)$, we take the derivative of the log-likelihood function with respect to each β_j, $j = 0, 1, 2, \ldots, p$ and θ and set equal to zero, and solve for the β_j's and θ. This process is typically performed numerically because the equations do not have closed-form solutions. In general, there are no closed-form solutions, so the ML estimates are obtained by using iterative algorithms such as NR and IRWLS by using any the statistical software's like R, Python, and SPSS or STATA. By substituting the values of the parameters $(\beta_0, \hat{\beta}_1, \hat{\beta}_2, \ldots, \hat{\beta}_p)$ obtained through MLE in equation (12.17), we get the fitted model as:

$$\log(\hat{\mu}_i) = \hat{\beta}_0 + \hat{\beta}_1 x_{1i} + \hat{\beta}_2 x_{2i} + \ldots + \hat{\beta}_p x_{pi} \quad (12.22)$$

Or, equivalently, the expected count for the ith observation as:

$$\hat{\mu}_i = e^{\hat{\beta}_0 + \hat{\beta}_1 x_{1i} + \hat{\beta}_2 x_{2i} + \ldots + \hat{\beta}_p x_{pi}} \quad (12.23)$$

12.2.14 Interpreting parameter estimates

The parameter estimates $\beta_0, \hat{\beta}_1, \hat{\beta}_2, \ldots, \hat{\beta}_p$ of NB regression obtained through MLE are used to interpret the relationship between predictors and the dependent variable similar to Poisson regression. The intercept β_0 represents the expected log count when all predictors are zero. The exponentiated coefficients (e^{β_j}, $j = 0, 1, 2, \ldots p$) represent the rate ratio (relative risk) associated with a one-unit increase in the corresponding predictor x_j, holding other predictors constant. If $e^{\beta_j} > 1$, then x_j is associated with an increase in the expected count. If $e^{\beta_j} < 1$, then x_j is associated with a decrease in the expected count. The dispersion parameter $\theta = \frac{1}{r}$, quantifies overdispersion, if $\theta = 0$, the model reduces to Poisson model and if $\theta > 0$, overdispersion is present and the NB regression is a better choice than Poisson regression.

12.2.15 Hypothesis testing for significance of predictors in negative binomial regression

NB regression hypothesis testing helps us in determining whether a predictor (independent) variable has a significant impact on the response variable similar as in Poisson regression. To test whether each predictor's regression coefficient is significantly different from zero, we use Wald test statistic [7]. The steps that are typically included in the hypothesis testing process are same as discussed above in section 12.2.7.

12.2.16 Goodness-of-fit assessment

NB regression model's goodness of fit is evaluated using a number of measures that evaluate how well the model handles the data. These include deviance, information criteria (AIC and BIC), residual analysis, which collectively helps to assess if the model is appropriate and identify potential issues such as lack of fit.

12.2.16.1 Deviance

Deviance measures the difference between the fitted model and the saturated model (a model that perfectly fits the data). It is often used to evaluate how much of the variation in the response (dependent) variable remains unexplained by the fitted model. The formula for deviance (D) in NB regression is given by:

$$D = 2\sum_{i=1}^{n}\left[y_i \ln\left(\frac{y_i}{\hat{\mu}_i}\right) - (y_i + r)\ln\left(\frac{y_i + r}{\hat{\mu}_i + r}\right)\right] \tag{12.24}$$

where

y_i is the observed count for the ith observation.

$\hat{\mu}_i$ is the fitted (predicted) count for the ith observation from the model.

$r = \frac{1}{\theta}$ is the dispersion parameter.

n is the number of observations.

Similar to Poisson regression model, a smaller deviance measure of NB regression indicates better fit. Deviance measure can also be compared with a chi-squared distribution with $(n-p)$, degrees of freedom, where p is the number of model parameters. A large deviance relative to chi-squared distribution indicates the lack of fit.

12.2.16.2 Pearson chi-square

Another goodness-of-fit measure that examines how well the observed data fit the model's predictions is the Pearson chi-square. The formula for Pearson chi-square (χ^2) in NB regression is given by:

$$\chi^2 = \sum_{i=1}^{n} \frac{\left(y_i - \hat{\mu}_i\right)^2}{\hat{\mu}_i}$$

where

y_i is the observed count for the ith observation.

$\hat{\mu}_i$ is the fitted (predicted) count for the ith observation from the model.

n is the number of observations.

Similar to Poisson regression model if the Pearson chi-square (χ^2) value of NB regression is approximately equal to the degrees of freedom (df), the model indicates better fit. However, if the Pearson chi-square (χ^2) value is significantly larger than the (df), it suggests lack of fit, which could be caused by overdispersion or model misspecification. However, if the Pearson chi-square is much smaller than the (df), it may indicate that the model is overfitting the data.

12.2.16.3 Akaike information criterion (AIC)

The AIC [3] is a widely used metric that evaluates a model's goodness of fit while penalizing for model complexity to prevent overfitting. The formula for AIC in NB regression is given by:

$$AIC = 2p - 2\ln L(\beta, \theta) \tag{12.25}$$

where p is the number of parameters in the model and $\ln L(\beta, \theta)$ is log-likelihood of the fitted model.

Lower AIC values indicate better model fit, and AIC is useful for comparing multiple models; the model with the lowest AIC is preferred.

12.2.16.4 Bayesian information criterion (BIC)

The BIC [4] is similar to AIC but imposes a stronger penalty on model complexity, particularly when the sample size is large. The formula for BIC in NB regression is given by:

$$BIC = p\ln(n) - 2\ln L(\beta, \theta) \tag{12.26}$$

where p is the number of parameters in the model, $\ln L(\beta,\theta)$ is log-likelihood of the fitted model and n is the number of observations in the dataset.

Like AIC, smaller BIC values indicate better fit, and BIC is more conservative than AIC when the sample size is large. BIC is especially useful when comparing multiple models; the model with the lowest BIC is preferred as it adequately fits the data.

12.2.16.5 Residual analysis

Similar to Poisson regression, residuals in NB regression help diagnose potential model misspecification, lack of fit, or presence of outliers. They quantify the difference between observed and predicted values and should ideally follow a standard normal distribution if the model is correctly specified. In NB regression, two commonly used residuals are:

i. **Deviance Residuals:** The deviance residuals are derived from the deviance function, which measures the discrepancy between the observed and expected values based on the fitted model. The formula for the deviance residual for each observation i in the case of NB regression is given by:

$$r_i^{dev} = sign(y_i - \hat{\mu}_i)\sqrt{2\left[y_i \ln\left(\frac{y_i}{\hat{\mu}_i}\right) - (y_i + r)\ln\left(\frac{y_i + r}{\hat{\mu}_i + r}\right)\right]} \tag{12.27}$$

where

y_i is the observed count for the ith observation.

$\hat{\mu}_i$ is the fitted (predicted) count for the ith observation from the model.

$r = \frac{1}{\theta}$ is the dispersion parameter.

$sign(y_i - \hat{\mu}_i)$ is the sign function, which returns +1 if $y_i > \hat{\mu}_i$, −1 if $y_i < \hat{\mu}_i$ and 0 if $y_i = \hat{\mu}_i$.

Values close to 0 suggest a good fit for that observation; large residuals (in magnitude) indicate potential outliers or poor model fit. A random spread of residuals without a pattern indicates an appropriate model.

ii. **Pearson Residuals:** Pearson residuals quantify the standardized difference between observed and expected counts, making them a useful diagnostic tool for model fit assessment. The formula for the Pearson residual for each observation i in NB regression is given by:

$$r_i^{pear} = \frac{(y_i - \hat{\mu}_i)}{\sqrt{\hat{\mu}_i + \theta\hat{\mu}_i^2}} \tag{12.28}$$

where

y_i is the observed count for the ith observation.

$\hat{\mu}_i$ is the fitted (predicted) count for the ith observation from the model.

$r = \frac{1}{\theta}$ is the dispersion parameter.

If the NB is well-fitted, Pearson residuals approximately follow standard normal distribution, i.e., $r_i^{pear} \sim N(0,1)$ and residuals with large magnitudes, i.e., $\left(\left|r_i^{pear}\right| > 2\right)$ indicate potential lack of fit or outliers. If the residuals mentioned above still indicate a lack of fit, it may be necessary to utilize zero-inflated models in order to improve the accuracy.

iii. **Application of Poisson and Negative Binomial Regression:** Poisson and NB regression models are commonly used to analyze count data, where the response variable indicates the number of occurrences (count) of an event. In this section, we will fit these two regression models to real data to illustrate their practical use and compare their performance. To perform all the computations related to these models, R-Software [9] will be used.

The dataset utilized for the above-mentioned regression models comprises multiple variables; the names of these variables, along with their description, type, sources, and their respective roles in the analysis, are presented in Tables 12.1 and 12.2.

Based on the dataset collected for this study, the following table presents the descriptive analysis to see the characteristics of each variable used in this study.

12.2.17 Fitting of Poisson regression model

In this section, we will first fit the Poisson regression model to the dataset discussed above, as the response (dependent) variable VL Cases Y_t is count data, and all independent (predictors) $(T_{ct}, X_{1t}, X_{2t}, X_{3t}, X_{4t}, X_{5t}, X_{6t}, X_{7t}, X_{8t})$ are continuous numeric. Suppose that the number of VL Cases (Y_t) follows a Poisson distribution with parameter λ_t (expected number of VL Cases in year t), then the Poisson regression model can be written as:

$$\log(\lambda_t) = \beta_0 + \gamma \boldsymbol{T}_{ct} + \beta_1 \boldsymbol{X}_{1t} + \beta_2 \boldsymbol{X}_{2t} + \beta_3 \boldsymbol{X}_{3t} + \beta_4 \boldsymbol{X}_{4t} + \beta_5 \boldsymbol{X}_{5t} + \beta_6 \boldsymbol{X}_{6t} + \beta_7 \boldsymbol{X}_{7t} + \beta_8 \boldsymbol{X}_{8t}$$

Or, equivalently, we can write the above model as:

$$\lambda_t = e^{\beta_0 + \gamma T_{ct} + \beta_1 X_{1t} + \beta_2 X_{2t} + \beta_3 X_{3t} + \beta_4 X_{4t} + \beta_5 X_{5t} + \beta_6 X_{6t} + \beta_7 X_{7t} + \beta_8 X_{8t}}$$

Based on the dataset collected for this study, the parameter estimation of the Poisson regression model, along with the standard error, Wald test

Table 12.1 Description of variables used in count data models for predicting visceral leishmaniasis cases in Bihar (2001–2013)

Variable and notation	*Description*	*Type*	*Role*	*Source*
Year (T)	Year of observation of VL cases during 2001–2013.	Continuous numeric	Identifier	Yearly VL cases of Bihar were collected by the National Vector Borne Disease Control Programme (NVBDCP), Government of India. [10]
VL cases ($\boldsymbol{Y}_t$)	Number of reported VL cases during 2001–2013.	Count integer	Response (dependent) variable	
Centered time ($\boldsymbol{T}_{ct}$)	Time centered at the base year 2001 ($T - 2001$)	Continuous numeric	Predictor (independent) variable	Derived from Year (T) as ($T - 2001$)
Annual average temperature ($\boldsymbol{X}_{1t}$)	Annual average temperature in degrees Celsius (°C) corresponding to the given years.	Continuous numeric	Predictor (independent) variable	The data for the variables mentioned at S. No's 3–10 were accessed from an open-source repository OSF dataset [11], [12]
Average monsoon temperature ($\boldsymbol{X}_{2t}$)	Average temperature in degrees Celsius (°C) during monsoon corresponding to the given years.	Continuous numeric	Predictor (independent) variable	
Average temperature sandfly peak ($\boldsymbol{X}_{3t}$)	Average temperature in degrees Celsius (°C) during sandfly peak season corresponding to the given years.	Continuous numeric	Predictor (independent) variable	
Average monsoon humidity ($\boldsymbol{X}_{4t}$)	Average specific humidity in (kg/kg) during Monsoon corresponding to the given years.	Continuous numeric	Predictor (independent) variable	
Sandfly peak average humidity ($\boldsymbol{X}_{5t}$)	Average specific humidity in kg/kg during sandfly peak corresponding to the given years.	Continuous numeric	Predictor (independent) variable	
Annual rainfall ($\boldsymbol{X}_{6t}$)	Annual rainfall in (cm) corresponding to the given years	Continuous numeric	Predictor (independent) variable	
Annual monsoon rainfall ($\boldsymbol{X}_{7t}$)	Annual rainfall in (cm) during monsoon corresponding to the given years.	Continuous numeric	Predictor (independent) variable	
Sandfly peak annual rainfall ($\boldsymbol{X}_{8t}$)	Annual Rainfall in (cm) during sandfly peak corresponding to the given years.	Continuous numeric	Predictor (independent) variable	

Table 12.2 Descriptive statistics for key variables in the study (2001–2013)

Variables	*N*	*Mean*	*Standard deviation*	*Minimum*	*Maximum*
VL cases (Y_t)	13	20,546.07	8649.608	9505	37,822
Annual average temperature (X_{1t})	13	21.894	0.325	21.480	22.520
Average monsoon temperature (X_{2t})	13	25.348	0.433	24.890	26.460
Average temperature sandfly peak (X_{3t})	13	23.753	0.506	23.260	24.880
Average monsoon humidity (X_{4t})	13	19.692	0.480	19.000	20.000
Sandfly peak average humidity (X_{5t})	13	12.231	0.599	11.000	13.000
Annual rainfall (X_{6t})	13	94.852	19.132	72.325	132.542
Annual monsoon rainfall (X_{7t})	13	231.929	54.444	180.975	343.225
Sandfly peak annual rainfall (X_{8t})	13	60.731	21.221	26.650	94.033

statistic, p-value, and 95% confidence interval limits for each parameter, are presented in Table 12.3.

Based on the results of Table 12.2, the fitted Poisson regression can be written as:

$$log\left(\hat{\lambda}_t\right) = 3.0111 + 0.0469T_{ct} + 0.7939X_{1t} + 0.0142X_{2t} - 0.6904X_{3t} + 0.0507X_{4t} + 0.3221X_{5t} - 0.0177X_{6t} + 0.0097X_{7t} - 0.0047X_{8t}$$

Or, equivalently, we can write the fitted model as:

$$\hat{\lambda}_t = e^{3.0111+0.0469T_{ct}+0.7939X_{1t}+0.0142X_{2t}-0.6904X_{3t}+0.0507X_{4t}+0.3221X_{5t}-0.0177X_{6t}+0.0097X_{7t}-0.0047X_{8t}}$$

This is the fitted Poisson regression model to the defined data set. Based on this fitted Poisson regression model, we can predict the expected number of

Table 12.3 Parameter estimation of Poisson regression model

Predictors	*Regression coefficients*	*S.E.*	*Wald test statistic*	*p*	*Lower limit*	*Upper limit*
(Intercept)	3.0111	0.3169	9.5014	<0.001	2.3899	3.6322
T_{ct}	0.0469	0.0011	44.6711	<0.001	0.0449	0.0490
X_{1t}	0.7939	0.0158	50.0991	<0.001	0.7629	0.8250
X_{2t}	0.0143	0.0240	0.5954	0.552	−0.0327	0.0613
X_{3t}	−0.6904	0.0193	−35.8042	<0.001	−0.7282	−0.6526
X_{4t}	0.0507	0.0083	6.1131	<0.001	0.0345	0.0670
X_{5t}	0.3221	0.0149	21.6715	<0.001	0.2930	0.3513
X_{6t}	−0.0177	0.0008	−22.4477	<0.001	−0.0192	−0.0161
X_{7t}	0.0097	0.0002	44.0696	<0.001	0.0092	0.0101
X_{8t}	−0.0047	0.0004	−12.2546	<0.001	−0.0055	−0.0040

VL cases by substituting the values of the predictor variables corresponding to that year in the model.

Now, we will discuss which independent (predictor) variables are significant in the Poisson regression model. Based on the results of Table 12.3, we observe that the p-values corresponding to the intercept and the independent (predictors) variables centered time (T_{ct}), annual average temperature (X_{1t}), average temperature sandfly peak (X_{3t}), average monsoon humidity (X_{4t}), sandfly peak average humidity (X_{5t}), annual rainfall (X_{6t}), annual monsoon rainfall (X_{7t}), and sandfly peak annual rainfall (X_{8t}) are less than 0.05, so with 95% confidence they are statistically significant in the model and the p-value for the predictor average monsoon temperature (X_{2t}) is greater than 0.05, so it is not statistically significant in the model for this dataset.

Now, based on the dataset defined for this study, the goodness of fit measures like deviance, Pearson chi-square, AIC, and BIC of the fitted Poisson regression model are presented in Table 12.4.

From the results of Table 12.4, we observe that the deviance and Pearson chi-square values are significantly greater than the degrees of freedom, suggesting that the Poisson model does not provide an adequate fit to the dataset.

Also, by comparing the deviance and Pearson chi-square values with the theoretical chi-square value for 3 degrees of freedom at 5% level of significance, we observe that both deviance and Pearson chi-square values are significantly greater than ($\chi^2_{(0.05,3)} = 7.815$), suggesting that the Poisson model does not provide an adequate fit to the dataset.

Also, from the results of Table 12.4, we observe that the overdispersion measure of the Poisson regression model based on deviance and Pearson chi-square are ($\Phi = 1574.567 > 1$) and ($\Phi = 1514.781 > 1$), respectively, which are significantly higher than 1, so we conclude that overdispersion is present in the Poisson regression model and the Poisson model is not appropriate fit to the dataset, so the alternative models like NB regression should be used to handle the overdispersion. Next, we fit the NB regression to the dataset, as the Poisson regression analysis shows that the model is not an appropriate fit to the dataset.

Table 12.4 Goodness of fit statistical measures for Poisson Regression Model

Goodness of fit measures	*Value*	*Degrees of freedom (df)*	*Overdispersion* (Φ)
Deviance	4723.702	3	1574.567
Pearson chi-square	4544.344	3	1514.781
Log-likelihood	2437.776		
Akaike information criterion (AIC)	4895.551		
Bayesian information criterion (BIC)	4901.201		

12.2.18 Fitting of negative binomial regression model

In this section, we will fit the NB regression model as the Poisson regression to the dataset is not appropriate due to the presence of overdispersion in the dataset. Suppose that the number of VL Cases (Y_t) follows a NB distribution with parameters μ_t (expected number of VL Cases in year t) and dispersion (scale) parameter $r = \frac{1}{\theta}$, then the NB regression model can be written as:

$$\log(\mu_t) = \beta_0 + \gamma T_{ct} + \beta_1 X_{1t} + \beta_2 X_{2t} + \beta_3 X_{3t} + \beta_4 X_{4t} + \beta_5 X_{5t} + \beta_6 X_{6t} + \beta_7 X_{7t} + \beta_8 X_{8t}$$

Or, equivalently, we can write the above model as:

$$\mu_t = e^{\beta_0 + \gamma T_{ct} + \beta_1 X_{1t} + \beta_2 X_{2t} + \beta_3 X_{3t} + \beta_4 X_{4t} + \beta_5 X_{5t} + \beta_6 X_{6t} + \beta_7 X_{7t} + \beta_8 X_{8t}}$$

Based on the dataset collected for this study, the parameter estimation of the NB regression model, along with the standard error, Wald test statistic, p-value, and 95% confidence interval limits for each parameter, is presented in Table 12.5.

Based on the results of Table 12.5, the fitted NB regression can be written as:

$$\log(\hat{\mu}_t) = 2.2634 + 0.0456T_{ct} + 0.8476X_{1t} - 0.0123X_{2t} - 0.6248X_{3t} - 0.0055X_{4t} + 0.2937X_{5t} - 0.0275X_{6t} + 0.0.0134X_{7t} - 0.0013X_{8t}$$

Or, equivalently, we can write the fitted model as:

$$\hat{\mu}_t = e^{2.2634+0.0456T_{ct}+0.8476X_{1t}-0.0123X_{2t}-0.6248X_{3t}-0.0055X_{4t}+0.2937X_{5t}-0.0275X_{6t}+0.0.0134X_{7t}-0.0013X_{8t}}$$

This is the fitted NB regression model to the defined data set. Based on this fitted NB regression model, we can predict the expected number of VL cases

Table 12.5 Parameter estimation of negative binomial regression model

Predictors	*Regression coefficients*	*S.E.*	*Wald test statistic*	*p*	*Lower limit*	*Upper limit*
(Intercept)	2.2634	6.5063	0.3479	0.7279	−10.8931	15.3600
T_{ct}	0.0456	0.0214	2.1338	0.0329	0.0057	0.0866
X_{1t}	0.8476	0.3567	2.3763	0.0175	0.1538	1.5606
X_{2t}	−0.0123	0.4166	−0.0294	0.9765	−0.8319	0.8062
X_{3t}	−0.6248	0.3652	−1.7106	0.0872	−1.3309	0.0734
X_{4t}	−0.0055	0.1690	−0.0328	0.9738	−0.3345	0.3212
X_{5t}	0.2937	0.2559	1.1478	0.2510	−0.1981	0.7872
X_{6t}	−0.0275	0.0165	−1.6630	0.0963	−0.0600	0.0057
X_{7t}	0.0134	0.0046	2.9097	0.0036	0.0041	0.0227
X_{8t}	−0.0013	0.0078	−0.1684	0.8663	−0.0168	0.0139

Table 12.6 Goodness of fit statistical measures for Negative Binomial Regression Model

Goodness of fit measures	*Value*	*Degrees of freedom (df)*
Deviance	13.05	2
Pearson chi-square	12.09	2
Log-likelihood	121.611	
Akaike information criterion (AIC)	265.22	
Bayesian information criterion (BIC)	271.436	
Dispersion (scale) parameter (r)	45.801	

by substituting the values of the predictor variables corresponding to that year in the model.

Now, we will discuss which independent (predictors) variables are significant in the NB regression model. Based on the results of Table 12.5, we observe that the p-values corresponding to the independent (predictors) variables centered time (T_{ct}), annual average temperature (X_{1t}), and annual monsoon rainfall (X_{7t}) are less than 0.05, so with 95% confidence they are statistically significant in the model and the p-values corresponding to the intercept and the independent (predictors) variables average monsoon temperature (X_{2t}), average temperature sandfly peak (X_{3t}), average monsoon humidity (X_{4t}), sandfly peak average humidity (X_{5t}), annual rainfall (X_{6t}), and sandfly peak annual rainfall (X_{8t}) are greater than 0.05, so they are not statistically significant in the model for this dataset.

Now, based on the dataset defined for this study, the goodness of fit measures like deviance, Pearson chi-square, AIC, and BIC of the fitted NB regression model are presented in Table 12.6.

From the results of Table 12.6, we observe that the deviance value is 13.05 with 2 degrees of freedom (df) and the Pearson chi-square is 12.09 with 2 df, resulting in a deviance-to-df ratio of 6.525 and a chi-square-to-df ratio of 6.045, these ratios suggesting that the NB regression model may not provide an adequate fit to the dataset, potentially due to the lack of important predictors or factors or the presence of greater variability in the dataset than the model assumptions allows. This is further supported by the high dispersion parameter $r = 45.801$, which indicates that the variability in the dataset exceeds what the NB regression model assumes. These findings suggest that adjustments, such as including additional predictors/factors, interaction terms, or considering other available regression models for count data, may be necessary to better capture the data's structure and reduce overdispersion.

12.3 CONCLUSION

Based on the findings of Poisson and NB regression models, the findings suggest that the NB regression model provides a significantly better fit to the dataset compared to the Poisson regression model, as evidenced by its lower deviance, Pearson chi-square, AIC, and BIC values. The Poisson model suffers from extreme overdispersion, making it an unsuitable choice, whereas the NB model effectively accounts for this overdispersion. However, the high dispersion parameter (r = 45.801) indicated that some variability remains unexplained, suggesting the need for further adjustments. Future adjustments such as incorporating additional predictors, interaction terms, or exploring alternative regression models for count data, may be necessary to improve model fit and reduce overdispersion. Overall, the NB model may be the preferred choice; further adjustments may enhance its accuracy and improve its ability to capture the underlying data structure more accurately.

ACKNOWLEDGMENT

Authors are thankful to the ICMR-RMRIMS, Patna, for providing necessary research support.

REFERENCES

1. P. Saini *et al*, "Visceral Leishmaniasis Caused by *Leishmania donovani* Zymodeme MON-37, Western Ghats, India," *Emerg. Infect. Dis*, vol. 26, no. 8, p. 1956, Aug. 2020. doi: 10.3201/EID2608.200557.
2. A. C. Cameron and P. K. Trivedi, "Regression Analysis of Count Data," *Regres. Anal. Count Data*, 1998, doi: 10.1017/CBO9780511814365.
3. H. Akaike, "A New Look at the Statistical Model Identification," *IEEE Trans. Automat. Contr*, vol. 19, no. 6, pp. 716–723, 1974. doi: 10.1109/TAC.1974.1100705.
4. G. Schwarz, "Estimating the Dimension of a Model," vol. 6, no. 2, pp. 461–464, Mar. 1978, doi: 10.1214/AOS/1176344136.
5. P. McCullagh and J. A. Nelder, "Generalized Linear Models," Jan. 2019, doi: 10.1201/9780203753736.
6. R. A. Fisher. "On the mathematical foundations of theoretical statistics," *Philosophical transactions of the Royal Society of London. Series A, containing papers of a mathematical or physical character*, 222(594-604), pp. 309–368, 1922.
7. X. Liu, "Linear Mixed-Effects Models," *Methods Appl. Longitud. Data Anal.*, pp. 61–94, 2016, doi: 10.1016/B978-0-12-801342-7.00003-4.
8. G. H. Dunteman and M.-H. R. Ho, "An Introduction to Generalized Linear Models," *An Introd. to Gen. Linear Model.*, Jul. 2006. doi: 10.4135/9781412983273.
9. R Core Team (2020) R: A Language and Environment for Statistical Computing. R Foundation for Statistical Computing, Vienna, Austria. https://www.r-project.org/.
10. National Center for Vector Borne Diseases Control (NCVBDC). [Online]. Available: https://ncvbdc.mohfw.gov.in/index.php
11. R. M. Deb *et al.*, "Visceral Leishmaniasis Cyclical Trends in Bihar, India – Implications for the Elimination Programme.," 2018, doi: 10.17605/OSF.IO/2T9DX.
12. "OSF | Visceral Leishmaniasis Cyclical Trends in Bihar, India – Implications for the Elimination Programme." Accessed: Feb. 19, 2025. [Online]. Available: https://osf.io/2t9dx/

Chapter 13

An analysis of 2001–2017 time series data of kala-azar for Bihar, India, and its prediction for the year 2018–2020

*Niyamat Ali Siddiqui**

Department of Biostatistics, ICMR-Rajendra Memorial Research Institute of Medical Sciences, Patna, Bihar, India

Krishna Pandey

Department of Clinical Medicine, ICMR-Rajendra Memorial Research Institute of Medical Sciences, Patna, Bihar, India

13.1 INTRODUCTION

Visceral leishmaniasis (VL), also known as kala-azar (KA), is a vector-borne neglected disease ranked by the WHO as the infectious disease with the ninth highest burden worldwide [1]. More than 70% of the cases worldwide occur in India, Bangladesh, and Nepal [2]. The incidence of kala-azar in India is among the highest in the world [3]. Kala-azar is a major public health problem in four states of India, namely, Bihar, Jharkhand, Uttar Pradesh, and West Bengal, with a major proportion of cases (>70%) reported alone from Bihar. The total population of VL endemic states in India is 1339.2 million, out of which 165.4 million are at risk. The population at risk of VL infection alone in Bihar is 110 million, which accounts for 66% of the population at risk in India. Bihar state is in the eastern part of India, which carries an estimated 40% of the world's VL burden [4]. In this region, *Leishmania donovani* is the only species causing VL, the female sandfly *Phlebotomus argentipes* is the only vector, and humans are the only known reservoir [5]. VL is characterized by frequent episodes of fever, hepatosplenomegaly, pancytopenia, weight loss, and anemia. VL primarily affects the undernourished, socially marginalized population, who are far from socio-economic development. The World Health Organization (WHO)'s new neglected tropical diseases (NTDs) roadmap 2021–2030 has targeted the VL elimination as a public health problem in the South-East Asia Region (SEARO) countries and its validation by 2030, and reducing the case fatality rate to less than 1% due to primary VL [6]. In a similar line, the Ministry of Health, Government of India, has set the goal

* Corresponding author: niyamatalisiddiqui@yahoo.com

 DOI: 10.1201/9781003596455-13

for eliminating kala-azar from India by the year 2023 [7]. India has a comprehensive elimination strategy and has made great strides to eliminate kala-azar as a public health problem among the estimated 133 million population at risk in 54 districts of four states, where the disease is endemic [7]. In spite of the availability of effective drugs, other technical advances, and various elimination interventions, VL attracts global concerns as the disease periodically re-emerges in newer areas. Kala-azar is considered eliminated if the known case prevalence is <1 per 10,000 inhabitants. Despite a remarkable decline in overall kala-azar cases and deaths, India's commitment toward the revised target for eliminating kala-azar in India by 2030, as set by the WHO NTD Road Map, requires several additional measures for its achievement. With a progressive reduction of cases, it is important to maintain robust surveillance.

After many years of use of anti-kala-azar drugs, there is no evidence of their impact on transmission. Better knowledge of transmission mechanisms is still needed. Many PHCs of the state still present high kala-azar incidence. The higher incidence represents the flow of cases from the hidden prevalence to the known prevalence through easily available diagnostics. It is then a function of the size of the hidden prevalence and the access to the diagnosis. Elimination needs permanent efforts to ensure early diagnosis and treatment. A report published by Pieter Feenstra on leprosy documented that a strong decrease in registered prevalence is not based on a decreasing incidence [8]. Another study on leprosy reported that a drop in prevalence was due to operational change, as a reduction in diagnosis or a modification in the routine reporting, more likely [9]. A recent study on malaria in Orissa revealed that the state's annual malaria incidence declined from 10.82/1000 to 5.28/1000 during 2003–2013. Furthermore, it observed almost a 200% greater decline in the post-intensification period. Thereby, it made the achievement of malaria elimination (incidence <1/1000) unlikely by 2017 [10]. Recent reports from the NCVBDC suggest that the kala-azar situation in Bihar has improved considerably over the years, notwithstanding the various pitfalls in the anti-kala-azar activities mounted by the public health system of the country. Although Bihar is the largest contributor to kala-azar burden in India, and historically, the VL disease trend in India has been regarded as cyclical, with case resurgence characteristically occurring every 15 years. Understanding this pattern is essential if the VL elimination gains are to be sustained [11]. On the other hand, seasonal variation of kala-azar had already been reported from different parts of the world; however, no definite and consistent pattern was documented. To better understand the cyclical trend, a clear understanding of the spatio-temporal distribution of VL may improve surveillance and control implementation. For a vector-borne disease, it is important to recognize the spatial and temporal characteristics of its transmission.

Despite tremendous progress, last-mile challenges to eliminate kala-azar still remain. Therefore, the present study aimed to investigate spatio-temporal patterns of kala-azar cases in the state for the period of 2001–2017 and to predict the future trends related to the disease for the period 2018–2020, in the light of the various anti-kala-azar measures undertaken. Predicting the incidence of kala-azar may play an important role in planning KA control strategies for the future, developing intervention programs, and allocating resources. Additionally, the study evaluated the disease trend in the state, before and after the interventions initiated under the elimination program. Such studies are highly cost-effective, and findings may provide evidence for developing intervention programs and allocating resources for the VL in the country.

13.2 MATERIALS AND METHODS

The present study was designed as a mixed methods study design with a convergent design. The quantitative programmatic data (NCVBDC) were utilized to compare the kala-azar occurrence trends before and after elimination intervention, and the qualitative data were simultaneously collected to assess the details of the implemented intervention packages. The district-wise (districts) monthly kala-azar surveillance data from 2001 to 2017 were obtained from the NCVBDC for Bihar and converted into the annualized rates of kala-azar incidence. After that, these two rates were structured as multiple month-wise time-series data. The mid-year district-wise population was used for computing these rates derived from the national censuses (2011) to account for the growth of the population.

13.2.1 Qualitative information

Following informed consent, semi-structured key-informant interviews were conducted on anti-kala-azar strategies adopted by the state during 2001–2017 with stakeholders of the state to explore the details of the kala-azar control program.

13.2.2 Data analysis

13.2.2.1 Estimation of trend and seasonality

The time-series data were decomposed to describe their trends and seasonality. The trends were graphically examined using moving average, centered moving averages, least square method, semi-average and exponential semi-average, raw monthly time series data, and linear regression lines, including time as the only explanatory variable [12,13]. The decomposed seasonality was graphically examined and presented using the seasonal variation index, moving average, link relative method, and ratio to trend method. The decomposed seasonality was further explored to determine monthly fluctuations from the mean.

13.2.2.2 Analysis of the pre- and post-intensification periods

Significant strategic changes in the national and state anti-kala-azar program, roll-out, scaling-up of new interventions under the elimination program, along with the strengthening of some of the ongoing strategies in the state, were recorded. Kala-azar time-series data was analyzed after dividing them into two periods, i.e., interrupting the time series into two segments: the first segment ranging from 2001 to 2008, denoting the pre-intensification period, and the second segment ranging from 2009 to 2017, denote the post-intensification period. The slopes of the time-trends of kala-azar incidence between these two segments were graphically presented and compared using the unadjusted and adjusted trend estimates of VL incidence. Furthermore, a segmented regression model of the interrupted time-series, which estimated the quantum of change in the slopes of these two segments, was tested. The statistical significance of that change was evaluated, as it was done elsewhere [14,15]. To confirm the robustness of the segmented regression model and identify the optimum time point at which the significant changes occurred after the intensification of anti-kala-azar activities, a change-point analysis was conducted, as it was done elsewhere, by using the R change-point package that employs the Hinkley algorithm [16,17].

13.2.2.3 Analysis of the districts stratified by the baseline disease burden

The districts were stratified as per their annual kala-azar incidence in 2001. The district's annual incidence rate was stratified into the following three clusters: high (incidence >4 per 10,000 populations), moderate (1–4 per 10,000), and low (<1 per 10,000) as per the criteria [18]. Then the trends in kala-azar incidence in these three clusters were estimated using the same process.

13.2.2.4 Spatial distribution of kala-azar

The districts stratified as per the aforesaid cut-off points were color-coded in the state map once at baseline (2001) and then for 2009 and 2017 to display the progress of each district with regard to their disease burden.

13.2.2.5 Prediction

The kala-azar incidence was predicted for the next three years (2018–2020), using historical disease incidence rates of the period 2001–2017 and the Holt-Winters exponential smoothing forecasting models. The appropriateness of the predictive model was checked using the Box–Ljung test.

13.2.2.6 Analysis of the qualitative data

The information from the key-informants was noted during the interview by one researcher while the other conducted the interview. After each interview, the researcher debriefed the data and adopted a new guide based on the previous findings. The qualitative data were analyzed using content analysis, as used elsewhere [17]. Meaning units were identified and coded. Similar codes were clustered together and merged into categories. The main theme-emergent concepts were developed based on the interpretation of codes and categories.

13.2.3 Statistical tools and analysis

Centered moving averages, linear regression, generalized least square regression models, and moving averages model, segmented regression model for interrupted time-series, and change-point analysis (by using the R change-point package that employs the Hinkley algorithm) were used to perform statistical analysis. Additionally, Holt–Winters exponential smoothing forecasting models were used. The appropriateness of the predictive model was checked using the Box–Ljung test. Additionally, descriptive statistics (percentage, mean, 95% CI, IQR, etc.) were utilized to describe the important findings. Appropriate inferential statistics, such as the t-test (for comparing means between two groups), ANOVA (analysis of variance) for comparing means across multiple groups, and the Pearson correlation coefficient for assessing the linear relationship between two quantitative variables, and the Kruskal–Wallis test, deemed fit were also used to see the difference (statistical significance) between variables. However, the qualitative data were analyzed using content analysis. Data entry and analysis was done using the Epi Info v.7.1.4 and R v3.0.1 software, respectively. A p-value <0.05 was considered statistically significant.

13.2.4 Ethics statement

The data used in our analysis were district-aggregate programmatic data of NCVBDC, collected in de-identified form from the state program. For that, prior departmental permission from the NCVBDC was obtained prior to the start of the study. These were routinely collected data by the program information system. No individual data was analyzed under this study. Hence, there was no scope for anonymization or de-identification of the data. Thus, the ethical consideration is deemed to be inapplicable to the quantitative data. However, for the stakeholders' interviews, ethical approval was obtained from the Institutional Ethics Committee. Informed and written consents were obtained from all the participants before the interview.

13.3 RESULTS

13.3.1 Trends

In Bihar, a total of 290,725 KA cases and 1583 deaths were reported during 2001–2017. Incidence was assessed per 10,000 and presented in Table 13.1. The annual KA incidence of Bihar decreased from 1.2 in 2001 to 0.3 per 10,000 population in 2017, and this decline was found statistically significant ($p<0.001$). The adjusted decline in KA incidence (2001–2017), 0.61 (1.5–1.9; $p<0.001$), was statistically highly significant.

Trends of KA cases using the central moving average and the moving average were presented in Figures 13.1 and 13.2.

However, trends of KA cases and their semi-average were presented in Figure 13.3.

Table 13.1 Trends of kala-azar incidence and cause-specific death rate of Bihar

Year	Population	*No. of KA Cases*	*No. of death Cases*	*Annual KA incidence (per 10,000)*	*Cause-specific death rate (per 100,000)*
2001	82,878,796	10,029	203	1.2	0.24
2002	85,250,787	9505	158	1.1	0.19
2003	87,622,778	13,960	189	1.6	0.22
2004	89,994,769	17,324	115	1.9	0.13
2005	92,366,761	23,383	128	2.5	0.14
2006	94,738,752	29,711	162	3.1	0.17
2007	97,110,743	37,822	172	3.9	0.18
2008	99,482,734	29,094	147	2.9	0.15
2009	101,854,725	21,318	80	2.1	0.08
2010	104,226,716	23,084	95	2.2	0.09
2011	104,099,452	25,222	76	2.4	0.07
2012	106,712,348	16,036	27	1.5	0.03
2013	109,325,244	10,611	17	1	0.02
2014	111,938,141	8028	9	0.7	0.01
2015	114,551,037	6782	5	0.6	0
2016	117,163,933	4773	0	0.4	0
2017	119,776,829	4043	0	0.3	0

Adjusted decline: 0.61 (1.5–1.9; p<0.001).

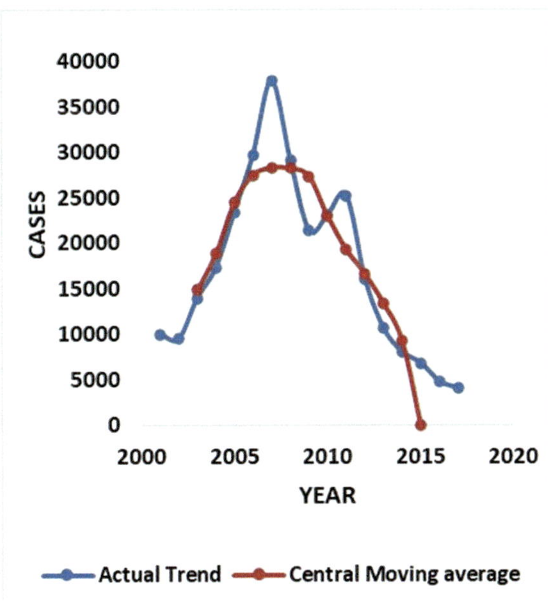

Figure 13.1 Trends of kala-azar cases using central moving average.

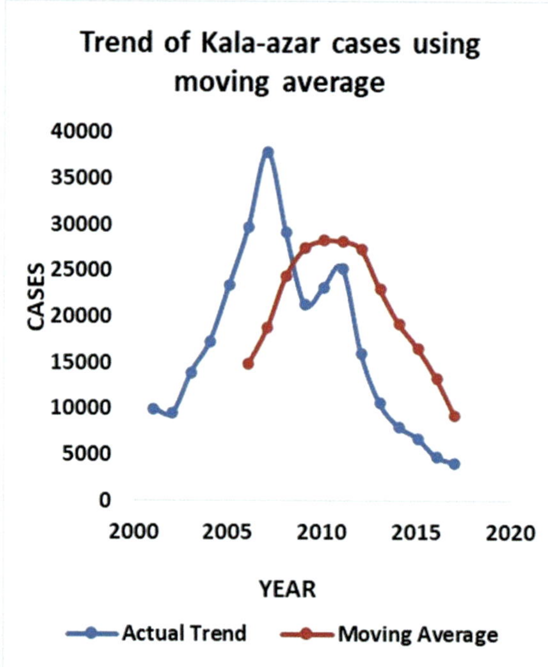

Figure 13.2 Trends of kala-azar cases using moving average.

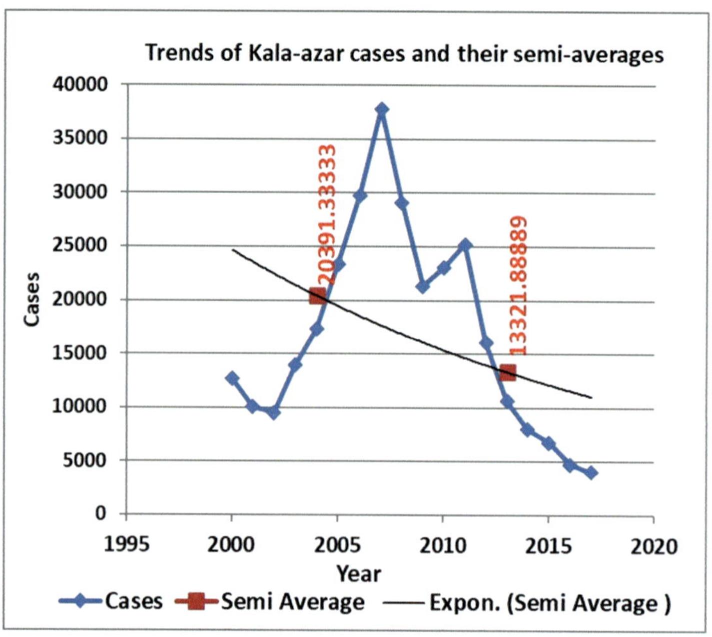

Figure 13.3 Trends of KA cases and their semi-average.

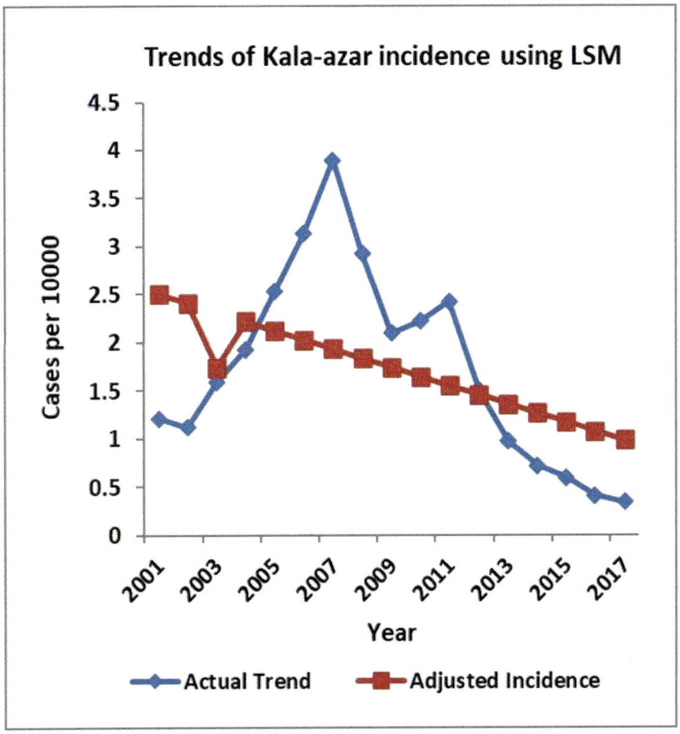

Figure 13.4 Trends of kala-azar incidence using LSM.

Additionally, trends of kala-azar incidence using the generalized least square method were presented in Figure 13.4.

The annualized rate of VL incidence, using monthly time-series data, central moving average, and linear regression line, was presented in Figure 13.5.

The overall linear slope of decline in the KA incidence was seen from 2007 to 2017, which was considerably flatter for 2012–2017.

13.3.2 Trends of KA incidence as per pre- and post-intensification periods

The slopes of the time-trends of kala-azar incidence between these two segments were graphically presented and compared using the unadjusted and adjusted trend estimates of VL incidence in Figure 13.6. Both unadjusted and adjusted incidence rates showed a consistent increase during the pre-intensification period (2001–2008) with peaks around 2007 and 2008. The year 2009 is considered the intensification year, marking a transition point where the rates begin to level off and then decline in the following years. The post-intensification period (2009–2017) revealed a notable decrease in both unadjusted and adjusted kala-azar incidence rates, particularly after the year

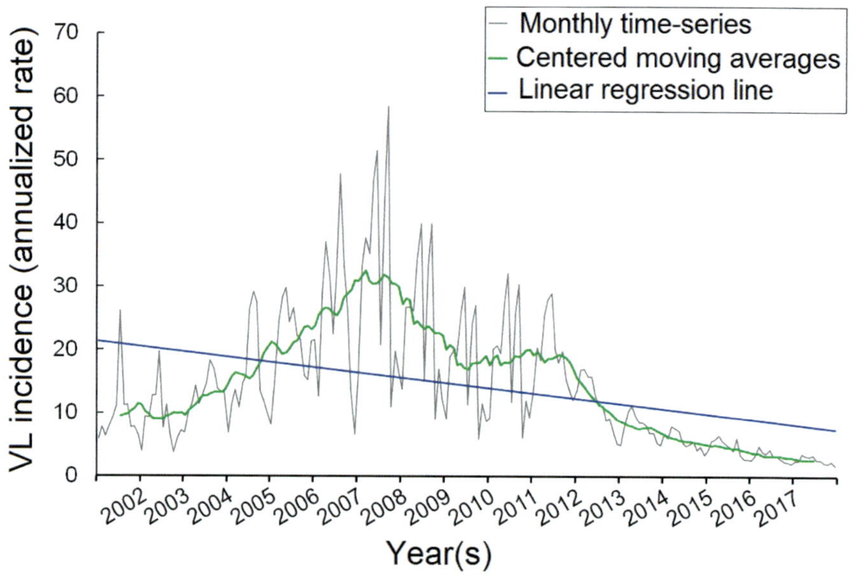

Figure 13.5 Annualized rate of VL incidence, using monthly time-series data, central moving average, and linear regression line.

2013. The adjusted incidence rate dropped significantly after the year 2017. The possible explanations for that may be elimination initiatives implemented effectively at grassroots level.

The unadjusted and adjusted trend estimate of VL incidence was presented in Table 13.2. There was as average decline of 5% in the adjusted estimate

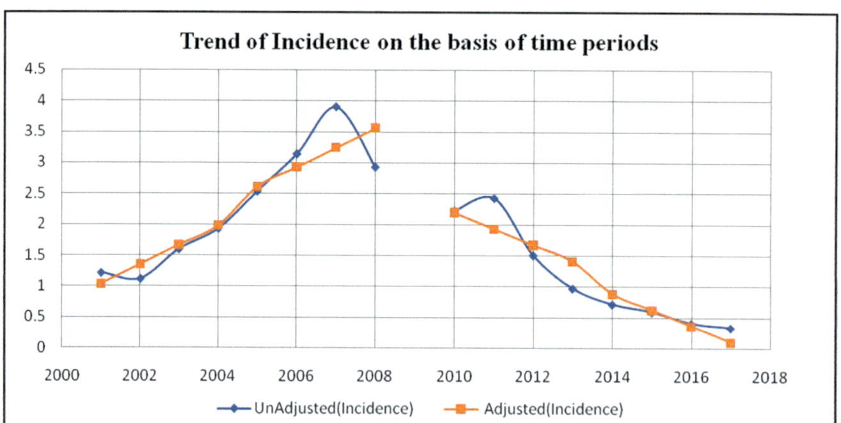

Figure 13.6 Trends of KA incidence as per pre- (2001–2008) and post-intensification (2009–2017) periods.

Table 13.2 Unadjusted and adjusted trends of VL incidence of Bihar

Unadjusted trend estimate of VL incidence	
Bihar, 2001–2017	Avg. declined=4% (1.252–2.229)
Pre-intensification period (2001–2008)	Avg. increase=15% (1.602–2.980)
Post-intensification period (2009–2017)	Avg. declined=21% (0.688–1.604)
Adjusted trend estimate of VL incidence	
Bihar, 2001–2017	Avg. declined=5% (1.490–1.924)
Pre-intensification period (2001–2008)	Avg. increase=19% (1.652–2.930)
Post-intensification period (2009–2017)	**Avg. declined=31% (0.617–1.675)**

of VL incidence for the period 2001–2017. However, an increase of 19% in adjusted estimate of VL incidence was observed for the period 2001–2008 (pre-intensification period), whereas a significant decline (31%) in the adjusted estimate of VL incidence was observed for 2009–2017 (post-intensification period).

13.3.3 Seasonality

Majority (62%) of the kala-azar incidence occurred in only two quarters (April to June: Q2 and July to September: Q3). The highest peak of this disease in almost all the years was observed during April to June, quarter (Q2). Seasonal variation index was found to be highest (127.10) in Q2 (April to June), followed by (120.71) in Q3 (July to September), presented in Figure 13.7.

However, seasonal variation using moving average, link relative method, and ratio to trend method were presented in Figures 13.8–13.10. In every

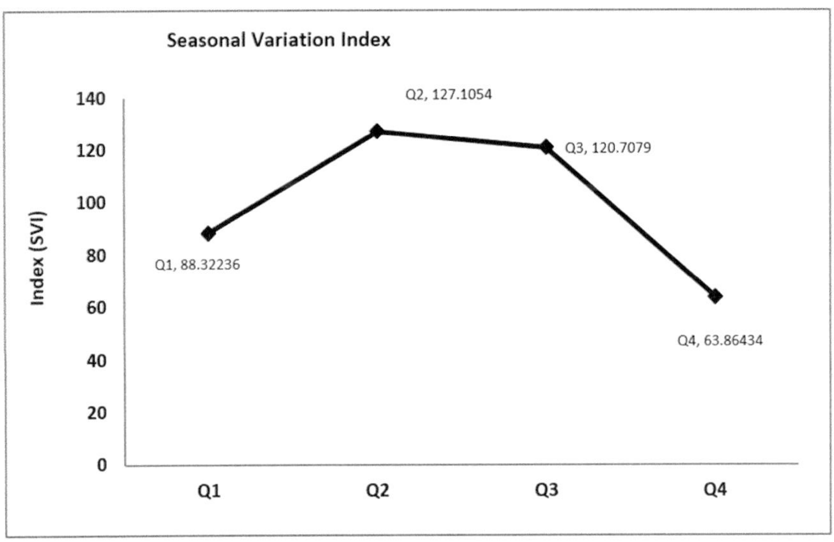

Figure 13.7 Seasonal variation index.

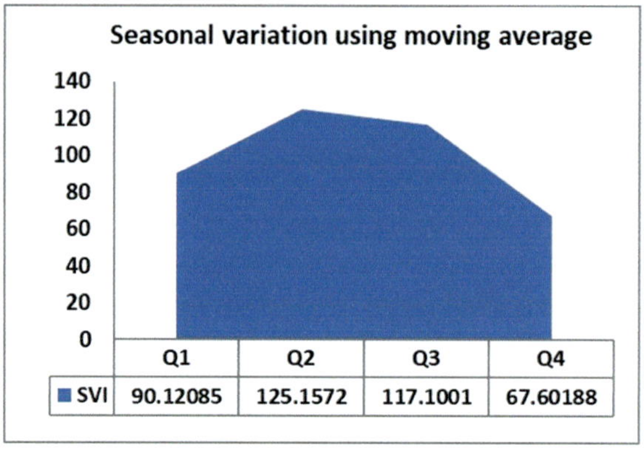

Figure 13.8 Seasonal variation using moving average.

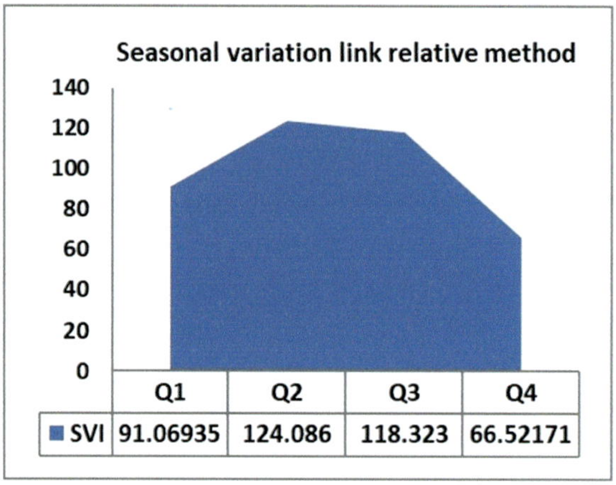

Figure 13.9 Seasonal variation using the link relative method.

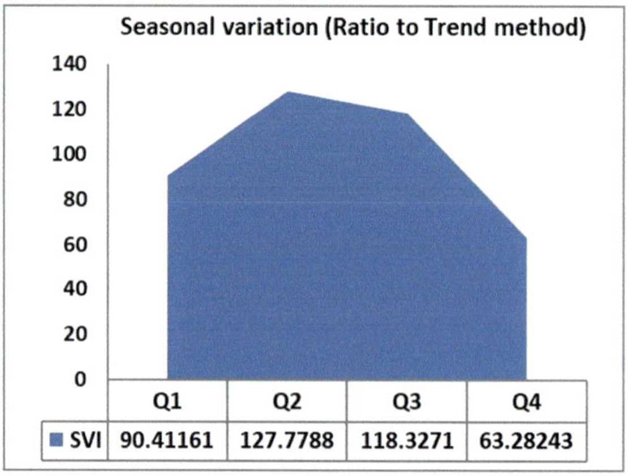

Figure 13.10 Seasonal variation using the ratio to trend method.

method, Q2 showed the maximum value, followed by Q3. A statistically highly significant difference ($p<0.0001$) was observed for kala-azar incidence between Q2 and Q1 and Q4, respectively. Similarly, kala-azar incidence noted for Q3 was also significantly different when compared to Q1 and Q4, respectively. All these measures of seasonality concluded that seasonality really exists for kala-azar disease.

13.3.4 Spatial distribution

Bihar map showing districts in each zone (North, South, West, and East) was segregated by a line drawn on the map (Figure 13.11).

Maximum KA cases (56%) were from the North zone (12 districts in total), followed by 40% from the East zone (11 districts in total) of Bihar, when compared to the West (7 districts in total) and South zone (8 districts in total) areas with 1.5% and 2.5%, respectively. The average annual incidence was significantly higher (1.0/10,000) in the North zone as compared to the West and South zones ($p<0.001$). Furthermore, spatial analysis revealed that the numbers of district from the North zones showed more than 20,000 KA cases, while only one district of the East zone showed up to 15,000 KA cases, and the rest of the districts were below 10,000 KA cases. It may be concluded that the North zone districts are crucial to handle. Accordingly, additional interventions are specifically needed for these areas.

Figure 13.11 Bihar map showing districts in each zone. (Source: https:www.mapsofindia.com>maps>bihar-district.)

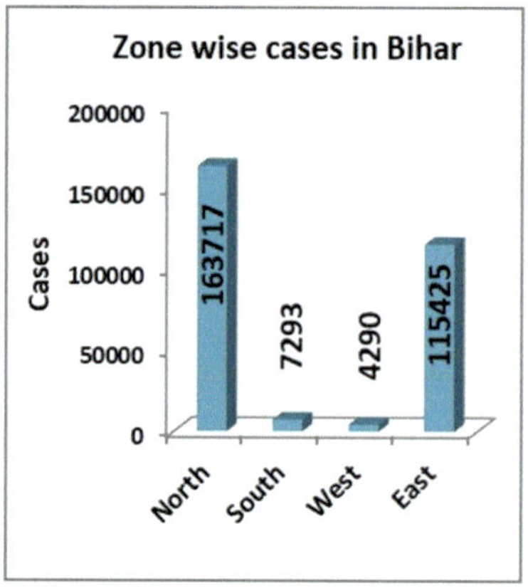

Figure 13.12 Zone-wise kala-azar cases of Bihar.

Spatial distribution of KA cases and deaths in Bihar was presented in Figures 13.12 and 13.13, and the districts stratified as per the aforesaid cut-off points were color-coded in the state map once at baseline 2001 and then for 2009 and 2017, to display the progress of each district with regard to their disease burden (Figure 13.14).

The zone-wise adjusted and unadjusted KA incidence was presented in Figure 13.15.

The unadjusted data reflects raw data, while adjusted data has been modified to account for predictable and external factors, providing a clearer view

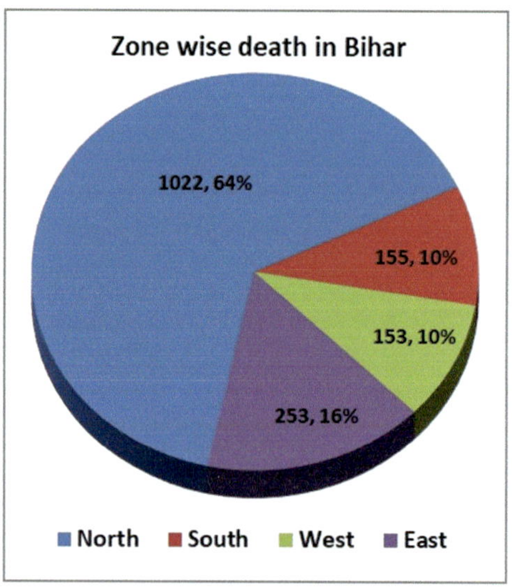

Figure 13.13 Zone-wise deaths due to kala-azar of Bihar.

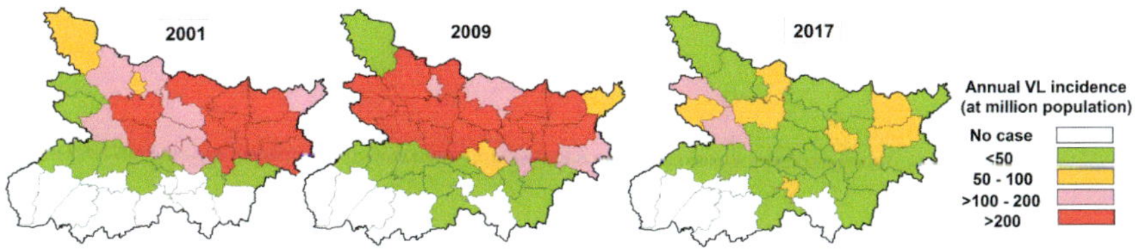

Figure 13.14 Status of endemic districts of Bihar in 2001 (pre-intensification), 2009 (post-intensification), and 2017, showing annual VL incidence.

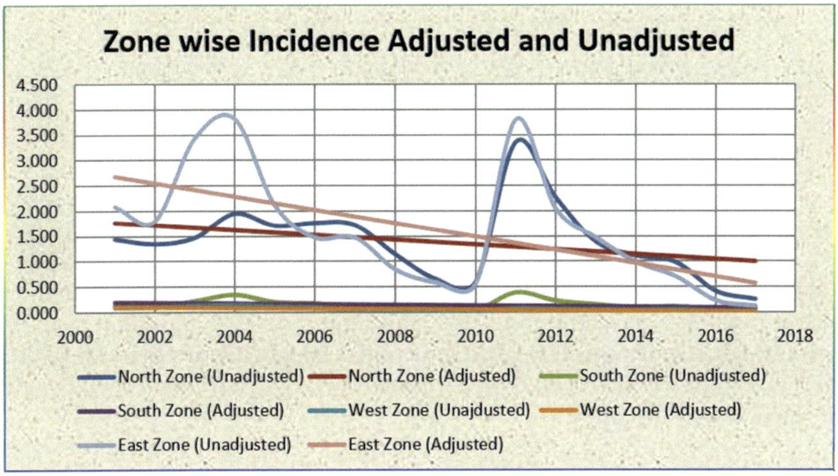

Figure 13.15 The zone-wise adjusted and unadjusted KA incidence of Bihar.

of long-term trends or patterns. A statistically highly significant difference ($p<0.001$) was observed between the unadjusted incidence and the adjusted incidence of these zones. Overall, the North and East zones were consistently in the very high KA incidence zone from 2003 to 2017. Stratification of districts according to disease burden was described in Table 13.3.

Table 13.3 Stratification of districts on the basis of five-year average incidence (2013–2017), Bihar

Strata	
High endemic district (>1.5) **(Saran, Saharsa, and Gopalgunj)**	1.796 (1.021–2.571), p=0.848
Moderate endemic districts (1.5–1) **(Muzaffarpur, Purnia, and Araria)**	1.311 (1.019–1.602), p=0.655
Low endemic districts (<1) **(Siwan, Sitamarhi, and Khagaria)**	0.884 (0.608–1.161), p=0.539

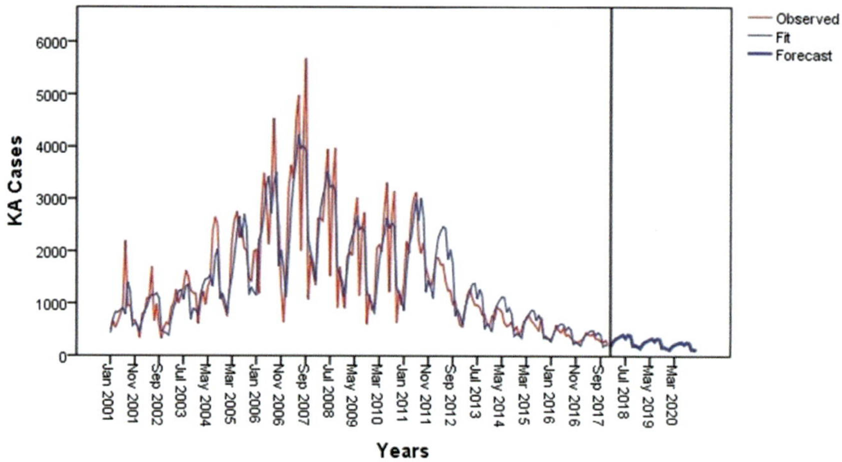

Figure 13.16 Forecasting KA cases for 2018–2020 based on the trends from 2001 to 2017, Bihar.

It was surprising to note that the difference in KA incidence between high-burden districts was not found statistically significant (P=0.848). Thus, each high-burden district is unique, and the same degree of resources needs to be in place to handle the situation. Furthermore, these districts may be challenging for the national program.

13.3.5 Forecast

The forecast for the next three years (2018–2020) showed that, as compared to the early phases of intensified interventions, the decline was likely to slow down, as presented in Figure 13.16. This implies that the elimination target of KA incidence (<1 case per 10,000 population) at the block level might not be achieved by 2020 for the whole state, which is the national goal. However, the disease will be under good control and much lower than the average KA incidence (2001–2017) if the existing interventions are continued. Each high-burden district is unique, and the same degree of resources needs to be in place to handle the situation.

13.4 DISCUSSION

The VL program in Bihar has made significant progress in adopting best practices for improved treatment and vector control, with the aim of achieving VL elimination. But still VL in Bihar State (India) continues to be endemic, despite the existence of effective treatment and a vector control program to control disease morbidity. This is most probably the first study to attempt to utilize spatio-temporal techniques to explore the trends of the historical data of 17 years on VL incidence from Bihar, India, with detailed analysis. In the present study, trend estimate of VL incidence of Bihar was performed for the pre-intensification period (2001–2008) and the post-intensification period (2009–2017) separately. The annual KA incidence of Bihar decreased from 1.2 in 2001 to 0.3 per 10,000 population in 2017, and this decline was found statistically significant ($p<0.001$). The adjusted decline in KA incidence (2001–2017): 0.61 (1.5–1.9; $p<0.001$), which was statistically highly significant. The adjusted trend estimate of VL incidence in Bihar revealed that the average decline was 5%, with a 95% CI of (1.49–1.92) for the period 2001–2017. An average increase in VL incidence was 19% with 95% CI (1.65–2.93) for the pre-intensification period (2001–2008). On the other side, an average decline in VL incidence was 31% with 95% CI (0.62–1.70) for the post-intensification period (2009–2017). The overall linear slope of decline in the KA incidence was seen from 2007 to 2017, which was considerably flatter for the year 2012–2017. These findings were consistent with other studies conducted in India, which documented that the lowest incidence recorded was seen in 2014 (7615 cases – 0.705 per 10,000) and 2013 (1.005 per 10,000 (10,730 cases) [11]. Another study conducted by Malaviya et al. (2011) in India reported average annual VL incidence rate 2.2/10,000, 95% CI (2.1–2.4) for the year 1999–2004, and average annual VL incidence 5.8/10,000, 95% CI (5.5–6.0) for the year 2005–2008 in Muzaffarpur district of Bihar [19]. Similar observations were noted by other studies, which concluded that kala-azar showed epidemic cycles taking place almost regularly every 15–20 years [20,21]. Another study conducted by Rakesh Mandal et al. (2018) documented the spatio-temporal analysis of village's VL endemicity, stated that 624 villages (43.9% of all inhabited villages) were endemic during 2012–2014, of which 232 (37.2%) remained endemic for more than one year, of which 154 (66.4%) for two years (56.5% of 2012–2013; 43.5% of 2013–2014), and 78 (33.6%) for three years (2012–2014) [22].

The decline in the incidence of VL cases from 2001 to 2017 is encouraging. However, this drop may be related to the regional efforts to eliminate VL in general and to the use of improved diagnostic tools (i.e., rK39 dipstick) and anti-*Leishmania* drugs, indoor residual spray (IRS), etc., in particular. Despite this improvement, efforts should continue to ensure a clear picture of the situation. The reporting system should be improved so VL incidence per age, gender, or population at risk groups (i.e., HIV, tuberculosis) as well as PKDL cases can be monitored.

Majority (62%) of the kala-azar incidence occurred in only two quarters (April to June: Q2 and July to September: Q3). The highest peak of this disease in almost all the years was observed during April to June, quarter (Q2). Seasonal variation index was found to be highest (127.10) in Q2 (April to June), followed by (120.71) in Q3 (July to September). Additionally, using moving average, link relative method, and ratio to trend method, the highest peak of the disease was observed for Q2, followed by Q3. All these measures of seasonality concluded that seasonality really exists for kala-azar disease in Bihar. These findings were identical to those of a study conducted in India on tuberculosis, which also documented a seasonal pattern and a declining trend, with variable amplitudes of fluctuation observed in the incidence of tuberculosis [23]. Another study by Rinki M Deb et al. (2018) also concluded statistically significant associations ($p<0.05$) for VL incidence and maximum temperature, and average temperature, when considering annual and pre-sandfly peak time periods [11]. Another study conducted by Paritosh Malaviya et al. (2011) in India concluded the highest peak of kala-azar incidence in March to April, followed by another peak in July, and inferred that VL incidence was associated positively with rainfall and negatively with relative humidity [19].

Results of the study clearly revealed that seasonality really exists for kala-azar disease in Bihar. This may be another potential implication of these study findings, given that seasonal variation was observed, clinicians in Bihar public health facilities should have a higher-than-usual clinical suspicion for kala-azar in the hot (April to June) and rainy (July to September) months. These observations are suggestive of the presence of a seasonal disease-modifying factor. This regularity of peak seasonality in the KA case detection may prove useful to initiate measures that warrant a better implementation of control measures. This information may also be useful for administration and managers to take extra care to arrange and provide extra facilities during the peak seasons.

Maximum KA cases (56%) were from the North zone (12 districts in total), followed by 40% from the East zone (11 districts in total) of Bihar, when compared to the West (7 districts in total) and the South zones (8 districts in total) areas with 1.5% and 2.5%, respectively. The average annual incidence was significantly higher (1.0/10,000) in the North zone as compared to the West and South zones ($p<0.001$). Furthermore, spatial analysis revealed that the number of districts from the North zone showed more than 20,000 KA cases, while only one district of the East zone showed up to 15,000 KA cases, and the rest of the districts were below 10,000 KA cases. It may be concluded that the North zone districts are crucial to handle. These results were in line with other study findings, which also concluded spatially clustered patterns with significant differences by village. The hotspots showed the spatial trend of kala-azar diffusion ($p<0.01$) [24]. Spatial heterogeneity in the incidence of VL was also recorded by another study conducted elsewhere in the world [25]. Another scientist, Rakesh Mandal et al. (2018), concluded that high CIR plays a major role in the spatial dispersion of the VL cases, and the spatial distribution of CIR among the endemic villages was found to be clustered with positive Moran's I (between 0.10 and 0.17) and significant (at $p<0.05$) [22].

Overall, the North and East zones were consistently in the very high KA incidence zone from 2003 to 2017. It was surprising to note that the difference in KA incidence between high-burden districts was not found statistically significant ($p=0.848$). Thus, each high-burden district is unique, and the same degree of resources needs to be in place to handle the situation. Furthermore, these districts may be challenging for the national program. This finding coincided with other study findings, which concluded that some blocks had constantly higher (Sahebganj and Bochahan) or lower (Sakra) SIR over time [19].

Spatial analysis demonstrated that the North zone districts are crucial to handle. These zones may be challenging for the national program. Accordingly, additional interventions are specifically needed for these areas. Each high-burden district is unique, and the same degree of resources needs to be in place to handle the situation. The results pointed to the usefulness of a spatial statistical approach to improve our understanding of the spatio-temporal dynamics and control of kala-azar. The study also showed that the North zone (12 districts in total) and the East zone (11 districts in total) are most likely higher endemic cluster regions. To employ exact and geographically suitable risk-reduction programs, the application of such spatial analysis tools should be a vital constituent in epidemiology research and risk evaluation of kala-azar. The spatial statistical approaches can expansively recognize the high-risk areas, timely in the epidemic based on their integrated spatial-temporal profiles.

The forecast for the next three years (2018–2020) showed that, as compared to the early phases of intensified interventions, the decline was likely to slow down. This implies that the elimination target of KA incidence (<1 case per 10,000 population) at the block level might not be achieved by 2020 for the whole state, which is the national goal. However, the disease will be under good control and much lower than the average KA incidence (2001–2017) if the existing interventions are continued. Each high-burden district is unique, and the same degree of resources needs to be in place to handle the situation. A similar observation was made by a study conducted by Varun Kumar et al. (2014) on tuberculosis, which reported that in the near future, the reported annual TB incidence numbers in India will decrease only slightly [23]. On the contrary, another study on tuberculosis, using Box–Jenkins and SARIMA models, documented that prediction results show an increasing trend of TB cases in Iran [26]. Another study conducted by Fathelrhman EL Guma on VL elsewhere in the world concluded that the LSTM approach may be better suited for predicting VL incidents and has the potential to contribute to the formulation of efficient preventive measures [27]. Other study conducted by Emily S. Nightingale et al. (2020) demonstrated a framework for forecasting VL incidence at sub-district level in India, which achieves good predictive performance based on the available routinely collected surveillance data and concluded that this framework could be used to make short-term forecasts to provide an early indication of where case numbers are higher (or lower) than expected [28]. Rakesh Mandal et al. (2018) concluded that an elevated CIR increased the risk of VL cases dispersal between endemic and non-endemic villages [22].

13.5 CONCLUSIONS

The VL program in Bihar has made significant progress in adopting best practices for improved treatment and vector control, with the aim of achieving

VL elimination. But still VL in Bihar state (India) continues to be endemic, despite the existence of effective treatment and a vector control program to control disease morbidity. This study explored the trends in spatio-temporal dynamics of VL endemicity in Bihar, India. The spatio-temporal techniques not only provide an improved understanding of the distribution pattern of the disease but also help to optimize the control resources more effectively. The cyclic pattern of kala-azar recorded cases, the Box–Jenkins model is suitable for predicting its incidence in the future. Predicting the incidence of kala-azar plays an important role in planning health control strategies for the future, developing intervention programs, and allocating resources. The result of this study shows that epidemiological measures can be used to make our high- and low-risk areas more efficient, which aids in optimizing resources and minimizing VL cases. The present study provides public health planners with a more sophisticated tool to differentiate risk patterns of VL epidemic using spatial statistical approaches rather than relying on annual cumulative incidence alone, so that high-risk areas can be expansively recognized early in the epidemic based on their integrated spatial-temporal profiles. Spatio-temporal diffusion patterns and hotspot detection may offer useful information to help epidemiologists to control and predict kala-azar spread over critical hotspot areas only. The results of this study may help public health scientists and researchers to design and implement control strategies in an advanced way to achieve the VL elimination target for the highly endemic region of Bihar, as well as others in the Indian subcontinent.

Considering the complex network of factors involved in the occurrence of VL in endemic areas, the results of this study showed that the development of a predictive model for VL might be feasible and useful for guiding interventions against the disease, but under the present study, model building was not attempted. It is still a challenge, as demonstrated by the unsatisfactory predictive performance of the model developed in various earlier studies. However, using time series prediction techniques for VL might offer valuable insights to aid public health professionals in strategizing and implementing effective measures for VL prevention. Furthermore, it is suggested that future studies should investigate the possibility of integrating SARIMA and LSTM techniques to improve the VL forecast's precision.

13.6 LIMITATIONS

The present study has its own limitations, with the major one being the lack of clinical and epidemiological data. A spatio-temporal approach may not offer useful information to sustain epidemiologists to control and predict kala-azar throughout the whole region. The study only used time series data of kala-azar cases from Bihar, and hence, the cause-and-effect relationship could not be established.

ACKNOWLEDGMENT

Authors are thankful to the ICMR for providing the necessary research support.

REFERENCES

1. World Health Organization. Kala-Azar Elimination in South-East Asia Region: Training Module. Document SEA-182. World Health Organization, SEARO, New Delhi, 2008.
2. Stauch A, Sarkar RR, Picado A, Ostyn B, Sundar S, Rijal S, Boelaert M, Dujardin JC, Duerr HP. Visceral leishmaniasis in the Indian subcontinent: Modelling epidemiology and control. PLoS Negl Trop Dis. 2011, 5(11):e1405.
3. Alvar J, Velez I, Bern C, Herrero M, Desjeux P, et al. Leishmaniasis worldwide and global estimates of its incidence. PLoS One. 2012, 7:e35671.
4. World Health Organization. Report on the consultative meeting on Leishmania/HIV co-infection. Report no. WHO/LEISH/95.35, pp. 1–14. World Health Organization, Geneva, 1995.
5. Joshi A, Narain JP, Prasittisuk C, Bhatia R, Hashim G, Jorge A, Banjara M, Kroeger A. Can visceral leishmaniasis be eliminated from Asia? J Vector Borne Dis. 2008, 45(2):105–111.
6. Ending the neglect to attain the Sustainable Development Goals; a road map for neglected tropical diseases 2021–2030, WHO; 28 January 2021/Global Strategy, ISBN 978-92-4-001035-2, pages: 196.
7. National road map FOR KALA-AZAR ELIMINATION. Directorate of National Vector Borne Disease Control Programme (NVBDCP) Delhi. KA_Road-Map-NVDBCP-Nov 2014.
8. Feenstra P. “Elimination” of leprosy and the need to sustain leprosy services, expectations, predictions and reality. Int J Leprosy Other Mycobact Dis. 2003, 71(3): 248–256.
9. Penna MLF, Penna GO. Trend of case detection and leprosy elimination in Brazil. Trop Med Int Health. 2007, 12(5):647–650.
10. Pradhan A, Anasuya A, Pradhan MM, Kavitha AK, Kar P, Sahoo KC, Panigrahi P, Dutta A. Trends in Malaria in Odisha, India—An analysis of the 2003–2013 time-series data from the National Vector Borne Disease Control Program. PLOS One. 2016. DOI: 10.1371/journal.pone.0149126.
11. Deb RM, Stanton MC, Foster GM, Das Gupta RK, Roy N, Das P, Dhariwal AC, Coleman M. Visceral leishmaniasis cyclical trends in Bihar, India – Implications for the elimination programme. Gates Open Res. 2018, 2:10 last updated: 01 OCT 2018.
12. Fox J. Time-Series Regression and Generalized Least Squares. Appendix to An R and S-PLUS Companion to Applied Regression [Internet].2002[cited 2014 May 3]: 1–8. Available from: http://cran.rproject.org/doc/contrib/Fox-Companion/appendix-timeseries-regression.pdf.
13. Dayer MJ, Jones S, Prendergast B, Baddour LM, Lockhart PB, Thornhill MH. Incidence of infective endocarditis in England, 2000–13: A secular trend, interrupted time-series analysis. Lancet [Internet]. 2014 Nov 28 [cited 2014 Nov 19]; 385(9974):1219–28. Available from: https://doi.org/10.1016/s0140-6736(14)62007-9
14. Taljaard M, McKenzie JE, Ramsay CR, Grimshaw JM. The use of segmented regression in analyzing interrupted time series studies: An example in pre-hospital

ambulance care. Implement Sci [Internet. 2014, Jan [cited 2014 Dec 5]; 9(1):77. Available from: http://www.implementationscience.com/content/9/1/77
15. Wagner AK, Soumerai SB, Zhang F, Ross-degnan D. Segmented regression analysis of interrupted time series studies in medication use research. J Clin Pharm Ther [Internet]. 2002; 27:299–309. Available from: http://isites.harvard.edu/fs/docs/icb.topic79832.files/L06_Program_Evaluation_2/Segmented_Regression.Wagner.2002.pdf PMID: 12174032.
16. Aregawi M, Lynch M, Bekele W, Kebede H, Jima D, Taffese HS, et al. Time series analysis of trends in malaria cases and deaths at hospitals and the effect of antimalarial interventions, 2001–2011, Ethiopia. PLoS One [Internet]. 2014, Jan [cited 2014 Nov 19]; 9(11):e106359. Available from: : 10.1371/journal. pone.0106359 PMID: 25406083
17. Graneheim UH, Lundman B. Qualitative content analysis in nursing research: Concepts, procedures and measures to achieve trustworthiness. Nurse Educ Today [Internet]. 2004 Feb 2 [cited 2014 Jul 9]; 24(2):105–12. Available from: http://www.nurseeducationtoday.com/article/S0260691703001515/fulltext PMID: 14769454.
18. Mamun Huda, M, Hirve, S, Siddiqui, NA, Malaviya, P, Banjara, MR, Das, P, Kansal, S, Gurung, CK, Naznin, E, Rijal, S, Arana, B, Kroeger, A, Mondal, D. Active case detection in national visceral leishmaniasis elimination programs in Bangladesh, India, and Nepal: Feasibility, performance and costs. BMC Public Health. 2012, 12:1001. http://www.biomedcentral.com/1471-2458/12/1001
19. Malaviya P, Picado A, Singh SP, Hasker E, Singh RP, Boelaert M, Sundar S. Visceral leishmaniasis in Muzaffarpur District, Bihar, India from 1990 to 2008. PLoS ONE | www.plosone.org 1 March 2011, 6(3):e14751.
20. Kumar, V., Mandal, R., Das, S., Kesari, S., Dinesh, D. S., Pandey, K., Das, V. R., Topno, R. K., Sharma, M. P., Dasgupta, R. K., & Das, P. (2020). Kala-azar elimination in a highly-endemic district of Bihar, India: A success story. PLoS Neglected Tropical Diseases, 14(5), e0008254. https://doi.org/10.1371/journal.pntd.0008254
21. Alvar J, Yactayo S, Bern C. Leishmaniasis and poverty. Trends Parasitol. 2006, 22(12):552–7.
22. Mandal R, Kesari S, Kumar V, Das P. Trends in spatio-temporal dynamics of visceral leishmaniasis cases in a highly endemic focus of Bihar, India: An investigation based on GIS tools. Parasites & Vectors. 2018, 11:220. DOI: 10.1186/s13071-018-2707-x.
23. Kumar V, Singh A, Adhikary M, Daral S, Khokhar A, Singh S. Seasonality of tuberculosis in Delhi, India: A time series analysis. Tuberc Res Treat. 2014, 2014:514093. DOI: 10.1155/2014/514093. Epub 2014 Mar 23.
24. Bhunia GS, Kesari S, Chatterjee N, Kumar V, Das P. Spatial and temporal variation and hotspot detection of kala-azar disease in vaishali district (Bihar), India. BMC Infect Dis. 2013 Feb 2, 13:64. DOI: 10.1186/1471-2334-13-64.
25. Almeida AS, Werneck GL. Prediction of high-risk areas for visceral leishmaniasis using socioeconomic indicators and remote sensing data. Int J Health Geograph. 20 May, 2014, 13(2014). DOI: 10.1186/1476-072X-13-13.
26. Moosazadeh M, Nasehi M, Bahrampour A, Khanjani N, Sharafi S, Ahmadi S. Forecasting tuberculosis incidence in Iran using Box–Jenkins models. Iran Red Crescent Med J. 2014 May 5, 16(5):e11779. DOI: https://doi.org/10.5812/ircmj.11779.
27. Guma FEL. Comparative analysis of time series prediction models for visceral leishmaniasis: Based on SARIMA and LSTM. Appl. Math. Inf. Sci. 2024, 18(1): 125–132.
28. Nightingale ES, Chapman LAC, Srikantiah S, Subramanian S, Jambulingam P, Bracher J, Cameron MM, Medley GF. A spatio-temporal approach to short-term prediction of visceral leishmaniasis diagnoses in India. PLOS Negl Trop Dis; July 9, 2020| https://doi.org/10.1371/journal.pntd.0008422.

Chapter 14

In silico resources and methods for the discovery and optimization of treatments for visceral leishmaniasis and post-kala-azar dermal leishmaniasis

*Saravanan Vijayakumar**
Department of Bioinformatics, ICMR-Rajendra Memorial Research Institute of Medical Sciences, Patna, India

Ravi Ranjan
Department of Bioinformatics, ICMR-Rajendra Memorial Research Institute of Medical Sciences, Patna, India
Department of Biochemistry, ICMR-Rajendra Memorial Research Institute of Medical Sciences, Patna, India
Department of Biochemistry, University of Calcutta, Kolkata, West Bengal, India

14.1 INTRODUCTION

A wide range of diseases have long threatened human health, including non-communicable chronic diseases like cancer and diabetes, as well as infectious diseases. As a neglected tropical disease, Leishmaniasis, caused by parasites of the genus *Leishmania*, is transmitted by phlebotomine sandflies that bite infected people. The disease is primarily endemic to impoverished populations worldwide. In a recent fact sheet published by the World Health Organization (WHO), it was estimated that 700,000–1 million new cases of leishmaniasis occur every year (https://www.who.int/news-room/fact-sheets/detail/leishmaniasis) (1). Clinically, leishmaniasis manifests three primary forms: cutaneous leishmaniasis (CL), mucocutaneous leishmaniasis (MCL), and visceral leishmaniasis (VL), with the latter being the most severe and potentially fatal. However, no effective vaccine has been approved for leishmaniasis, despite the fact that multiple studies suggest it may be a vaccine-preventable disease (2). Consequently, disease management remains reliant on chemotherapy (3). The most commonly prescribed medications for VL include paromomycin, amphotericin B, pentamidine, and miltefosine. While these

* Corresponding author: brsaran@gmail.com

DOI: 10.1201/9781003596455-14

medications have demonstrated clinical efficacy, they also present several challenges, such as substantial treatment costs, dose-limiting toxicity, and the development of drug resistance. These complications have incited interest in the development of innovative treatments that specifically target the key/vital enzymes of the parasite *Leishmania*. Identifying and refining such drug candidates is a crucial component of enhancing treatment outcomes (4). However, the process of discovering, developing, and approving novel chemicals is inherently arduous, expensive, and time-consuming.

A significant shift toward in silico approaches has been observed in recent years in the fight against VL and post-kala-azar dermal leishmaniasis (PKDL). As a result, we are able to study disease mechanisms, identify potential treatments, and evaluate drug candidates in a faster and more cost-effective manner than with traditional experiments. Molecular docking, virtual screening, and ADMET prediction techniques have made rapid assessment of a compound's safety and effectiveness possible, speeding up the discovery process (5). In other hand, network analysis has revolutionized our understanding of host–pathogen interactions, identifying critical pathways and molecular targets that could lead to new therapies. This has been further enhanced by developments in NGS-based bioinformatics, which enable large-scale genomic and transcriptomic data analysis, identifying novel targets and resistance mechanisms (6). Additionally, the identification of parasite-specific proteins that contribute to the development of disease and the survival of parasites has been made possible by proteomics-driven in silico studies (7, 8). When combined, these computational techniques produce a potent, data-driven method of drug development that enhances and rationalizes real-world results. It is not only novel but also essential for VL and PKDL research to use in silico methods. Thus, this chapter's goal is to investigate and comprehend in silico approaches for enhancing VL and PKDL treatment plans. In silico tools provide promising alternatives to traditional drug discovery, which is confronted with growing challenges such as drug resistance, treatment toxicity, and limited therapeutic options, by facilitating quick, affordable, and focused research (9) (Figure 14.1).

A thorough suite of in silico resources created especially for leishmaniasis and related kinetoplastid parasites is examined at the beginning of this chapter. The databases under discussion are crucial resources for researchers looking into different facets of leishmaniasis pathobiology because they offer insightful information on parasite genomics, proteomics, metabolomics, drug discovery, and immune response predictions. Building on this framework, the chapter explores the various computational methods used in leishmaniasis research and treatment, highlighting their contribution to the understanding of disease mechanisms, the discovery of new therapeutic targets, and the development of potent treatments and vaccines. The fundamental ideas, particular uses for *Leishmania* research, and related drawbacks of several important approaches are covered in detail, including molecular docking, molecular dynamics simulation, next-generation sequencing (NGS) analysis, pathway and network analysis, proteomic analysis, and immunoinformatics. A case study addressing miltefosine resistance in *Leishmania donovani* is also provided to highlight the usefulness

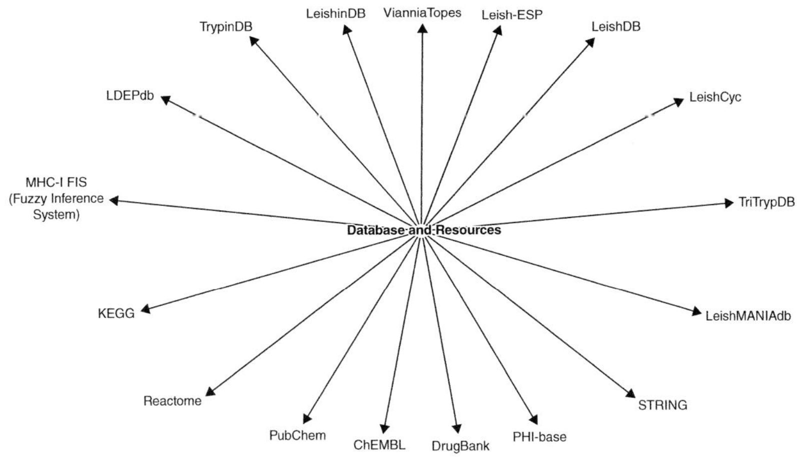

Figure 14.1 Resources and databases for *Leishmania* research.

of these tools in practice. It demonstrates how an integrated computational approach can guide vaccine development as well as therapeutic repurposing.

14.2 COMPREHENSIVE RESOURCES AND DATABASES FOR *LEISHMANIA* RESEARCH

To understand *Leishmania* spp.'s molecular and biochemical intricacies, it is necessary to use specialized resources that provide detailed genomic, proteomic, immunological, and chemical information. These curated databases have become indispensable tools for researchers in parasitology, immunoinformatics, drug discovery, and systems biology. Below is a detailed overview of key *Leishmania*-centric databases.

14.2.1 TriTrypDB

TriTrypDB is a central genomic database devoted to the study of kinetoplastid parasites, including *Trypanosoma brucei*, *Trypanosoma cruzi*, and *Leishmania* species. It is a component of the VEuPathDB Bioinformatics Resource Center and offers a complete collection of genome, transcriptome, proteome, metabolic pathway, and host–pathogen interaction datasets (10). The platform is specifically made to support comparative and functional genomics research, in addition to providing a number of important features.

- **Querying:** TriTrypDB enables users to effectively search and analyze genomic and proteomic data through a graphical user interface and customized workflows. Researchers can quickly locate relevant genes, transcripts, or protein sequences by using TriTrypDB.
- **Genome Annotation and Expression Data:** The database includes gene expression information for different life stages (including promastigote

and amastigote forms) and comprehensive genome annotations for several *Leishmania* species.
- **Pathway Enrichment Tools:** A pathway enrichment analysis tool in TriTrypDB aids in locating metabolic or signaling pathways essential to parasite pathogenesis and survival.
- **Cross-Species Comparative Tools:** Cross-species studies can be improved by comparing *Leishmania* and other kinetoplastids using data from EuPathDB resources.

TriTrypDB can be a useful tool for drug discovery as a resource for identifying critical genes, investigating patterns of gene expression, and identifying metabolic vulnerabilities.

14.2.2 LeishCyc

By manually curating its metabolic pathway database, this database offers comprehensive details on the main metabolic pathways of *Leishmania*, along with gene annotations, compound relationships, and pathway diagrams (11). This database provides a number of important features and is intended to assist systems biology research.

- **Visualization of Parasite-Specific Biochemical Pathways:** LeishCyc makes it easier to see essential metabolic pathways that are essential to *Leishmania* pathogenesis and survival.
- **Identification of Essential Enzymes and Chokepoints:** Researchers can find the enzymes and possible metabolic chokepoints that might be targets for medication development by mapping the metabolic network of *Leishmania*.
- **Exportable Data for Systems Biology Analysis:** Researchers can simulate metabolic flux and forecast the effects of metabolic changes by using LeishCyc's support for exporting pathway data for use in computational modeling, such as flux balance analysis.

LeishCyc is essential to understanding how *Leishmania* maintains infection in varying environments and adapts to changing conditions.

14.2.3 LeishDB

The non-coding RNAs (ncRNAs) of *Leishmania braziliensis*, which are extremely important for MCL, are the main focus of the LeishDB database. By storing small RNAs (sRNAs), long ncRNAs (lncRNAs), and their expression profiles based on transcriptomic data, LeishDB offers crucial tools for the following applications (12).

- **Exploration of Post-Transcriptional Regulation:** Researchers can investigate how ncRNAs control post-transcriptional gene expression using

LeishDB, especially in connection to *Leishmania*'s adaptation during infection.
- **Identification of Regulatory RNAs:** The database aids in the identification of ncRNAs that could aid in immune evasion, stage transitions, or the regulation of virulence factors.
- **Comparative Genomics:** Comparative genomics data from various *Leishmania* species are included in the LeishDB, which enables researchers to evaluate the conservation and functions of ncRNAs.

In order to understand how *Leishmania* regulates gene expression at the RNA level and how this contributes to its pathogenicity, it is vital to access this resource.

14.2.4 Leish-ESP

Protein expression information for various *Leishmania* species and developmental stages is compiled in Leish-ESP. Among Leish-ESP's salient characteristics are combined experimental proteomic data, such as iTRAQ, LC-MS/MS, and 2D gel electrophoresis results (13).

- **Cross-Species Proteomic Comparisons:** By comparing protein expression across various *Leishmania* species, Leish-ESP enables users to gain the understanding of their virulence and adaptations.
- **Post-Translational Modifications:** This database has proteins that have been changed by phosphorylation, glycosylation, and other post-translational modifications. It shows how these changes affect protein function and parasite survival.
- **Differential Expression Across Developmental Stages:** One can learn how protein expression changes during the life cycle of Leishmania, from promastigote to amastigote, and how these changes affect pathogenicity.

Leish-ESP can be used to find new biomarkers, treatment targets, and possible vaccine candidates based on differential protein expression.

14.2.5 ViannaTopes

A freely available online database called ViannaTopes was developed especially to help with the logical development of vaccines to prevent leishmaniasis, which is brought on by the parasite *Leishmania* (*Viannia*). The proteomes of several important Viannia species are used to identify CD4+ and CD8+ T cell epitopes, which is a comprehensive collection of computationally predicted immunogenic peptides (14).

- **A Curated Collection of Predicted T Cell Epitopes:** The database provides a searchable and easily accessible collection of CD4+ and CD8+ T cell epitopes that have been computationally predicted from the protein sequences of several *Leishmania* (*Viannia*) species. This makes it unnecessary for lone researchers to carry out intricate and time-consuming in silico analyses.

- **A User-Friendly Interface for Data Exploration:** Through the database's web interface, researchers can quickly query and filter the predicted epitope data according to a number of different criteria. This allows for the targeted identification of epitopes of interest for particular research questions by searching by parasite species, protein, MHC allele, and predicted binding affinity.

It is an essential resource for researchers studying *Leishmania* species with unique clinical manifestations and differential responses to drugs.

14.2.6 LeishInDB

More than 7000 small molecule inhibitors with proven efficacy against *Leishmania* species are accessible through this database (15). In order to tackle the problem of drug resistance, it provides a variety of features, such as the following:

- **Detailed Activity Data:** In order to evaluate the effectiveness of possible medications, each compound in LeishInDB is annotated with important data points like IC_{50}, EC_{50}, Ki, and binding energy.
- **Target Species and Developmental Forms:** The database contains details about the *Leishmania* species and the different stages of development (such as promastigote and amastigote) that the inhibitors target.
- **Protein Targets Linked to TriTrypDB:** Finding possible drug targets is made easier by LeishInDB's integration of data with TriTrypDB, which enables researchers to connect small molecules to their protein targets.
- **Chemical Descriptors and Toxicity Predictions:** Researchers can evaluate the safety profiles of compounds by using the database's toxicity predictions using admetSAR and chemical properties like logP, rotatable bonds, and molecular weight.
- **Advanced Search and Download Tools:** LeishInDB is a priceless tool for drug discovery because it provides tools for bulk molecular structure downloads, dynamic filtering, and substructure searching.

The database is used in screening, virtual screening, and QSAR modeling for anti-leishmanial drug development.

14.2.7 TrypinDB

TrypinDB has been created as an extensive, searchable database of small molecule inhibitors for Trypanosoma species. This database includes information from 700 research articles and more than 14,000 small molecules with different inhibitory activities (16). In addition to providing extensive information, the database can also be used to repurpose existing drugs. Key features include the following:

- **General Information:** Essential information like activity values, enzyme targets, and the inhibitor's source are linked to each inhibitor. This enables scientists to evaluate each compound's possible usefulness and applicability.
- **Structural Information:** Every inhibitor in the database has comprehensive structural data that can be utilized to optimize current compounds and create new ones.
- **Toxicity Information:** TrypinDB helps find compounds with good safety profiles by offering toxicity data and predictions for safe drug development.
- **Literature Information:** Every inhibitor is supported by pertinent literature and citations, which add important context to the experimental findings.

For every inhibitor, the database contains about 25 distinct data points, many of which are expected to shed light on their pharmacological profiles. TrypinDB's multi-field filters are user-friendly and allow researchers to conduct comprehensive searches, including substructure searches and FDA-approved drug similarity searches. The platform offers a variety of formats for exporting query results and chemical structures for additional analysis.

Additionally, TrypinDB can be used in studies on other kinetoplastid diseases like leishmaniasis because it is integrated with LeishInDB. This integration will enable TrypinDB to investigate repurposed inhibitors and new therapeutic approaches, as well as perform cross-disease studies.

14.2.8 LDEPdb

A thorough analysis of the proteomics literature and a methodically structured DMP database were combined to create LDEPdb. Researchers can learn a great deal about the disease at the molecular level from a review and database of articles published between 2000 and 2019 (17). The review and database are divided into five sections:

- **DMP in the Host After *Leishmania* Infection:** Proteins that change in the host after infection and provide insight into how the disease progresses.
- **DMP Between Different Strains of *Leishmania*:** Comparative proteomics data from various strains of *Leishmania* that help identify therapeutic targets and biomarkers unique to each strain.
- **DMP in Drug-Resistant *Leishmania*:** Proteomic alterations in drug-resistant strains are essential for creating plans to get past resistance and improve the effectiveness of treatment.

- **DMP in *Leishmania* Under Stress:** *Leishmania* adapts to harsh conditions within the host by modulating proteins in response to environmental stressors.
- **DMP in Different Life Stages of *Leishmania*:** Targeting distinct infection stages requires an understanding of how protein expression varies over the course of the parasite's life cycle.

This database enables more focused proteomics research on leishmaniasis, revealing possible biomarkers and treatment targets. The LDEPdb offers valuable resources for furthering research on *Leishmania* and its host interactions by fostering the development of improved diagnostic and treatment approaches.

14.2.9 MHC-I Peptide Screening using fuzzy inference system

The MHC-I Peptide Screening tool predicts MHC-I binding peptides in *L. donovani* proteomes by employing a fuzzy inference system (FIS). Fuzzy logic-based methods imitate expert reasoning and enhance prediction sensitivity and accuracy (18). Key features include the following:

- **Multi-Feature Scoring System:** To predict peptide binding to MHC-I molecules, this platform uses a multi-feature scoring mechanism that takes into account physicochemical, immunological, and structural parameters.
- **Customizable Search Filters:** To find the most promising vaccine candidates, the peptides can be filtered using criteria like length, score thresholds, and epitope location.
- **Novel Scoring Engine:** This fuzzy logic engine increases the sensitivity of peptide prediction, enabling users to make more accurate peptide predictions.

In particular, this tool is useful for the design of epitope-based vaccines and for understanding the T-cell responses required for immunity against *Leishmania* infections. For predicting B-cell epitopes in *Leishmania*, general-purpose tools not specifically designed for this parasite might also prove valuable (19–21).

14.2.10 KEGG and Reactome

Two well-known bioinformatics databases, KEGG (Kyoto Encyclopedia of Genes and Genomes) and Reactome, provide curated maps of metabolic and signaling pathways, many of which are conserved in *Leishmania* (22, 23). For researchers looking into the molecular interactions and cellular processes that occur within parasites and their host cells, these resources are priceless. Important characteristics include the following:

- **Comparative Pathway Mapping:** By offering comprehensive pathway maps, KEGG and Reactome allow scientists to contrast Leishmania's metabolic and signaling networks with those of other organisms, such as the host. This makes it easier to find species-specific variations and conserved pathways.
- **Integration with Host–Parasite Interaction Studies:** Data from host–parasite interactions can be integrated into both databases, which improves our knowledge of how *Leishmania* alters host cell functions to create infection.
- **Drug Target Visualization:** Researchers can find new therapeutic targets for leishmaniasis by using KEGG and Reactome to visualize possible drug targets within metabolic and signaling pathways.

By offering more comprehensive systems biology insights into the metabolic and signaling processes crucial to parasite survival and pathogenesis, these databases supplement specialized resources like LeishCyc and TriTrypDB.

14.2.11 PubChem, ChEMBL, and DrugBank

Three important chemical compound repositories that are essential to the development of anti-leishmanial drugs are PubChem, ChEMBL, and DrugBank:

- **PubChem:** For a wide variety of chemical compounds, PubChem offers comprehensive structural details, activity tests, and bioactivity data. It is frequently utilized for compound optimization and in silico screening in the search for anti-leishmanial drugs (10).
- **ChEMBL:** This carefully curated database of bioactivity information, which includes inhibition constants and clinical trial stages for different compounds, is derived from scientific literature. When evaluating the pharmacological characteristics of possible anti-leishmanial agents, it is especially helpful (24).
- **DrugBank:** DrugBank provides extensive information on approved medications and experimental compounds, including molecular properties, pharmacological details, and ADMET (absorption, distribution, metabolism, excretion, and toxicity) data. It is an essential tool for leishmaniasis medication candidate optimization (25).

These repositories provide a wealth of information about the pharmacology, safety profiles, and bioactivity of chemical compounds, making them invaluable for researchers engaged in compound optimization, drug discovery, and in silico screening.

14.2.12 PHI-Base and STRING

Two crucial databases that concentrate on host–pathogen and protein–protein interactions, respectively, are PHI-base and STRING:

- **PHI-Base:** This carefully curated database on host–pathogen interactions provides useful details on effector proteins, virulence factors, and mutant phenotypes in a range of pathogens, including *Leishmania*. It is an essential tool for figuring out possible therapeutic targets and comprehending the molecular mechanisms underlying parasite–host interactions (26).
- **STRING:** This database of protein–protein interactions offers *Leishmania* and other organisms' interaction networks. Researchers can discover important hubs in protein interactions, learn more about cellular functions, and investigate possible therapeutic targets by examining these networks (27).

14.2.13 LeishMANIAdb

A database called LeishMANIAdb was created expressly to look into potential interference between host proteins and *Leishmania* virulence factors. The database creates a unified, easily navigable web resource by combining different experimental data and computational predictions to identify proteins secreted by *Leishmania* (28). It gives users access to all available data on predicted and experimental secretomes. Furthermore, 3764 transmembrane proteins of various *Leishmania* species have their localization functions and 211 proteins' host–pathogen interactions manually annotated in the database. The database might help find new treatment targets for this underappreciated illness and offer fresh perspectives on *Leishmania* host–pathogen interactions.

By offering thorough insights into the molecular networks and interactions that underlie immune evasion, parasite virulence, and survival tactics, these databases enhance one another. Such knowledge is crucial for the development of new therapeutic strategies.

14.3 COMPUTATIONAL APPROACHES IN LEISHMANIASIS RESEARCH

Understanding the biology of *Leishmania* parasites, figuring out host–pathogen interactions, finding new therapeutic targets, and creating efficient treatments or vaccines all depend heavily on computational techniques. An extensive examination of the main computational techniques used in leishmaniasis research is given in this chapter. The fundamental scientific tenet of the approach is explained in detail in each section, along with its practical application in the context of the illness and any limitations to take into

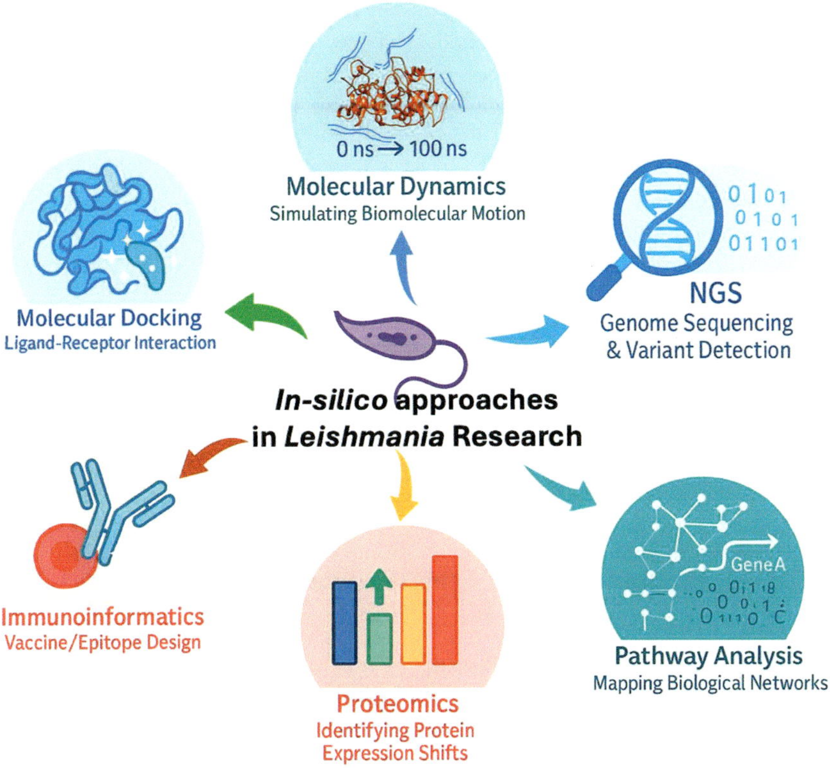

Figure 14.2 In silico approaches for *Leishmania* research.

account when interpreting the findings. The chapter concludes with a real-world case scenario illustrating how these tools and databases can work together to solve pressing clinical problems (Figure 14.2).

14.3.1 Molecular docking

- **Basic Principle:**
  A computational method for simulating the interaction between two molecules—usually a small ligand, such as a medication, and a target protein—is called molecular docking. By using scoring functions that take into account steric, electrostatic, and hydrogen-bonding contributions, the goal is to predict the most stable configuration (binding pose) of the ligand within the protein's binding site and to estimate the strength of this interaction.
- **Application in Leishmaniasis:**
  Research on leishmaniasis has made extensive use of molecular docking to find putative inhibitors of vital enzymes involved in the parasite's

pathogenicity and survival. To assess their potential as therapeutic candidates, scientists have docked a variety of synthetic and natural substances to *Leishmania* proteins, including 3-MST, O-acetylserine (S) synthase (OASS), trypanothione reductase, pteridine reductase, and cysteine proteases (7, 29–32). Prior to experimental validation, these docking studies are frequently used as a first-line screening. In this regard, databases such as LeishInDB (33) and TrypinDB (34) are useful tools that provide details on protein targets and recognized inhibitors pertinent to *Leishmania* species (35, 36).

- **Limitations:**
  - Relies on high-quality 3D structures of target proteins, which may not always be available.
  - Scoring functions may not always correlate with experimental binding affinities.
  - Typically treats the protein as rigid, which may overlook important conformational changes.

14.3.2 Molecular dynamics (MD) simulation

- **Basic Principle:**
  By resolving Newton's equations of motion, molecular dynamics simulation simulates how atoms and molecules behave over time. MD determines the position and velocity of every atom in a system subject to physical forces using parameters from force fields. On a nanosecond to microsecond scale, this enables researchers to see how biomolecular complexes move and react to their surroundings.
- **Application in Leishmaniasis:**
  MD simulations are widely used in the field of leishmaniasis to evaluate the dynamics and stability of drug–protein complexes that were first predicted using molecular docking. For example, MD can be used to model the behavior of a complex in a solvated, physiological environment after docking a compound into the active site of a *Leishmania* enzyme. This aids in determining whether the binding is likely to dissociate or is stable. Additionally, MD has been used to investigate conformational changes in parasite proteins that may result from mutations that cause drug resistance (37). Important information for targeted specificity is also provided by the comparative structural conformational changes (38).
- **Limitations:**
  - Computationally intensive, requiring significant time and hardware resources.
  - Limited by the accuracy of the force field used.
  - May not capture long-term conformational changes beyond the simulation timescale.

14.3.3 Next-generation sequencing (NGS) and genomic analysis

- **Basic Principle:**
  NGS encompasses a suite of high-throughput technologies that enable rapid sequencing of nucleic acids. By aligning millions of short reads to a reference genome, it becomes possible to identify mutations, assess gene expression levels (via RNA-seq), and examine genomic diversity. Bioinformatics tools are employed to preprocess, map, quantify, and statistically analyze the data.
- **Application in Leishmaniasis:**
  NGS technologies have significantly enhanced our ability to study *Leishmania* genomes and transcriptomes. In resistant vs. sensitive *L. donovani* strains, RNA-seq has been used to pinpoint differentially expressed genes linked to drug resistance mechanisms, such as ABC transporters or oxidative stress enzymes (39, 40). Genomic sequencing can also uncover point mutations, deletions, or gene amplifications that contribute to virulence or treatment failure. Platforms like TriTrypDB offer curated genomic data and tools to facilitate such analyses, allowing for comparative genomics and integration of transcriptomic datasets.
- **Limitations:**
  - Requires deep sequencing coverage and high-quality data to avoid noise.
  - Dependent on comprehensive and accurate reference genome annotations.
  - Computational analysis is complex and may yield false positives without experimental validation.

14.3.4 Pathway and network analysis

- **Basic Principle:**
  Pathway analysis involves mapping biological molecules such as genes or proteins onto known biological pathways to identify functional patterns or dysregulations. Network analysis, on the other hand, constructs graphs of molecular interactions—such as protein–protein interactions or gene regulatory networks—to identify key hubs, bottlenecks, or clusters associated with particular biological functions.
- **Application in Leishmaniasis:**
  To understand *Leishmania's* metabolic and regulatory systems, researchers have employed network and pathway analysis. For instance, scientists can determine which metabolic pathways are upregulated during infection or drug exposure by incorporating transcriptomic or proteomic data into programs like KEGG, Reactome, or LeishCyc. Critical proteins involved in host–pathogen interactions have been found through network analyses using platforms such as STRING or

PHI-base. These proteins could be targets for future medications or vaccines. This method works particularly well for comprehending intricate system-level reactions to infection or therapy.

- **Limitations:**
  - Dependent on the quality and completeness of interaction databases.
  - Pathways may be inferred from other organisms and not validated in *Leishmania*.
  - Static networks often ignore time-dependent or context-specific interactions.

14.3.5 Proteomics and Differential Modulation Analysis

- **Basic Principle:**
  Mass spectrometry is used in proteomic analysis to detect and measure proteins in biological samples. Data analysis includes identifying proteins that are differentially expressed under various experimental conditions, quantifying the abundance of detected peptide fragments, and matching them to protein databases. These proteins can be categorized and their functional roles deduced with the aid of computational tools.
- **Application in Leishmaniasis:**
  At the protein level, proteomics offers insight into how *Leishmania* and its host react to infection or therapy. For instance, researchers can find proteins that are upregulated in response to drug pressure by comparing the proteomes of sensitive and resistant parasite strains (41). Finding biomarkers or new therapeutic targets is facilitated by the LDEPdb database, which compiles data on differentially modulated proteins from various studies (42). Additionally, proteomic analysis helps validate transcriptomic study gene expression data.
- **Limitations:**
  - Detection bias toward high-abundance proteins.
  - Proteins with similar mass/charge ratios may be difficult to distinguish.
  - Post-translational modifications are not always captured without targeted analysis.

14.3.6 Immunoinformatics and epitope prediction

- **Basic Principle:**
  Immunoinformatics predicts epitopes, which are tiny antigenic protein segments that the immune system recognizes, using algorithms. The ability of the peptide to bind to MHC molecules, a crucial step for T-cell activation, is the main focus of T-cell epitope prediction. Tools include fuzzy logic systems that integrate several parameters (e.g., binding affinity, antigenicity, conservation), motif-based models, and machine learning (ML) models.

- **Application in Leishmaniasis:**
  Immunoinformatics plays an essential role in rational vaccine design for leishmaniasis. Using tools like the **MHC-I Fuzzy Peptide Screening System**, researchers can predict epitopes from *Leishmania* proteins that are likely to elicit protective immune responses. These predicted epitopes can be evaluated across diverse HLA alleles to maximize population coverage (8). The approach is especially valuable for designing multi-epitope subunit vaccines that can be produced synthetically and tested in preclinical models.
- **Limitations:**
  - Predictions may not account for antigen processing or in vivo immunogenicity.
  - High variability in HLA types complicates universal vaccine development.
  - Requires validation in animal models and clinical settings.

14.4 CASE SCENARIO: OVERCOMING RESISTANCE IN *LEISHMANIA DONOVANI*

The following case scenario is entirely hypothetical and does not represent real events or data. The results presented in each section are illustrative and not derived from actual database searches or experimental work. This is intended solely to demonstrate a potential workflow.

14.4.1 Background

In the Indian state of Bihar, an alarming number of visceral leishmaniasis patients are experiencing relapse after treatment with the frontline oral anti-leishmanial drug. With resistance suspected, a multidisciplinary team embarks on a computational investigation to identify resistance mechanisms and alternative therapies.

14.4.1.1 Step 1: Transcriptomic investigation

RNA samples from resistant and sensitive *L. donovani* strains are sequenced. Analysis using tools from TriTrypDB reveals overexpression of ABC transporters, suggesting drug efflux as a resistance mechanism. Downregulation of lipid metabolism pathways is also observed.

14.4.1.2 Step 2: Proteomic validation

Mass spectrometry data from the same strains are processed using LDEPdb and other proteomics pipelines. Differential expression confirms upregulation of proteins involved in redox balance and membrane transport, supporting the hypothesis that resistant parasites expel the drug and counter oxidative stress.

14.4.1.3 Step 3: Pathway analysis for target selection

Pathway mapping using LeishCyc and KEGG identifies the trypanothione metabolism pathway as crucial for survival under drug-induced stress. Trypanothione reductase is highlighted as a potential drug target due to its centrality and differential expression.

14.4.1.4 Step 4: Docking-based drug repurposing

Using LeishInDB and TrypinDB, researchers identify several known inhibitors of trypanothione reductase. Virtual screening and docking experiments are performed using AutoDock Vina, with posaconazole—an antifungal drug—emerging as a promising repurposed candidate based on high binding affinity and favorable interactions.

14.4.1.5 Step 5: Validation via MD simulation

To test binding stability, the drug–target complex undergoes 100 ns of MD simulation. The complex remains stable throughout, with consistent hydrogen bonding and minimal deviation in the active site—indicating strong, sustained interaction.

14.4.1.6 Step 6: Epitope prediction for long-term vaccine design

The team uses the LDEPdb and MHC-I Fuzzy Peptide Screening System to analyze proteins upregulated during infection. Conserved 9-mer epitopes are identified across multiple strains and are predicted to bind a wide range of HLA alleles (43). These epitopes are prioritized for preclinical testing in a multi-epitope vaccine strategy (44).

14.5 CONCLUSION

Through the coordinated use of transcriptomic, proteomic, docking, MD, and immunoinformatic tools, the team proposes a dual strategy:

- Short-term: Repurpose posaconazole for treating drug-resistant VL cases.
- Long-term: Develop a subunit vaccine using predicted immunogenic epitopes.

This integrative approach showcases the power of computational biology in rapidly responding to evolving parasitic threats.

The integration of computational tools and in silico resources has revolutionized the landscape of leishmaniasis research, offering powerful avenues to understand the complex biology of the parasite and its interaction with the host. Throughout this chapter, we have examined a wide array of dedicated databases such as TriTrypDB, LeishCyc, LeishDB, Leish-ESP, ViannìaDB, LeishInDB, TrypinDB, and LDEPdb, each contributing uniquely to aspects of genomics, proteomics, drug discovery, and vaccine design. These resources not only facilitate data retrieval but also empower researchers with tools for advanced analysis and hypothesis generation.

We also discussed the foundational principles and practical applications of computational methods such as molecular docking, molecular dynamics simulations, NGS analysis, pathway and network analysis, and proteomics-based investigations. Each of these approaches plays a critical role in uncovering mechanisms of pathogenesis, identifying druggable targets, and informing strategies for diagnostic and therapeutic development. Their collective utility is further underscored through a real-world case scenario of drug resistance, demonstrating the translational potential of computational insights.

Crucially, the way we investigate and treat any disease is changing as a result of the expanding use of Artificial Intelligence (AI) and ML techniques in research. These technologies improve the identification of vaccine candidates with high immunogenic potential, improve the prediction of drug–target interactions, and allow the automated analysis of complex omics datasets. Additionally, treatment outcome forecasting, systems-level host–pathogen dynamics research, and the discovery of hidden patterns that conventional analytical techniques might miss are all being done with AI-driven models. Combining curated databases and experimental data with AI and ML frameworks as they develop could speed up research and provide more accurate, individualized, and successful treatments to fight leishmaniasis.

Even with the impressive advancements, there are still issues, such as incomplete data, inaccurate predictions, and the requirement for experimental confirmation. But as biological datasets, AI/ML techniques, and computational technologies advance, their combination will deepen our knowledge of parasite biology and pathogenesis.

In conclusion, this chapter emphasizes the critical role that computational biology plays in contemporary parasitology by showing how carefully chosen materials, in silico techniques, and AI-driven methodologies all work together to provide a more thorough, systems-level understanding of leishmaniasis. These resources not only change our perspective on the illness but also open the door to creative and more successful methods of prevention, treatment, and control (Table 14.1).

Table 14.1 Key databases and resources for *Leishmania* research

Name	*URL*	*Focus area*	*Key features*	*Reference*
TriTrypDB	https://tritrypdb.org	Genomics, transcriptomics, proteomics	Genome-wide datasets, gene expression, SNPs, BLAST, metabolic pathways, integrated query interface	(10)
LeishCyc	http://www.leishcyc.org	Metabolic pathways (*Leishmania major*)	Manually curated pathways, enzyme functions, metabolic chokepoints, pathway maps	(11)
LeishDB	www.leishdb.com	Non-coding RNAs (*Leishmania braziliensis*)	lncRNAs, sRNAs, expression profiles, regulatory RNA identification	(12)
Leish-ESP	http://leish-esp.cbm.uam.es/	Proteomics	Protein expression in various species/stages, PTMs, differential analysis	(13)
VianniaTopes	https://doi.org/10.1155/2023/2274415	Epitope screening (Viannia subgenus)	B-cell/T-cell epitope data, antigenicity, MHC-binding, multi-epitope design	(14)
LeishInDB	http://leishindb.biomedinformri.com	Small molecule inhibitors	7000+ inhibitors, IC_{50}/EC_{50} data, target proteins (linked to TriTrypDB), chemical properties, admetSAR-based toxicity, substructure search	(15)
TrypinDB	https://www.trypindb.biomedinformri.com/	Protease inhibitors	Validated/predicted inhibitors, structural info, protease targets, virtual screening tools	(16)
LDEPdb	https://www.ldepdb.biomedinformri.com/	Epitope prediction (*L. donovani*)	B-cell/T-cell epitope data, antigenicity, MHC-binding, multi-epitope design	(17)
MHC-I FIS (fuzzy inference system)	https://www.mhcfisld.biomedinformri.com/	Epitope screening (*L. donovani*)	Fuzzy logic-based MHC-I peptide predictions, physicochemical and immunogenic filters, custom scoring	(18)
KEGG	https://www.kegg.jp/	Pathways	General pathways, host-parasite comparisons, drug-pathway maps	(22)

(*Continued*)

Table 14.1 (Continued)

Name	*URL*	*Focus area*	*Key features*	*Reference*
Reactome	https://reactome.org/	Pathways and interactions	Conserved pathway visualization, drug target mapping	(23)
PubChem	https://pubchem.ncbi.nlm.nih.gov/	Chemical information	Bioassays, structures, activity data	(45)
ChEMBL	https://www.ebi.ac.uk/chembl/	Bioactive molecules	Drug-likeness, bioactivity, clinical phases	(24)
DrugBank	https://www.drugbank.ca/	Drug data	Pharmacokinetics, approved drugs, repurposing info	(25)
PHI-base	https://www.phi-base.org/	Host–pathogen interactions	Virulence genes, phenotypes, pathogen strategies	(26)
STRING	https://string-db.org/	Protein–protein interactions	Interaction networks, functional clustering	(27)
LeishMANIAdb	https://leishmaniadb.ttk.hu/	Protein	Contains experimental, computational and annotated information about *Leishmania* proteins	(28)

ACKNOWLEDGMENT

The authors express their gratitude to the ICMR-Rajendra Memorial Research Institute of Medical Sciences, Patna, for the invaluable support they provided during the course of this study.

REFERENCES

1. WHO. https://www.who.int/news-room/fact-sheets/detail/leishmaniasis.
2. Dinc R. *Leishmania* vaccines: The current situation with its promising aspect for the future. Korean J Parasitol. 2022;60(6):379.
3. Uliana SRB, Trinconi CT, Coelho AC. Chemotherapy of leishmaniasis: Present challenges. Parasitology. 2018;145(4):464–80.
4. Ghorbani M, Farhoudi R. Leishmaniasis in humans: Drug or vaccine therapy? Drug Des Devel Ther. 2017;25–40.
5. Adelusi TI, Oyedele AQK, Boyenle ID, Ogunlana AT, Adeyemi RO, Ukachi CD, et al. Molecular modeling in drug discovery. Inform Med Unlocked [Internet]. 2022;29:100880. Available from: https://www.sciencedirect.com/science/article/pii/S235291482200034X
6. Dotolo S, Esposito Abate R, Roma C, Guido D, Preziosi A, Tropea B, et al. Bioinformatics: From NGS data to biological complexity in variant detection and oncological clinical practice. Biomedicines. 2022;10(9):2074.

7. Kant V, Kumar P, Ranjan R, Kumar P, Mandal D, Vijayakumar S. In silico screening, molecular dynamic simulations, and in vitro activity of selected natural compounds as an inhibitor of *Leishmania donovani* 3-mercaptopyruvate sulfurtransferase. Parasitol Res [Internet]. 2022 Jul 10;121(7):2093–109. Available from: https://link.springer.com/10.1007/s00436-022-07532-5
8. Kumari S, Vijaykumar S, Kumar V, Ranjan R, Alti D, Singh V, et al. In silico and in vitro evaluation of the immunogenic potential of *Leishmania donovani* ascorbate peroxidase and its derived peptides. Acta Trop. no. 260 (2024) Dec;107381.
9. Vijayakumar S, Das P. Recent progress in drug targets and inhibitors towards combating leishmaniasis. Acta Tropica. Elsevier B.V. 2018;181:95–104.
10. Shanmugasundram A, Starns D, Böhme U, Amos B, Wilkinson PA, Harb OS, et al. TriTrypDB: An integrated functional genomics resource for kinetoplastida. PLoS Negl Trop Dis. 2023;17(1):e0011058.
11. Doyle MA, MacRae JI, De Souza DP, Saunders EC, McConville MJ, Likić VA. LeishCyc: A biochemical pathways database for *Leishmania major*. BMC Syst Biol. 2009;3:1–12.
12. Torres F, Arias-Carrasco R, Caris-Maldonado JC, Barral A, Maracaja-Coutinho V, De Queiroz ATL. LeishDB: A database of coding gene annotation and non-coding RNAs in *Leishmania braziliensis*. Database. 2017;2017:bax047.
13. Das A, Biswas N, Chakrabarti S. Leish-ExP: A database of exclusive proteins from *Leishmania* parasite. bioRxiv. 2020;2020–5.
14. Llanes A, Restrepo CM, Lleonart R. ViannіaTopes: A database of predicted immunogenic peptides for *Leishmania* (Viannia) species. Database. 2018;2018:bay111.
15. Vijayakumar S, Kant V, Das P. LeishInDB: A web-accessible resource for small molecule inhibitors against *Leishmania* sp. Acta Trop. 2019;190:375–9.
16. Vijayakumar S, Ranjan R, Kumar R, Das P. TrypInDB: A searchable online resource of small molecule inhibitors against *Trypanosoma* sp. Parasitol Int. 2020; 78:102131.
17. Ranjan R, Das P, Vijayakumar S. Differentially modulated proteins associated with Leishmaniasis—A systematic review of in-vivo and in-vitro studies. Mol Biol Rep. 2020;47(11):9159–78.
18. Vijayakumar S. Harnessing fuzzy rule based system for screening major histocompatibility complex class I peptide epitopes from the whole proteome: An implementation on the proteome of *Leishmania donovani*. J Computat Biol. 2022;29(9):1045–58.
19. Saravanan V, Gautham N. BCIgEPRED—A dual-layer approach for predicting linear IgE epitopes. Mol Biol. 2018;52(2):285–93.
20. Carroll M, Rosenbaum E, Viswanathan R. Computational methods to predict conformational B-cell epitopes. Biomolecules. 2024;14(8):983.
21. Saravanan V, Gautham N. Harnessing computational biology for exact linear B-cell epitope prediction: A novel amino acid composition-based feature descriptor. OMICS. 2015;19(10):648–58.
22. Kanehisa M, Araki M, Goto S, Hattori M, Hirakawa M, Itoh M, et al. KEGG for linking genomes to life and the environment. Nucleic Acids Res. 2007; 36(suppl_1):D480–D484.
23. Fabregat A, Jupe S, Matthews L, Sidiropoulos K, Gillespie M, Garapati P, et al. The reactome pathway knowledgebase. Nucleic Acids Res. 2018;46(D1):D649–D655.
24. Gaulton A, Bellis LJ, Bento AP, Chambers J, Davies M, Hersey A, et al. ChEMBL: A large-scale bioactivity database for drug discovery. Nucleic Acids Res. 2012;40(D1):D1100–D1107.
25. Wishart DS, Knox C, Guo AC, Shrivastava S, Hassanali M, Stothard P, et al. DrugBank: A comprehensive resource for in silico drug discovery and exploration. Nucleic Acids Res. 2006;34(suppl_1):D668–D672.

26. Urban M, Irvine AG, Cuzick A, Hammond-Kosack KE. Using the pathogen-host interactions database (PHI-base) to investigate plant pathogen genomes and genes implicated in virulence. Front Plant Sci. 2015;6:605.
27. von Mering C, Huynen M, Jaeggi D, Schmidt S, Bork P, Snel B. STRING: a database of predicted functional associations between proteins. Nucleic Acids Res. 2003;31(1):258–61.
28. Tusnády GE, Zeke A, Kálmán ZE, Fatoux M, Ricard-Blum S, Gibson TJ et al. LeishMANIAdb: A comparative resource for *Leishmania* proteins. Database. 2023;2023:baad074.
29. Kant V, Vijayakumar S, Sahoo GC, Ali V, Singh K, Chaudhery SS, et al. In-silico screening and validation of high-affinity tetra-peptide inhibitor of *Leishmania donovani* O-acetyl serine sulfhydrylase (OASS). J Biomol Struct Dyn [Internet]. 2019;37(2):481–92. Available from: http://doi.org/10.1080/07391102.2018.1429315
30. Yadav S, Mandal H, Saravanan V, Das P, Singh SK. In vitro and in silico analysis of *L. donovani* enoyl acyl carrier protein reductase – A possible drug target. J Biomol Struct Dyn [Internet]. 2020;0(0):1–14. Available from: https://doi.org/10.1080/07391102.2020.1802337
31. Mandal H, Vijayakumar S, Yadav S, Kumar Singh S, Das P. Validation of NAD synthase inhibitors for inhibiting the cell viability of *Leishmania donovani*: In silico and in vitro approach. J Biomol Struct Dyn [Internet]. 2019;37(17):4481–93. Available from: https://doi.org/10.1080/07391102.2018.1552199
32. Pandey RK, Kumbhar BV, Sundar S, Kunwar A, Prajapati VK. Structure-based virtual screening, molecular docking, ADMET and molecular simulations to develop benzoxaborole analogs as potential inhibitor against *Leishmania donovani* trypanothione reductase. J Recept Signal Transduct. 2017;37(1):60–70.
33. Kwofie SK, Broni E, Dankwa B, Enninful KS, Kwarko GB, Darko L, et al. Outwitting an old neglected nemesis: A review on leveraging integrated data-driven approaches to aid in unraveling of leishmanicides of therapeutic potential. Curr Top Med Chem. 2020;20(5):349–66.
34. Ungogo MA, Ebiloma GU, Ichoron N, Igoli JO, de Koning HP, Balogun EO. A review of the antimalarial, antitrypanosomal, and antileishmanial activities of natural compounds isolated from Nigerian flora. Front Chem [Internet]. 2020;8:617448. Available from: http://www.ncbi.nlm.nih.gov/pubmed/33425860
35. Arya PK, Barik K, Singh AK, Kumar A. Databases and web resources for neglected tropical disease research. J Appl Pharm Sci [Internet]. 2023; Available from: https://japsonline.com/abstract.php?article_id=4024&sts=2
36. de Angelo RM, Nascimento LA, Encide JPP, Barbosa H, Lago JHG, da Silva Emery F, et al. Advances and challenges in molecular docking applied to neglected tropical diseases. Curr Med Chem. 2025;32(28):5939–5959.
37. Vijayakumar S, Das P. Structural, molecular motions, and free-energy landscape of *Leishmania* sterol-14α-demethylase wild type and drug resistant mutant: a comparative molecular dynamics study. J Biomol Struct Dyn [Internet]. 2019; 37(6):1477–93. Available from: http://doi.org/10.1080/07391102.2018.1461135
38. Ranjan R, Kumari R, Kumar A, Vijayakumar S. Harnessing computational methods for uncovering structural insights into *Leishmania donovani* 3-MST: implications for drug design and target specificity. In Silico Pharmacol. 2025;13(1):51.
39. Kumar V, Das S, Kumar A, Tiwari N, Kumar A, Abhishek K, et al. *Leishmania donovani* infection induce differential miRNA expression in CD4+ T cells. Sci Rep. 2020 Dec 1;10(1).
40. Kumar A, Vijayakumar S, Dikhit MR, Abhishek K, Mukherjee R, Sen A, et al. Differential regulation of miRNA profiles of human cells experimentally infected by *Leishmania donovani* isolated from Indian visceral leishmaniasis and post-kala-azar dermal leishmaniasis. Front Microbiol. 2020;11(July).

41. Reyaz E, Puri N, Selvapandiyan A. Global remodeling of host proteome in response to *Leishmania* infection. ACS Infect Dis. 2023;10(1):5–19.
42. Routaray CB, Choudhari V, Prakash D, Patil R, Jagtap S, Bai S, et al. Quantitative proteomic analysis reveals differential modulation of crucial stage specific proteins during promastigote to amastigote differentiation in *Leishmania donovani*. J Proteins Proteom. 2022;1–11.
43. Vijayakumar S. Harnessing fuzzy rule based system for screening major histocompatibility complex class I peptide epitopes from the whole proteome: An implementation on the proteome of *Leishmania donovani*. J Comput Biol. 2022; 29(9):1045–58.58.
44. Hashemzadeh P, Ghorbanzadeh V, Lashgarian HE, Kheirandish F, Dariushnejad H. Harnessing bioinformatic approaches to design novel multi-epitope subunit vaccine against *Leishmania infantum*. Int J Pept Res Ther [Internet]. 2019;(0123456789). Available from: https://doi.org/10.1007/s10989-019-09949-6
45. Wang Y, Xiao J, Suzek TO, Zhang J, Wang J, Bryant SH. PubChem: a public information system for analyzing bioactivities of small molecules. Nucleic Acids Res. 2009;37(suppl_2):W623–W633.

Chapter 15

Preventive measures for PKDL and major challenges

Saeme Asgari
Department of Biochemistry and Biophysics,
TeMS.C., Islamic Azad University, Tehran, Iran

Mojdeh Valizadeh
Department of Psychology, Islamic Azad University,
Rasht Branch, Rasht, Iran

Behrooz Ataei
Infectious Disease and Tropical Medicine Research Center,
Isfahan University of Medical Sciences, Isfahan, Iran

Hakimeh Karami
Information System and Technology Management, College of
Technological Innovation, Zayed University, Dubai, UAE

Saeedeh Azaraeen and Ali Kamali
School of Medicine, Jiroft University of Medical Sciences, Jiroft, Iran
Universal Scientific Education and Research Network (USERN)
JMU Office, Jiroft University of Medical Sciences, Jiroft, Iran

Kumar Abhishek
Department of Health Research, Model Rural Health Research Unit,
Muzaffarpur, Bihar, India
ICMR-Rajendra Memorial Research Institute of Medical Science,
Patna, India

Moein Gholami
Universal Scientific Education and Research Network (USERN)
JMU Office, Jiroft University of Medical Sciences, Jiroft, Iran

*Amir Sasan Mozaffari Nejad**
Infectious Disease and Tropical Medicine Research Center,
Isfahan University of Medical Sciences, Isfahan, Iran

* Corresponding author: asmnejad@res.mui.ac.ir

DOI: 10.1201/9781003596455-15

15.1 INTRODUCTION

Post-kala-azar dermal leishmaniasis (PKDL) is a distressing and often neglected complication that can arise following the treatment of visceral leishmaniasis (VL), also known as kala-azar. This debilitating skin condition is characterized by the appearance of macular, papules, or nodular lesions on the body, typically occurring months to years after the initial kala-azar infection has been treated and cured. While PKDL is not immediately life-threatening, it can have significant social and psychological impacts on affected individuals, leading to stigmatization and reduced quality of life [1–4]. Caused by the same *Leishmania* parasite responsible for VL, PKDL is endemic to several regions of the world, including the many parts of India, Bangladesh, Nepal, eastern Africa (Kenya, Sudan, Ethiopia), and parts of the Middle East. In these areas, the disease is particularly prevalent, with reported rates of PKDL ranging from 2.5% to 20% of treated kala-azar patients. However, the true burden of PKDL is likely underestimated, as many cases go undiagnosed or unreported, especially in resource-limited settings [2, 5–8]. The development of PKDL is a complex and multifactorial process, involving a delicate interplay between the host's immune response, the *Leishmania* parasite, and environmental factors. While the exact mechanisms are not fully understood, it is believed that persistent low-level parasitemia, genetic predisposition, and altered immune regulation all contribute to the emergence of PKDL in some individuals [9–11]. Addressing the challenge of PKDL is crucial, not only for improving the well-being of affected patients but also for the broader goal of eliminating VL. PKDL patients can serve as a reservoir for the *Leishmania* parasite, potentially perpetuating transmission and undermining efforts to control and eliminate the disease. Therefore, the development of effective preventive measures and targeted interventions is of paramount importance [9, 12]. To this end, researchers and public health authorities such as the World Health Organization (WHO) have made significant strides in understanding the epidemiology, risk factors, and potential preventive strategies for PKDL. However, numerous challenges remain, including the development of reliable diagnostic tools, the optimization of treatment regimens, and the implementation of comprehensive surveillance and monitoring systems [13–16]. This review will delve into the current state of knowledge on PKDL, highlighting the latest advancements in preventive measures and the major challenges that must be addressed to effectively mitigate the burden of this neglected tropical disease. By identifying and addressing the key barriers to PKDL prevention, we can work toward improving the health and well-being of individuals affected by this debilitating condition, while also contributing to the broader goal of eliminating VL in endemic regions.

15.2 UNDERSTANDING KALA-AZAR AND ITS COMPLICATIONS: THE DEVELOPMENT OF PKDL

Kala-azar, also known as VL, is a neglected tropical disease caused by the *Leishmania* parasite and transmitted by the bite of infected sandflies mainly of the genera *Phlebotomus* and *Lutzomyia*. Although sandflies are primarily recognized as vectors for *Leishmania*, they also transmit other pathogens, including bacteria and viruses. For instance, Carrion's disease is a sandfly-transmitted biphasic illness caused by *Bartonella bacilliformis*, prevalent in Central and South America. This disease manifests as either intermittent febrile states (Oroya fever), which can be life-threatening without timely treatment, or as cutaneous lesions known as Peruvian warts in individuals previously exposed to the bacteria [17, 18]. While the primary visceral form of kala-azar can be life-threatening if left untreated, an equally concerning manifestation is the development of PKDL. PKDL arises as a complication in some patients following treatment for the visceral disease and has become an increasingly pressing public health concern, particularly in the Indian subcontinent where it is endemic [1, 9, 19]. PKDL is characterized by the appearance of skin lesions, ranging from scattered nodules to diffuse reddish-brown discoloration of the skin. These lesions can persist for months or even years and serve as a reservoir for the *Leishmania* parasite, potentially leading to the continued transmission of the disease within affected communities. Individuals with PKDL, though not necessarily symptomatic, can act as a source of infection, perpetuating the cycle of transmission and hampering efforts to control and eliminate kala-azar [20–22]. In addition to *Leishmania*, some arboviruses are also transmitted by sandflies, particularly those belonging to the *Phlebovirus* genus. These viruses can cause uncomplicated to moderate fever states, often referred to as "sand fly fever" or "pappataci fever". Notably, the Toscana virus has a marked tropism for the central and peripheral nervous systems, potentially causing neuro-invasive conditions such as meningitis or encephalitis [18, 23]. The development of PKDL following treatment for VL is a complex and not fully understood phenomenon. While the exact mechanisms are still being investigated, it is believed that a dysregulated immune response, genetic predisposition, and other host-related factors may contribute to the progression from the visceral to the dermal form of the disease. Complicating matters further, PKDL can manifest at varying time intervals after the initial kala-azar episode, ranging from a few months to several years [8, 9, 24, 25]. Addressing the challenge of PKDL is of paramount importance, as it poses a significant obstacle to the goal of eliminating kala-azar in the Indian subcontinent, which has been a major public health priority in recent years. The WHO has set a target of eliminating kala-azar as a public health problem in the region by 2030, defined as reducing the annual incidence to less than 1 case per 10,000 population at the district level. However, the presence of PKDL cases, which can serve as a constant reservoir for the *Leishmania* parasite, threatens to undermine these efforts [26, 27]. To achieve the WHO's elimination goal, a multifaceted approach addressing both the visceral and dermal forms of the disease is crucial. This includes improving early detection and treatment of PKDL cases, implementing effective vector control measures, and strengthening surveillance and reporting systems. Additionally, research to better understand the pathogenesis of PKDL and develop novel preventive and therapeutic strategies is of paramount importance [9, 28, 29].

15.3 UNDERSTANDING THE TRANSMISSION AND RISK FACTORS OF PKDL

15.3.1 Feeding mechanisms and disease transmission in sandflies

Approximately 1,000 sandfly species belong to the genera *Phlebotomus* and *Lutzomyia*, with some capable of transmitting phleboviruses during blood feeding. While they are primarily known for spreading diseases like Leishmaniasis, their feeding mechanism is more similar to that of ticks. Female sandflies use sharp mouthparts to create shallow wounds in the skin, allowing for blood pooling (Figure 15.1). Unlike ticks, sandflies extract blood quickly, and their saliva promotes vascular leakage. Their saliva has significantly higher protein content than that of non-blood-feeding males, and research is ongoing to study the sialome of different species. Their small size helps them evade detection by hosts, despite causing painful bites [30] (Figure 15.1).

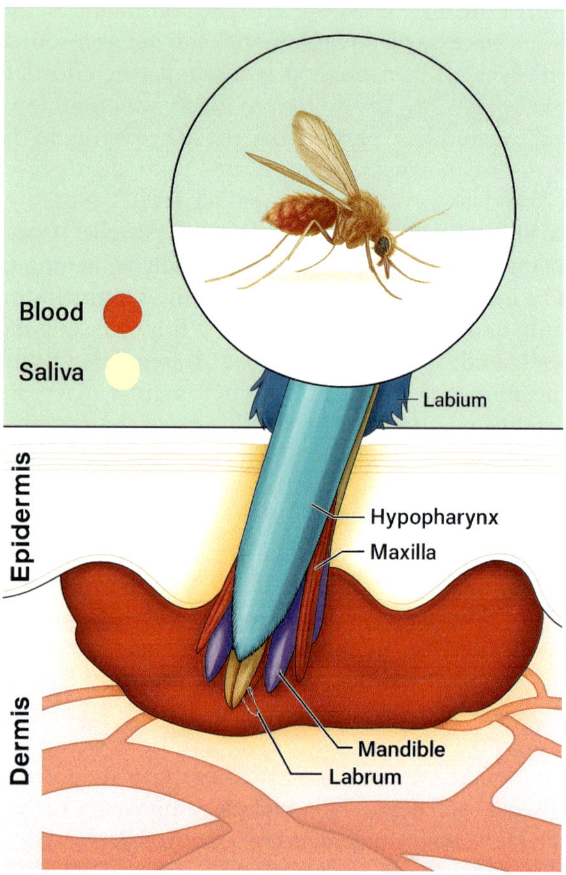

Figure 15.1 Sandfly feeding method and mouthpart.

15.3.2 Transmission cycle of *Leishmania* parasites

Leishmaniasis is a disease transmitted through the bite of female phlebotomine sandflies infected with the *Leishmania* parasite, as illustrated in Figure 15.2. During feeding, these sandflies inject the infective stage of the parasite, known as promastigotes, into the host's bloodstream [19, 31]. Once inside the host, the promastigotes are engulfed by macrophages and other types of mononuclear phagocytic cells, where they transform into the tissue stage of the parasite, called amastigotes. The amastigotes undergo replication via a process of simple division, subsequently infiltrating additional mononuclear phagocytic cells. The development and manifestation of the infection, whether it becomes symptomatic and leads to cutaneous or VL, depend on various factors, including the parasite, the host, and other environmental factors. The sandflies themselves become infected by ingesting these infected cells during their blood meals. Inside the sandflies, the amastigotes undergo a transformation process, reverting back into promastigotes. Subsequently, these promastigotes undergo development within the gut of the sandflies. Finally, they migrate toward the proboscis of the sandflies. Specifically, for leishmanial organisms in the *Viannia* subgenus, the promastigotes develop in the hindgut, while for organisms in the *Leishmania* subgenus, they develop in the midgut. Understanding the transmission and life cycle of the *Leishmania* parasite is crucial for developing effective prevention and control strategies for this disease. Further research is needed to elucidate the complex interactions between the parasite, the sandfly vector, and the host immune response, which will ultimately aid in the development of effective treatments and vaccines [18, 31–33] (Figure 15.2).

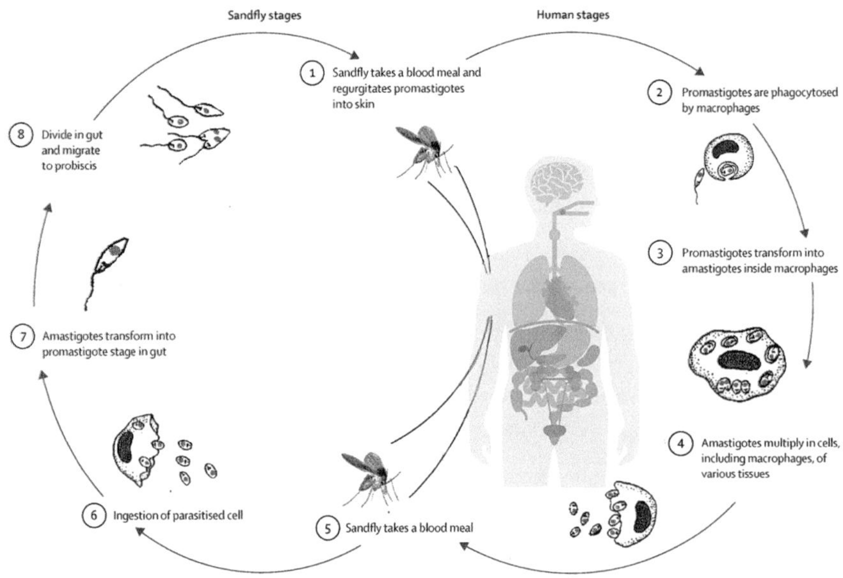

Figure 15.2 Life cycle of *Leishmania* parasite.

15.3.3 Risk factors contributing to the development of PKDL

PKDL is a dermatological condition that often occurs as a consequence of VL, which is caused by the *Leishmania* parasite. Several factors contribute to the development of PKDL.

15.3.3.1 Inadequate treatment of VL

One of the most significant risk factors is the insufficient or incomplete treatment of VL. If VL is not properly treated, the parasites responsible for the disease may persist in the body, increasing the likelihood of PKDL development [21].

15.3.3.2 Immune status

Individuals with compromised immune systems, such as those with HIV (human immunodeficiency virus)/acquired immune deficiency syndrome (AIDS) or malnutrition, are at a higher risk of developing PKDL. A weakened immune system is less effective at controlling *Leishmania* parasites, allowing them to multiply [25, 34].

15.3.3.3 Age and gender

PKDL tends to be more prevalent in young males, particularly those between the ages of 10 and 25. The reasons for this gender and age preference are not fully understood but may be related to hormonal or genetic factors [35–37].

15.3.3.4 Genetic susceptibility

Genetic factors play a role in determining susceptibility to PKDL. Some individuals may have genetic predispositions that make them more prone to developing PKDL after a VL infection [9, 38].

15.3.3.5 Geographical and environmental factors

Geographical factors can influence the prevalence of PKDL. It is more common in certain endemic regions, especially in the Indian subcontinent and East Africa. Environmental factors, such as climate and the presence of the sandfly vector, can also contribute to transmission [5].

15.3.3.6 Host immune response

The nature of an individual's immune response to *Leishmania* infection can influence PKDL development. An exaggerated or dysregulated immune response may contribute to the development of PKDL [21, 34].

15.4 EPIDEMIOLOGY OF PKDL

The occurrence of PKDL is primarily observed in regions that are endemic for VL. The prevalence of PKDL exhibits variation across different geographical areas, with high-endemicity regions being identified in parts of the Indian subcontinent, East Africa, and Sudan [5, 39]. PKDL affects both males and females, although there is a higher incidence observed in children and young adults. This age distribution pattern can be attributed to the increased exposure to sandfly vectors in regions endemic for the disease [12, 40]. The transmission dynamics of PKDL involve the involvement of sandfly vectors, particularly *Phlebotomus argentipes* and other species, which acquire the infection by feeding on infected individuals. However, the precise mechanisms underlying the transmission of PKDL remain incompletely understood and necessitate further research [41, 42].

15.5 PREVENTION OF PKDL

PKDL is a skin condition that frequently develops following VL, commonly referred to as kala-azar. It primarily affects the skin and mucous membranes and presents a significant challenge in the control and elimination of leishmaniasis. To effectively prevent PKDL, a comprehensive approach is necessary, incorporating various strategies and interventions supported by a growing body of research [1–4].

15.5.1 Vector control

The sandfly, particularly the species *P. argentipes*, serves as the primary vector for transmitting leishmaniasis [43]. Implementing effective vector control measures plays a crucial role in preventing the spread of the disease [44]. Some of the strategies include the following:

Insecticide-treated bed nets (ITNs): Distributing ITNs can reduce sandfly bites, especially during the nighttime when sandflies are most active [45].

Indoor residual spraying (IRS): Applying insecticides on the interior walls of houses can eliminate sandflies that rest inside dwellings [46].

Environmental management: Improving housing conditions and implementing proper waste management can reduce breeding sites for sandflies, effectively decreasing the vector population [47].

15.5.2 Early diagnosis and treatment

The WHO offers guidelines for the diagnosis of cutaneous and mucocutaneous leishmaniasis. Recommended diagnostic methods include parasitological techniques such as microscopy for visualizing *Leishmania* parasites in stained

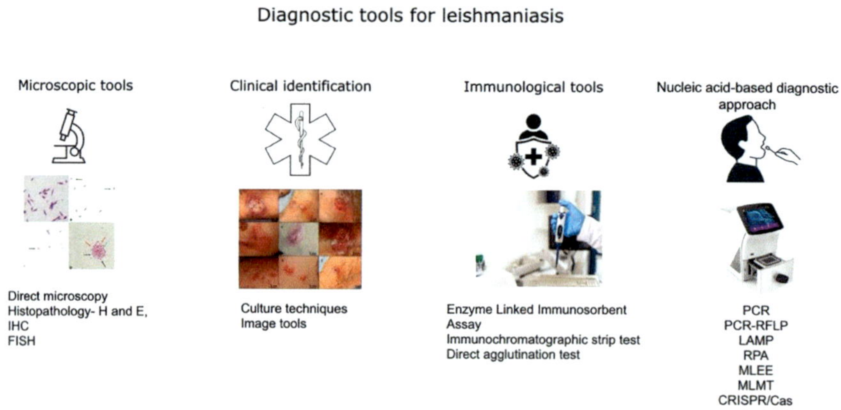

Figure 15.3 Graphical representation of diagnostic methods of leishmaniasis. PCR: polymerase chain reaction, RFLP: restriction fragment length polymorphism, LAMP: loop-mediated isothermal amplification, RPA: recombinase polymerase amplification assay, MLEE: multi-locus enzyme electrophoresis, MLMT: multi-locus microsatellite typing.

smears, culture methods for the growth and identification of the parasites, and polymerase chain reaction (PCR) for detecting *Leishmania* deoxyribonucleic acid (DNA) in lesion specimens. Serological techniques, including the direct agglutination test (DAT) and enzyme-linked immunosorbent assay (ELISA), are employed to identify antibodies against *Leishmania* antigens. Molecular methods, particularly PCR, facilitate the detection of *Leishmania* DNA in skin or blood samples. Additionally, immunological approaches, such as the immunofluorescent antibody test (IFAT), are used to identify *Leishmania* antibodies through fluorescence microscopy [48] (Figure 15.3).

Prompt diagnosis and treatment of VL are crucial in preventing the development of PKDL [49]. Timely intervention can prevent the persistence of parasites in the body, thereby reducing the risk of PKDL. Diagnostic tools and treatment regimens must be widely accessible and available to affected populations. Recent advancements in diagnostics include the utilization of rapid tests that detect antibodies or antigens associated with leishmaniasis, facilitating quick diagnosis of VL [9, 48, 50]. These tests are particularly valuable in resource-limited settings where more sophisticated diagnostic tools may be unavailable. Treatment options are continuously evolving, with a focus on improving drug regimens and therapies [48, 50]. Miltefosine, an oral medication, has shown promise in treating VL and preventing the development of PKDL [9]. Continued research and development in this area are essential for enhancing treatment outcomes and preventing the spread of PKDL. Laboratory tests are crucial for the diagnosis and management of PKDL, as timely intervention relies on precise diagnostic techniques. Co-infection with leishmaniasis and other infectious diseases, such as HIV, leprosy, tuberculosis, and schistosomiasis, is common in endemic areas and is influenced by environmental factors, vector availability, and host-parasite interactions. Leishmaniasis, being immunosuppressive, facilitates the invasion of opportunistic infections, often resulting in increased disease severity and mortality. PKDL, a dermal complication of VL, frequently mimics leprosy, complicating differentiation due to overlapping clinical presentations, including hypo-pigmented macular lesions and neural involvement [51–53].

15.5.2.1 Parasitological tests

a. **Skin smear:** One of the most commonly employed techniques for diagnosing PKDL is the utilization of a skin smear. This procedure entails the collection of a minute sample of the affected skin lesion, which is subsequently subjected to staining. The identification of *Leishmania* amastigotes within the sample serves as a definitive confirmation of the diagnosis. Nevertheless, it is important to note that this diagnostic method may not exhibit optimal sensitivity in certain instances [48, 50].
b. **Biopsy:** Skin biopsy, involving the extraction of a small fragment of the affected skin tissue for meticulous examination, represents an alternative diagnostic approach. This technique offers a more precise diagnosis by facilitating a histopathological evaluation, enabling the visualization of amastigotes. Biopsies are typically conducted in cases that are more severe or exhibit atypical characteristics [2, 53].

15.5.2.2 Molecular tests

a. **PCR:** PCR is a molecular technique widely used for the detection of *Leishmania* DNA in skin tissue samples. This method is known for its high sensitivity and specificity, making it a valuable tool in confirming the diagnosis of PKDL, particularly in cases where the parasite load is low or when other diagnostic tests yield inconclusive results [50, 54].
b. **Loop-mediated isothermal amplification (LAMP):** LAMP is a newer molecular diagnostic method that offers a rapid (1 hour until a result), high specificity, and simpler alternative to PCR. Similar to PCR, LAMP is capable of detecting *Leishmania* DNA in skin samples with high sensitivity and specificity. This method has shown promising results in the field of Leishmaniasis diagnosis and has the potential to be a valuable tool in clinical settings [55, 56].

15.5.2.3 Serological tests

a. **rK39 antigen-based ELISA:** This is a serological test that employs the rK39 antigen to identify antibodies specific to *Leishmania*. This diagnostic tool is highly valuable in the detection of VL, although its sensitivity may be somewhat restricted in PKDL cases [57, 58].

b. **DAT:** DAT is a serological assay designed to detect antibodies against *Leishmania* in the blood of patients. While it may not possess the same level of specificity as PCR or histopathology, it can still contribute to the diagnosis of PKDL and the monitoring of treatment response [41, 53].

15.5.2.4 Immunochromatographic test

rK39 rapid diagnostic test (RDT): Similar to the rK39 ELISA, this expedited assay identifies the presence of anti-*Leishmania* antibodies within the bloodstream. This diagnostic tool proves advantageous in the field due to its swift screening capabilities; however, it may not exhibit the same level of sensitivity as molecular tests [59].

15.5.2.5 Blood chemistry test

Routine blood tests, including complete blood counts (CBC) and liver function tests, play a crucial role in evaluating the overall well-being of patients with PKDL and tracking the disease's effects on their physiological state [9, 60].

Regular follow-up and monitoring of patients with PKDL are imperative subsequent to diagnosis in order to assess the advancement of treatment. Table 15.1 presents a comprehensive overview of the advantages and disadvantages associated with various diagnostic methodologies employed in the detection and management of *Leishmania* infections.

15.5.3 Surveillance and monitoring

Surveillance systems play a vital role in the monitoring of VL cases and the early detection of PKDL. This encompasses two main approaches.

15.5.3.1 Active case detection

Conducting periodic house-to-house surveys and actively searching for cases in areas endemic to leishmaniasis aids in the identification of new cases [6].

15.5.3.2 Passive case detection

Healthcare facilities should be adequately equipped to promptly diagnose and treat individuals with leishmaniasis. Regular monitoring of disease incidence and distribution is of utmost importance in identifying outbreaks and implementing effective control measures [61, 62].

15.5.4 Health education and awareness

Community awareness and education play a pivotal role in the prevention of PKDL [63]. Key components of this strategy include the following.

Table 15.1 Comparison of Diagnostic Methods for Leishmaniasis: Advantages and Disadvantages

Method	*Advantages*	*Disadvantages*
1. Microscopy	• Relatively simple and cost-effective. • Can be performed on various clinical samples such as splenic aspirates, bone marrow, or lymph nodes.	• Requires high-quality staining and trained personnel to ensure accurate results. • Sensitivity may be reduced in individuals with compromised immune systems.
2. Serological tests	• Non-invasive and can be conducted using blood samples. • Can serve as a screening tool. • Detects antibodies against the parasite.	• False-positive results can occur due to cross-reactions with other diseases. • Cannot differentiate between active and past infections. • Performance may vary in different geographic regions due to variations in parasite species.
3. Molecular tests (PCR)	• Highly sensitive and specific. • Can differentiate between different species of *Leishmania*. • Detects parasite DNA directly, even in low concentrations.	• Requires well-equipped laboratories and trained technicians. • Comparatively expensive compared to other diagnostic methods. • Limited availability in certain areas.
4. Rapid diagnostic tests	• Provides simple and rapid results (typically within 15–30 minutes). • Requires minimal training and equipment.	• Less sensitive than microscopy or PCR. • False-negative results can occur in early or mild infections. • Cross-reactions with other diseases can lead to false-positive results.

15.5.4.1 Educational campaigns

Raising awareness within affected communities about the causes, symptoms, and prevention of leishmaniasis [64].

15.5.4.2 Promoting hygiene

Encouraging practices such as maintaining clean living environments is essential in reducing the risk of PKDL. Additionally, promoting awareness about the importance of seeking early treatment and adhering to complete treatment courses is vital in reducing the incidence of PKDL [65].

15.5.5 Case management

Proper management of VL cases, including the completion of treatment courses and follow-up, is crucial in reducing the incidence of PKDL [29]. Ongoing research into treatment regimens and optimal drug combinations continues to enhance patient outcomes and decrease the likelihood of PKDL development [66].

15.6 MAJOR CHALLENGES IN PKDL PREVENTION

15.6.1 Limited healthcare infrastructure

In resource-limited settings, including diagnostic facilities, healthcare personnel, and treatment centers, can hinder early detection and timely treatment of VL cases, which may contribute to the development of PKDL [49].

15.6.2 Drug resistance

The emergence of drug-resistant strains of *Leishmania* parasites poses a challenge in the treatment of VL, indirectly impacting PKDL prevention. Sustaining long-term vector control interventions can be challenging due to resource constraints, logistical difficulties, and the need for community participation [67].

15.6.3 Sociocultural factors

Sociocultural beliefs, practices, and stigma associated with VL and PKDL can influence treatment-seeking behavior, compliance with treatment regimens, and community participation in prevention efforts [68].

15.6.4 Surveillance and reporting

Weak surveillance systems and underreporting of VL cases can hinder the identification of areas at risk for PKDL. Strengthening surveillance systems, improving reporting mechanisms, and enhancing data collection and analysis are crucial for targeted prevention strategies [69].

15.7 THE PSYCHOLOGICAL AND PSYCHOSOCIAL IMPACT OF PKDL

PKDL is a complex and multifaceted condition that extends far beyond its physical manifestations. The distinctive skin lesions and disfigurement associated with PKDL can have a profound impact on an individual's mental well-being, leading to a range of emotional and psychosocial challenges. One of the most significant effects of PKDL on mental health is the stigma and social isolation experienced by patients. The visible skin lesions often lead to societal discrimination, with PKDL patients facing rejection, ostracization, and even exclusion from community activities and social gatherings. This can result in feelings of shame, low self-esteem, and a diminished sense of self-worth, further exacerbating the psychological burden of the disease. Moreover, the chronic and unpredictable nature of PKDL can also contribute to high levels of anxiety, depression, and even suicidal ideation among affected individuals. The fear of disease recurrence, the uncertainty surrounding treatment outcomes, and the physical discomfort associated with the skin lesions can all take a significant toll on the mental health of PKDL patients [63, 70, 71]. Therefore, the integrated roles of psychologists and psychiatrists in the care of patients with PKDL are essential.

15.7.1 The role of psychologists in PKDL patient care

PKDL is a chronic condition that may develop as a dermal sequel to VL, commonly referred to as kala-azar. Although PKDL is not fatal, it is associated with significant cosmetic stigma due to visible skin lesions, which can manifest as macules, papules, and nodules [6]. This stigma often results in underreporting and reluctance among patients to seek treatment. The lesions characteristic of PKDL can be persistent, appearing months to years after the initial VL infection. Given its chronic nature, PKDL can last for several months to years, underscoring the need for ongoing management and comprehensive support for affected individuals. In this context, the role of psychologists and mental health professionals is crucial in providing holistic care and support for chronic disease especially PKDL patients, as they collaborate with healthcare providers to address both the physical and psychological aspects of the disease. Psychologists offer counseling and psychosocial support to patients and their families, including individual therapy sessions, group counseling, and the development of coping strategies to navigate the emotional and social challenges associated with PKDL. Additionally, they work to combat the stigma and discrimination faced by PKDL patients by partnering with community organizations and healthcare providers to raise awareness and promote social inclusion. Psychologists also play a vital role in assessing and managing mental health comorbidities, such as depression and anxiety, which may arise in PKDL patients, thereby enhancing their overall well-being and quality of life. Furthermore, they contribute to designing and implementing patient-centered interventions, such as support groups, educational programs, and vocational rehabilitation services, to empower PKDL patients and facilitate their reintegration into their communities [12, 72–77].

15.7.2 The role of psychiatrists in PKDL patient care

Psychiatrists play a vital role in the comprehensive care of patients with PKDL by addressing mental health comorbidities that often accompany the disease, such as depression, anxiety, and post-traumatic stress disorder. Their medical expertise enables them to conduct thorough assessments, make accurate diagnoses, and implement evidence-based pharmacological and psychotherapeutic interventions tailored to the needs of PKDL patients. Additionally, psychiatrists are essential in managing neuropsychiatric side effects from medications used to treat PKDL, such as sodium stibogluconate and miltefosine, which can cause adverse effects like confusion and psychosis. By closely monitoring these side effects and adjusting treatment regimens, psychiatrists help mitigate the impact on patients' mental well-being. Furthermore, psychiatrists contribute to research and advocacy efforts, collaborating with other healthcare professionals to explore the interplay between physical, psychological, and social factors affecting PKDL, thereby informing the development of holistic, patient-centered interventions [70, 77–80].

15.8 CONCLUSION

The prevention and management of PKDL pose significant public health challenges, particularly in endemic regions such as the Indian subcontinent and East Africa. As a debilitating complication of VL, PKDL imposes a substantial burden on affected individuals and threatens efforts to control and eliminate the disease. Psychologists and psychiatrists play a vital role in enhancing the well-being and quality of life for those affected by PKDL. A comprehensive, multidisciplinary approach to patient care is essential for addressing the psychological and social challenges these patients face, empowering them to navigate their condition with resilience and support. Despite advances in understanding PKDL's epidemiology and risk factors, challenges remain, including unreliable diagnostic tools, the need for improved treatment regimens, and the complexities of diverse social, economic, and environmental drivers. Addressing these challenges requires collaborative efforts among researchers, public health authorities, and affected communities. Strengthening surveillance and monitoring systems, improving access to healthcare services, and raising awareness about PKDL among healthcare providers and the general population are critical steps toward mitigating the burden of this neglected tropical disease. Ongoing research is necessary to elucidate the pathogenic mechanisms underlying PKDL and develop novel interventions for effective prevention and management. A multifaceted strategy that combines vector control, early diagnosis, treatment, and health education is essential for reducing the disease's burden. Through sustained collaboration, we can alleviate the suffering of PKDL patients and contribute to the goal of eliminating VL in endemic regions.

REFERENCES

1. Zijlstra, E., et al., *Post-kala-azar dermal leishmaniasis.* The Lancet Infectious Diseases, 2003. 3(2): p. 87–98.
2. Gedda, M.R., et al., *Post kala-azar dermal leishmaniasis: A threat to elimination program.* PLoS Neglected Tropical Diseases, 2020. **14**(7): p. e0008221.
3. Grifferty, G., et al., *Vulnerabilities to and the socioeconomic and psychosocial impacts of the leishmaniases: A review.* Research and Reports in Tropical Medicine, 2021. 12: p. 135–151.

4. Salam, M.A., et al., *Post-kala-azar dermal leishmaniasis with mucosal involvement: An unusual case presentation including successful treatment with miltefosine.* Journal of Health, Population, and Nutrition, 2013. **31**(2): p. 294.
5. Ghosh, P., et al., *Epidemiology of post-kala-azar dermal leishmaniasis.* Indian Journal of Dermatology, 2021. **66**(1): p. 12–23.
6. Desjeux, P., et al., *Report of the post kala-azar dermal leishmaniasis (PKDL) consortium meeting, New Delhi, India, 27–29 June 2012.* Parasites & Vectors, 2013. **6**(1): p. 196.
7. Roy, S., et al., *Monitoring the long-term effectiveness of miltefosine in Indian post-kala-azar dermal leishmaniasis.* The American Journal of Tropical Medicine and Hygiene, 2024. **110**(4): p. 656.
8. Kumar, P., Chatterjee, M., and Das, N.K., *Post kala-azar dermal leishmaniasis: Clinical features and differential diagnosis.* Indian Journal of Dermatology, 2021. **66**(1): p. 24–33.
9. Kumar, A., et al., *Post kala-azar dermal leishmaniasis in the Indian sub-continent: Challenges and strategies for elimination.* Frontiers in Immunology, 2023. **14**: p. 1236952.
10. Chatterjee, M., et al., *Immune responses in post kala-azar dermal leishmaniasis.* Indian Journal of Dermatology, 2020. **65**(6): p. 452–460.
11. Hartley, M.-A., et al., *The immunological, environmental, and phylogenetic perpetrators of metastatic leishmaniasis.* Trends in Parasitology, 2014. **30**(8): p. 412–422.
12. Joshi, A.B., et al., *Prevalence of post kala-azar dermal leishmaniasis (PKDL) and treatment seeking behavior of PKDL patients in Nepal.* PLoS Neglected Tropical Diseases, 2023. **17**(2): p. e0011138.
13. NTD Modelling Consortium Visceral Leishmaniasis Group, *Insights from mathematical modelling and quantitative analysis on the proposed WHO 2030 targets for visceral leishmaniasis on the Indian subcontinent.* Gates Open Research, 2019. **3**.
14. World Health Organization, *Regional Strategic Framework for accelerating and sustaining elimination of kala-azar in the South-East Asia Region.* 2022.
15. Mo, A.X., Pesce, J., and Hall, B.F., *Visceral leishmaniasis control and elimination: Is there a role for vaccines in achieving regional and global goals?* The American Journal of Tropical Medicine and Hygiene, 2016. **95**(3): p. 514.
16. Mondal, D., et al., *Relationship between treatment regimens for visceral leishmaniasis and development of post-kala-azar dermal leishmaniasis and visceral leishmaniasis relapse: A cohort study from Bangladesh.* PLoS Neglected Tropical Diseases, 2019. **13**(8): p. e0007653.
17. Telleria, E.L., et al., *Leishmania, microbiota and sand fly immunity.* Parasitology, 2018. **145**(10): p. 1336–1353.
18. Cecílio, P., Cordeiro-da-Silva, A., and Oliveira, F., *Sand flies: Basic information on the vectors of leishmaniasis and their interactions with leishmania parasites.* Communications Biology, 2022. **5**(1): p. 305.
19. Torres-Guerrero, E., et al., *Leishmaniasis: A review.* F1000Research, 2017. **6**.
20. Medley, G.F., et al., *Health-seeking behaviour, diagnostics and transmission dynamics in the control of visceral leishmaniasis in the Indian subcontinent.* Nature, 2015. **528**(7580): p. S102–S108.
21. Zijlstra, E.E., *The immunology of post-kala-azar dermal leishmaniasis (PKDL).* Parasites & Vectors, 2016. **9**(1): p. 464.
22. Kumar, P., Das, A., and Barkat, R., *Nodular post kala-azar dermal leishmaniasis: A report on two cases and brief review of literature.* Indian Dermatology Online Journal, 2020. **11**(5): p. 856–858.
23. Msellemu, D., et al., *Occupational exposure to malaria, leishmaniasis and arbovirus vectors in endemic regions: A systematic review.* Current Research in Parasitology & Vector-Borne Diseases, 2024. 6: p. 100185.

24. Burza, S., et al., *Post kala-azar dermal leishmaniasis following treatment with 20 mg/kg liposomal amphotericin B (Ambisome) for primary visceral leishmaniasis in Bihar, India*. PLoS Neglected Tropical Diseases, 2014. **8**(1): p. e2611.
25. Elmahallawy, E.K., Alkhaldi, A.A., and Saleh, A.A., *Host immune response against leishmaniasis and parasite persistence strategies: A review and assessment of recent research*. Biomedicine & Pharmacotherapy, 2021. **139**: p. 111671.
26. Subramanian, S., et al., *Modelling spatiotemporal patterns of visceral leishmaniasis incidence in two endemic states in India using environment, bioclimatic and demographic data, 2013–2022*. PLoS Neglected Tropical Diseases, 2024. **18**(2): p. e0011946.
27. Joshi, A.B., et al., *Follow-up assessment of visceral leishmaniasis treated patients and the impact of COVID-19 on control services in Nepal*. Tropical Medicine and Health, 2023. **51**(1): p. 57.
28. Zijlstra, E.E., et al., *Post-kala-azar dermal leishmaniasis in the Indian subcontinent: A threat to the South-East Asia Region kala-azar elimination programme*. PLoS Neglected Tropical Diseases, 2017. **11**(11): p. e0005877.
29. Maruf, S., et al., *Assessment of treatment outcomes of visceral leishmaniasis (VL) treated cases and impact of COVID-19 on VL management and control services in Bangladesh*. Journal of Infection and Public Health, 2023. **16**(11): p. 1716–1721.
30. Schneider, C.A., Calvo, E., and Peterson, K.E., *Arboviruses: How saliva impacts the journey from vector to host*. International Journal of Molecular Sciences, 2021. **22**(17): p. 9173.
31. Mans, D., et al., *Epidemiological, biological and clinical aspects of leishmaniasis with special emphasis on Busi Yasi in Suriname*. Journal of Clinical & Experimental Dermatology Research, 2017. 8(2): p. e1000388.
32. El-SayedAzab, M., Leishmaniasis, in *Textbook of Parasitic Zoonoses*, S.C. Parija and A. Chaudhury, Editors. 2022, Springer Nature Singapore: Singapore. p. 107–124.
33. Walker, D.M., et al., *Mechanisms of cellular invasion by intracellular parasites*. Cellular and Molecular Life Sciences, 2014. **71**: p. 1245–1263.
34. Zijlstra, E.E., *PKDL and other dermal lesions in HIV co-infected patients with leishmaniasis: Review of clinical presentation in relation to immune responses*. PLoS Neglected Tropical Diseases, 2014. 8(11): p. e3258.
35. Pace, D., *Leishmaniasis*. Journal of Infection, 2014. **69**: p. S10–S18.
36. Saurabh, S., et al, *Changing clinico-epidemiology of post-kala-azar dermal leishmaniasis (PKDL) in India: Results of a survey in four endemic states*. Journal of Vector Borne Diseases, 2020. 57(2): p. 161–169.
37. Kennedy, G.C., et al., *Visceral leishmaniasis follow-up and treatment outcomes in tiaty East and West sub-counties, Kenya: Cure, relapse, and post kala-azar dermal leishmaniasis*. PLoS One, 2024. **19**(6): p. e0306067.
38. Blackwell, J.M., Mohamed, H.S., and Ibrahim, M.E., *Genetics and visceral leishmaniasis in the Sudan: Seeking a link*. Trends in Parasitology, 2004. **20**(6): p. 268–274.
39. Yadav, P., et al., *Unusual observations in leishmaniasis—An overview*. Pathogens, 2023. **12**(2): p. 297.
40. Al-Dhafiri, M., et al., *Cutaneous leishmaniasis prevalence and clinical overview: A single center study from Saudi Arabia, Eastern Region, Al-Ahsa*. Tropical Medicine and Infectious Disease, 2023. 8(12): p. 507.
41. Zijlstra, E.E., *Biomarkers in post-kala-azar dermal leishmaniasis*. Frontiers in Cellular and Infection Microbiology, 2019. **9**: p. 228.
42. Calderon, A., et al., *Modeling post-kala-azar dermal leishmaniasis as an infection reservoir for visceral leishmaniasis*. Revista de Matemática: Teoría y Aplicaciones, 2020. **27**(1): p. 221–239.

43. Tripathi, L.K., and Nailwal, T.K., *Leishmaniasis: An overview of evolution, classification, distribution, and historical aspects of parasite and its vector.* Pathogenesis, treatment and prevention of leishmaniasis, 2021: p. 1–25.
44. Kumar, G., et al., *Addressing challenges in vector control: A review of current strategies and the imperative for novel tools in India's combat against vector-borne diseases.* BMJ Public Health, 2024. **2**(1).
45. Showler, A.J., et al., *Protozoan diseases: Malaria clinical features, management, and prevention.* 2023.
46. Deb, R., et al., *Impact of IRS: Four-years of entomological surveillance of the Indian visceral leishmaniases elimination programme.* PLoS Neglected tropical diseases, 2021. **15**(8): p. e0009101.
47. Dinesh, D., et al., *Efficacy and evaluation of environmental management system to control sandfly vector of kala-azar.* Journal of Environmental Management, 2017. **201**: p. 366–368.
48. Piyasiri, S.B., et al., *Diagnostic tools for cutaneous leishmaniasis caused by Leishmania donovani: A narrative review.* Diagnostics, 2023. **13**(18): p. 2989.
49. Pal, B., et al., *Barriers to the effective management and prevention of post kala-azar dermal leishmaniasis (PKDL) in the Indian subcontinent.* Medical Journal Armed Forces India, 2023. **79**(5): p. 500–505.
50. Thakur, S., Joshi, J., and Kaur, S., *Leishmaniasis diagnosis: An update on the use of parasitological, immunological and molecular methods.* Journal of Parasitic Diseases, 2020. **44**: p. 253–272.
51. Trindade, M.A.B., et al., *Post-kala-azar dermal leishmaniasis and leprosy: Case report and literature review.* BMC Infectious Diseases, 2015. **15**: p. 1–8.
52. Saini, I., Joshi, J., and Kaur, S., *Unwelcome prevalence of leishmaniasis with several other infectious diseases.* International Immunopharmacology, 2022. **110**: p. 109059.
53. Dixit, K.K., Singh, R., and Salotra, P., *Advancement in molecular diagnosis of post kala-azar dermal leishmaniasis..* Indian Journal of Dermatology, 2020. **65**(6): p. 465–472.
54. Ranasinghe, S., et al., *Polymerase chain reaction detection of leishmaniaDNA in skin biopsy samples in Sri Lanka where the causative agent of cutaneous leishmaniasis is Leishmania donovani.* Memórias do Instituto Oswaldo Cruz, 2015. **110**: p. 1017–1023.
55. Puri, M., et al., *Rapid diagnosis of Leishmania infection with a portable loop-mediated isothermal amplification device.* Journal of Biosciences, 2021. **46**: p. 1–15.
56. Verma, S., et al., *Development of a rapid loop-mediated isothermal amplification assay for diagnosis and assessment of cure of Leishmania infection.* BMC Infectious Diseases, 2017. **17**: p. 1–9.
57. Zijlstra, E., et al., *rK39 enzyme-linked immunosorbent assay for diagnosis of Leishmania donovani infection.* Clinical Diagnostic Laboratory Immunology, 1998. **5**(5): p. 717–720.
58. Zhang, W.-W., et al., *Development of a sandwich ELISA to detect leishmania 40S ribosomal protein S12 antigen from blood samples of visceral leishmaniasis patients.* BMC Infectious Diseases, 2018. **18**: p. 1–11.
59. Molinet, F.J.L., et al., *Specificity of the rapid rK39 antigen-based immunochromatographic test kalazar detect (r) in patients with cutaneous leishmaniasis in Brazil.* Memórias do Instituto Oswaldo Cruz, 2013. **108**(3): p. 293–296.
60. Topno, R.K., et al., *Noninvasive sweat-based diagnosis of visceral leishmaniasis and post kala-azar dermal leishmaniasis.* The American Journal of Tropical Medicine and Hygiene, 2018. **99**(5): p. 1162.
61. Rios-Zertuche, D., et al., *Performance of passive case detection for malaria surveillance: Results from nine countries in Mesoamerica and the Dominican Republic.* Malaria Journal, 2021. **20**(1): p. 208.

62. Sengupta, R., et al., *Active surveillance identified a neglected burden of macular cases of post kala-azar dermal leishmaniasis in West Bengal.* PLoS Neglected Tropical Diseases, 2019. **13**(3): p. e0007249.
63. Pal, B., et al., *Assessment of quality of life in patients with post kalaazar dermal leishmaniasis.* Health and Quality of Life Outcomes, 2017. **15**: p. 1–7.
64. Khan, W., et al., *Cutaneous leishmaniasis–Awareness, knowledge and practices among general population in rural and urban areas in Malakand region, Pakistan.* Brazilian Journal of Biology, 2021. **82**: p. e238665.
65. Srinivasan, R., et al., *Positive influence of behavior change communication on knowledge, attitudes, and practices for visceral Leishmaniasis/Kala-azar in India.* Global Health: Science and Practice, 2018. **6**(1): p. 192–209.
66. Singh-Phulgenda, S., et al., *Post-kala-azar dermal leishmaniasis (PKDL) drug efficacy study landscape: A systematic scoping review of clinical trials and observational studies to assess the feasibility of establishing an individual participant-level data (IPD) platform.* PLoS Neglected Tropical Diseases, 2024. **18**(4): p. e0011635.
67. Prakash Singh, O., et al., *Current challenges in treatment options for visceral leishmaniasis in India: A public health perspective.* Infectious Diseases of Poverty, 2016. **5**(02): p. 1–15.
68. Garapati, P., et al., *Knowledge, stigma, health seeking behaviour and its determinants among patients with post kalaazar dermal leishmaniasis, Bihar, India.* PLoS One, 2018. **13**(9): p. e0203407.
69. Chowdhury, R., et al., *How far are we from visceral leishmaniasis elimination in Bangladesh? An assessment of epidemiological surveillance data.* PLoS Neglected Tropical Diseases, 2014. **8**(8): p. e3020.
70. Pires, M., et al., *The impact of leishmaniasis on mental health and psychosocial well-being: A systematic review.* PLoS One, 2019. **14**(10): p. e0223313.
71. Duthie, M.S., et al., *Impact of sequelae of visceral leishmaniasis and their contribution to ongoing transmission of Leishmania donovani.* Pathogens and Disease, 2019. **77**(6): p. ftz057.
72. Cosio, D., and Demyan, A., *Behavioral medicine: How clinicians can reduce the stigma attached to chronic pain.* Practical Pain Management, 2021. **21**.
73. Wahass, S.H., *The role of psychologists in health care delivery.* Journal of Family and Community Medicine, 2005. **12**(2): p. 63–70.
74. Perugino, F., et al., *Stigma and chronic pain.* Pain and Therapy, 2022. **11**(4): p. 1085–1094.
75. Carr, E.R., and Miller, R., *Expanding our reach: Increasing the role of psychologists in public and community mental health..* Psychological Services, 2017. **14**(3): p. 352.
76. Conversano, C., *Common psychological factors in chronic diseases.* Frontiers in Psychology, 2019. **10**: p. 2727.
77. Bilgic-Temel, A., Murrell, D.F., and Uzun, S., *Cutaneous leishmaniasis: A neglected disfiguring disease for women.* International Journal of Women's Dermatology, 2019. **5**(3): p. 158–165.
78. Taheri, A.R., Rad, S.S., and Molkara, S., *Systemic treatments of leishmaniasis: A narrative review.* Reviews in Clinical Medicine, 2019. **6**(3).
79. Sundar, S., and Agarwal, D., *Visceral leishmaniasis—Optimum treatment options in children.* The Pediatric Infectious Disease Journal, 2018. **37**(5): p. 492–494.
80. Chahed, M.K., et al., *Psychological and psychosocial consequences of zoonotic cutaneous leishmaniasis among women in Tunisia: Preliminary findings from an exploratory study.* PLoS Neglected Tropical Diseases, 2016. **10**(10): p. e0005090.

Chapter 16

Conclusions and future perspectives

Research priorities and potential solutions

*Kumar Abhishek**
DHR- Model Rural Health Research Unit (MRHRU), Kurhani, Muzaffarpur, Bihar and ICMR-Rajendra Memorial Research Institute of Medical Science, Patna, India

Manas Ranjan Dikhit
Department of Bioinformatics, ICMR-Rajendra Memorial Research Institute of Medical Science, Patna, India

Abhik Sen
Department of Molecular Biology, ICMR-Rajendra Memorial Research Institute of Medical Science, Patna, India

Amir Sasan Mozaffari Nejad
Infectious Disease and Tropical Medicine Research Center, Isfahan University of Medical Sciences, Isfahan, Iran

Ashish Kumar
Department of Health Research, Model Rural Health Research Unit, Muzaffarpur, Bihar, India
ICMR-Rajendra Memorial Research Institute of Medical Science, Patna, India

16.1 INTRODUCTION

Leishmaniasis is endemic to more than 90 countries of the world in the forms of cutaneous leishmaniasis (CL), mucocutaneous leishmaniasis (ML), visceral leishmaniasis (VL) and as post kala-azar dermal leishmaniasis (PKDL), a sequel of VL in the areas where VL is prevalent (World Health Organization

* Corresponding author: abhisinghbhu41@gmail.com

DOI: 10.1201/9781003596455-16

(WHO), 2023). As estimated, around 700,000–1 million new cases of leishmaniasis occur annually in which the VL cases is between 50,000 and 90,000 (World Health Organization (WHO), 2023).

VL is the most severe form that affects the internal organs and is fatal if untreated. VL is endemic to Brazil, countries of East Africa, and Indian subcontinent; however, the better part is that the global burden of VL has shown a declining trend (Zhang et al., 2025). In India, a sharp decrease in the number of VL cases have been observed in the past decade with achieving the kala-azar elimination at block level in more than 98% of endemic blocks, as per the set standard of less than 1 case per 10,000 population (World Health Organization (WHO), 2025; NCVBDC, 2025). For achieving the elimination at such extent, intervention strategies like integrated vector management, creating awareness about the disease among the local endemic population, involvement of local bodies, public–private partnership to support health system have played a major role while research support in improved diagnosis and treatment options, identification and implementation of new insecticides have significant impact.

Leishmania parasites are known for greater adaptability and efficiency in modulating human host and are supposed to have co-evolved with the human host (Rossi and Fasel, 2018; Conceição-Silva and Morgado, 2019; Kumari et al., 2022). In such a scenario, to counter the parasite and control the disease in the future, as well as to sustain the achievements made in elimination, robust and continued research is the utmost requirement. Research in the following field is necessary for our preparedness against future encounters with leishmaniasis.

To strengthen our preparedness and response capacity, research must focus on several key areas. These include (A) identification of novel anti-leishmanial drugs, (B) development of an effective vaccine, (C) discovery of innovative insecticides for better vector control, (D) creation of robust molecular diagnostic tools tailored to endemic settings, and (E) in-depth understanding of the parasite's survival mechanisms within the host. A multidisciplinary approach will be essential to overcome remaining hurdles and sustain long-term success in combating leishmaniasis.

16.2 RESEARCH ON IDENTIFYING NEW DRUGS FOR LEISHMANIASIS

The primary therapeutic options against different forms of leishmaniasis are pentavalent antimonials (sodium stibogluconate and meglumine antimoniate), amphotericin B (AmBisome) (including liposomal formulations), miltefosine, paromomycin, and pentamidine (Pradhan et al., 2022). These antileishmanials have high therapeutic efficiency and efficacy but have limitations of adverse effects on patients, high cost, limited availability, need for multiple injections, and development of resistance. The pentavalent antimonials are considered first-line treatment but due to the emergence of resistance especially in India, it's not in use (Sundar et al., 2000). Amphotericin B is a certified treatment for VL with efficacy >95%, has drawbacks of long hospitalization time with major adverse effects of nephrotoxicity, hypokalemia, cardiotoxicity, etc. (Sundar et al., 2024). The liposomal formulation of Amphotericin B AmBisome is currently in use in the kala-azar elimination program which is single shot with low side effects. Miltefosine, which is basically an anti-cancer drug, is the only oral drug used for all forms of leishmaniasis (Bhattacharya et al., 2007; Soto et al., 2007; van Griensven et al., 2010), but in Indian subcontinent it is mainly included in the kala-azar elimination program for PKDL treatment (Khandpur et al., 2010; Ramesh et al., 2011). Again, miltefosine treatment has many side effects, including ocular complications (Saurabh and Mahabir, 2020). Paromomycin has differential efficacy in different regions and is generally preferred for combination therapy.

Limited therapeutic options and upcoming reports of resistance/unresponsiveness against antileishmanial drugs are matters of concern. Exploring new treatment options against leishmaniasis, research on multi-drug combination therapy, and other non-conventional therapies like immunotherapy, use of ant-microbial peptide (Alberola et al., 2004) should be the main focus.

16.3 RESEARCH ON DEVELOPMENT OF VACCINE

Vaccines have a key role in prevention and control of an infectious disease by boosting the immune system of the host. Immunization by vaccine administration stimulates the immune system to recognize and fight the invading pathogen and prevents establishment of disease.

The commercially available vaccines for canine VL are CaniLeish®, LetiFend® and Leish-Tec®, but their effectiveness has not been compared uniformly (Velez and Gállego, 2020).

Till date no vaccine has been registered for leishmaniasis for human use; some are under different phases of clinical trials. First-generation vaccine using killed parasite (*Leishmania major*) with BCG is in phase I and II clinical trials in healthy participants (Ayala et al., 2024). Second-generation recombinant antigen vaccines LEISH-F1 and LEISH-F2 that contain recombinant polyprotein of *L. major* and *Leishmania braziliensis* are in phase II trials (Chakravarty et al., 2011; Ayala et al., 2024) and LEISH-F3-containing components of *Leishmania donovani* and *Leishmania infantum* are under phase II clinical trials. The LEISH-F3 has demonstrated more promising results with the induction of a robust immune response (Coler et al., 2015). The live-attenuated parasite vaccine with *L. major centrin 1* gene knockout (Lm*Cen*$^{-/-}$) parasites prepared using CRISPR/Cas technology is in pre-clinical stage (Kaye et al., 2023). The ChAd63-KH vaccine that is a third-generation vaccine, containing *L. donovani* proteins, has been found efficient in eliciting a broad variety of CD8+ T cells specific to Leishmania antigens PKDL patients (Younis et al., 2021).

Several potential vaccine candidates against leishmaniasis have been identified, and few are under different stages of clinical trials, but still there is not a single vaccine for humans. This gap must be filled urgently with a prioritized focus on research in this field, as vaccines will help to properly control and prevent the spread of leishmaniasis to other non-endemic regions.

16.4 RESEARCH TO IDENTIFY NEW INSECTICIDES FOR VECTOR CONTROL

Leishmaniasis is a vector-borne disease transmitted from infected patient to uninfected individual from the bite of sandfly vector of *Phlebotomus* genus. Vector control is considered the best way to interrupt the transmission of disease in the endemic area (Balaska et al., 2021). For controlling vector population, use of insecticide-treated household items (like curtains) or bed nets (ITNs) and insecticide spray i.e. indoor residual spraying (IRS) are the common practices (Molina and Jiménez, 2022).

In Indian subcontinent, initially (dichloro-diphenyl-trichloroethane) DDT was used for IRS to control sandfly population but was replaced by synthetic pyrethroids (alpha-cypermethrin) (World Health Organization (WHO), 2015). The main reason for this replacement was decreased effectiveness and development of resistance in sandfly population against DDT, due to prolonged use for other vector control also (Dinesh et al., 2010; Kumar et al., 2015), while the pyrethroids exhibit low toxicity to mammals with high insecticide activity (Deb et al., 2021).

Finding the alternate insecticide option against sandfly is the utmost priority as resistances to pyrethroids (permethrin and deltamethrin) are also coming in reports from different geographical regions where it is in use from a long time (Amalraj et al., 1999; Molina and Jiménez, 2022). The widespread use of pyrethroids, including alpha-cypermethrin, is driving the emergence of resistance in sandflies. The mechanism of action of both pyrethroids and DDT is similar; both target the voltage-gated sodium channel, which means resistance to one can potentially lead to resistance to the other. Recent study explores that certain mutations in sandfly genome could present a threat to IRS-mediated vector control as it might provide resistance against currently utilized alpha-cypermethrin (Kristan et al., 2024). If resistance to pyrethroids develops, insecticides with different modes of action, such as organophosphates or carbamates, should be considered as alternatives.

16.5 RESEARCH ON DEVELOPMENT OF MOLECULAR DIAGNOSTIC TOOL FOR ENDEMIC SETTINGS

Accurate and early diagnosis of any infectious disease is crucial for its proper treatment, reduces morbidity, mortality and prevents its spreading to others.

Out of three main forms of leishmaniasis, diagnosis of VL or kala-azar and PKDL is challenging due to certain limitations. The clinical symptoms of VL like weight loss, splenomegaly, and anemia are common to many other diseases like malaria, typhoid, and tuberculosis. These overlapping symptoms lead to misdiagnosis or delayed diagnosis. In case of PKDL, the macular form is often misdiagnosed with other skin diseases like vitiligo having similar patterned patches.

Proper diagnosis of VL and PKDL starts with screening with rk39-based RDT kits, followed by confirmation with microscopic observation of LD bodies in bone marrow or splenic aspirate. Both bone marrow and splenic aspirate sample collection require much trained professional and specialized centers with adequate facilities. Again for PKDL diagnosis, confirmation of parasite in skin smear is necessary which is not available at endemic settings. These limitations hamper the accurate diagnosis and hence treatment course.

Detection of Leishmania DNA through polymerase chain reaction (PCR) or by quantitative PCR (qPCR) has emerged as a powerful tool offering higher sensitivity, specificity, and reliability. These molecular diagnostic techniques can be used to detect parasite DNA in skin smear, splenic or bone marrow aspirates and even in peripheral blood, with detection limit as low as less than 1 parasite/ml of blood (Mary et al., 2004; Galluzzi et al., 2018), makes it particularly valuable in diagnosing difficult or borderline cases. qPCR not only has its use in diagnosis, it can also quantitate parasitic load that is used to monitor treatment efficacy and detect relapses.

Limitations for molecular techniques are requirement of highly trained professionals, well-established laboratory, and a high cost per test compared to other conventional methods.

Research on other new molecular techniques like recombinase polymerase amplification (RPA) assay and loop-mediated isothermal amplification (LAMP) are ongoing (Puri et al., 2021). These techniques are under development, especially for endemic settings as these are less equipment-dependent and can be used in resource-limited kala-azar endemic regions. RPA technique is being under trial phases for VL and PKDL using both peripheral blood and tissue samples in a suitcase setup (Ghosh et al., 2021) and is providing more promising results. This innovation will make molecular testing more accessible to remote areas and practical for large-scale screening in kala-azar endemic areas and for control programs.

16.6 RESEARCH ON UNDERSTANDING THE MECHANISMS USED BY THE *LEISHMANIA* PARASITE FOR ITS SURVIVAL INSIDE HOST

Leishmania parasite survives as an intracellular parasite inside human host, thrives under harsh conditions of phagolysosomal compartment with an acidic pH (pH ~5.0). The dimorphic parasite exists as flagellated promastigote inside vector gut and as non-flagellated amastigote adapted to elevated temperature and acidic pH (37°C and pH—5.0). The parasite also copes the oxidative and nitrosative stress generated by the host cell by deploying a cascade of thiols and other anti-oxidants such as ascorbate peroxidase (LdApx), which is specifically present in the parasite (Kumar et al., 2014). LdApx is a suitable candidate for drug target and is now being studied for its therapeutic importance.

This oxidative stress management also enables the parasite to survive the drug pressure as the common anti-leishmanial drugs such as AmBisome, and miltefosine also acts by generating reactive oxygen species (ROS) to kill parasite (Purkait et al., 2012; Abhishek et al., 2017).

Leishmania employs a multi-faceted strategy to survive and replicate within host macrophages. By hacking the host cell signaling pathways (Akt-NF-κB), it delays the macrophage apoptosis that provides the time to survive (Abhishek et al., 2018). The parasite manipulates the host immune system by inhibiting IL-12 production and promoting IL-10 and TGF-β secretion, thus diverting the immune response toward a non-protective Th2 for its survival (Costa-da-Silva et al., 2022). Antimony unresponsive strains of *L. donovani* modulated the proteins cargo of the exosomal secretions of the host, enriching with immunomodulatory proteins which turns the host immune response toward parasite survival (Abhishek et al., 2022).

A better understanding of such metabolic pathways of parasite and pathways operative during host–pathogen interaction, which plays a crucial role in disease pathology, will help to find a better drug target, enables us to define better therapeutic procedures such as immune-chemoprophylaxis for complex infections and to design and develop an effective vaccine against this disease.

16.7 CONCLUSION

Leishmaniasis is one of the neglected tropical diseases, which is endemic to more than 90 countries of the world. A declining trend in global disease burden of VL especially in Indian subcontinent, with touching the elimination target, is a significant achievement; but sustenance of this achieved elimination and further eradication will be the goal. For this, along with implementation works, research in the field of new drugs and insecticides, development of vaccine, rapid field-deployable, and cost-effective diagnostic tools for endemic settings should be the major research focus.

ACKNOWLEDGMENT

Dr. Kumar Abhishek is thankful to the Department of Health Research, Model Rural Health Research Unit, Kurhani, Muzaffarpur, Bihar & ICMR-Rajendra Memorial Research Institute of Medical Science, Agamkuan, Patna for providing necessary research support.

REFERENCES

Abhishek K, Das S, Kumar A, Kumar A, Kumar V, Saini S, Mandal A, Verma S, Kumar M, Das P. *Leishmania donovani* induced Unfolded Protein Response delays host cell apoptosis in PERK dependent manner. PLoS Neglected Tropical Diseases. 2018 Jul 23;12(7):e0006646.

Abhishek K, Kumar A, Sardar AH, Vijayakumar S, Dikhit MR, Kumar A, Kumar V, Das S, Das P. Differential translational regulation of host exosomal proteins play key role in immunomodulation in antimony resistance in visceral leishmaniasis: A proteomic profiling study. Acta Tropica. 2022 Feb 1;226:106268.

Abhishek K, Sardar AH, Das S, Kumar A, Ghosh AK, Singh R, Saini S, Mandal A, Verma S, Kumar A, Purkait B. Phosphorylation of translation initiation factor 2-alpha in *Leishmania donovani* under stress is necessary for parasite survival. Molecular and Cellular Biology. 2017 Jan 1;37(1):e00344-16.

Alberola J, Rodriguez A, Francino O, et al. Safety and efficacy of antimicrobial peptides against naturally acquired leishmaniasis. Antimicrobial Agents and Chemotherapy. 2004;48(2):641–3. https://doi.org/10.1128/AAC.48.2.641-643.2004

Amalraj DD, Sivagnaname N, Srinivasan R. Susceptibility of *Phlebotomus argentipes* and *P. papatasi* (Diptera: Psychodidae) to insecticides. The Journal of Communicable Diseases. 1999 Sep 1;31(3):177–80.

Ayala A, Llanes A, Lleonart R, Restrepo CM. Advances in Leishmania vaccines: Current development and future prospects. Pathogens. 2024 Sep 20;13(9):812.

Balaska S, Fotakis EA, Chaskopoulou A, Vontas J. Chemical control and insecticide resistance status of sand fly vectors worldwide. PLoS Neglected Tropical Diseases. 2021;15(8): e0009586. https://doi.org/10.1371/journal.pntd.0009586

Bhattacharya SK, Sinha PK, Sundar S, Thakur CP, Jha TK, Pandey K, et al. Phase 4 trial of miltefosine for the treatment of Indian visceral leishmaniasis. The Journal of Infectious Diseases. 2007;196:591–8. https://doi.org/10.1086/519690.

Chakravarty, J, Kumar, S, Trivedi, S, Rai, VK, Singh, A, Ashman, JA, Laughlin, EM, Coler, RN, Kahn, SJ, Beckmann, AM, et al. A clinical trial to evaluate the safety and immunogenicity of the LEISH-F1+MPL-SE vaccine for use in the prevention of visceral leishmaniasis. Vaccine. 2011;29:3531–7.

Coler, RN, Duthie, MS, Hofmeyer, KA, Guderian, J, Jayashankar, L, Vergara, J, Rolf, T, Misquith, A, Laurance, JD, Raman, VS, et al. From mouse to man: Safety, immunogenicity and efficacy of a candidate leishmaniasis vaccine LEISH-F3+GLA-SE. Clinical & Translational Immunology. 2015;4:e35.

Conceição-Silva F, Morgado FN. *Leishmania* spp-host interaction: There is always an onset, but is there an end? Frontiers in Cellular and Infection Microbiology. 2019 Sep 19;9:330.

Costa-da-Silva AC, Nascimento DO, Ferreira JRM, Guimarães-Pinto K, Freire-de-Lima L, Morrot A, Decote-Ricardo D, Filardy AA, Freire-de-Lima CG. Immune responses in leishmaniasis: An overview. Tropical Medicine and Infectious Disease. 2022 Mar 31;7(4):54. https://doi.org/10.3390/tropicalmed7040054.

Deb R, Singh RP, Mishra PK, Hitchins L, Reid E, Barwa AM, et al. Impact of IRS: Four-years of entomological surveillance of the Indian visceral leishmaniases elimination programme. PLoS Neglected Tropical Diseases. 2021;15:e0009101.

Dinesh DS, Das ML, Picado A, Roy L, Rijal S, Singh SP, et al. Insecticide susceptibility of *Phlebotomus argentipes* in visceral leishmaniasis endemic districts in India and Nepal. PLoS Neglected Tropical Diseases. 2010;4:e859.

Galluzzi L, Ceccarelli M, Diotallevi A, Menotta M, Magnani M. Real-time PCR applications for diagnosis of leishmaniasis. Parasites & Vectors. 2018 May 2; 11(1):273. https://doi.org/10.1186/s13071-018-2859-8.

Ghosh P, Sharma A, Bhattarai NR, Abhishek K, Nisansala T, Kumar A, Böhlken-Fascher S, Chowdhury R, Khan MA, Faisal K, Hossain F. A multi-country, single-blinded, phase 2 study to evaluate a point-of-need system for rapid detection of leishmaniasis and its implementation in endemic settings. Microorganisms. 2021 Mar 12;9(3):588. https://ncvbdc.mohfw.gov.in/index1.php?lang=1&level=2&sublinkid=5945&lid=3750 https://www.who.int/india/health-topics/leishmaniasis

Kaye PM, Matlashewski G, Mohan S, Le Rutte E, Mondal D, Khamesipour A, Malvolti S. Vaccine value profile for leishmaniasis. Vaccine. 2023 Nov 3;41:S153–75.

Khandpur S, Chaturvedi P, Kumar U, Khaitan BK, Samantaray JC, Sharma VK. Oral miltefosine in post-kala-azar dermal leishmaniasis – Experience in three cases. International Journal of Dermatology. 2010;49:565–9. https://doi.org/10.1111/j.1365-4632.2010.04326.x.

Kristan M, Hazelgrove C, Kumar K, Kumar A, Kumar V, Das P, Collins E, Mark-Carew M, Campino S, Cameron M. Knockdown resistance mutations in *Phlebotomus argentipes* sand flies in Bihar, India. Parasites & Vectors. 2024 Aug 9;17(1):334.

Kumar A, Das S, Purkait B, Sardar AH, Ghosh AK, Dikhit MR, Abhishek K, Das P. Ascorbate peroxidase, a key molecule regulating amphotericin B resistance in clinical isolates of *Leishmania donovani*. Antimicrobial Agents and Chemotherapy. 2014 Oct;58(10):6172–84.

Kumar V, Shankar L, Kesari S, Bhunia GS, Dinesh DS, Mandal R, et al. Insecticide susceptibility of *Phlebotomus argentipes* & assessment of vector control in two districts of West Bengal, India. Indian Journal of Medical Research. 2015;142: 211–5.

Kumari D, Mahajan S, Kour P, Singh K. Virulence factors of Leishmania parasite: Their paramount importance in unraveling novel vaccine candidates and therapeutic targets. Life Sciences. 2022 Oct 1;306:120829. https://doi.org/10.1016/j.lfs.2022.120829.

Mary C, Faraut F, Lascombe L, Dumon H. Quantification of *Leishmania infantum* DNA by a real-time PCR assay with high sensitivity. Journal of Clinical Microbiology. 2004 Nov;42(11):5249–55. https://doi.org/10.1128/JCM.42.11.5249-5255.2004.

Molina R, Jiménez M. Assessing the susceptibility to permethrin and deltamethrin of two laboratory strains of *Phlebotomus perniciosus* from Madrid region, Spain. Acta Tropica. 2022 Jul;231:106453. https://doi.org/10.1016/j.actatropica.2022.106453.

National Centre for Vector Borne Disease Control: Diseases-Kala azar. Ministry of Health & Family Welfare. (2025). http://ncvbdc.mohfw.gov.in/index1.php?lang=1&level=1&sublinkid=5774&lid=3692"

Pradhan S, Schwartz RA, Patil A, Grabbe S, Goldust M. Treatment options for leishmaniasis. Clinical and Experimental Dermatology. 2022 Mar 1;47(3):516–21.

Puri, M, Brar, HK, Mittal, N. et al. Rapid diagnosis of Leishmania infection with a portable loop-mediated isothermal amplification device. Journal of Biosciences. 2021;46:92. https://doi.org/10.1007/s12038-021-00211-0

Purkait B, Kumar A, Nandi N, Sardar AH, Das S, Kumar S, Pandey K, Ravidas V, Kumar M, De T, Singh D. Mechanism of amphotericin B resistance in clinical isolates of *Leishmania donovani*. Antimicrobial Agents and Chemotherapy. 2012 Feb;56(2):1031–41.

Ramesh V, Katara GK, Verma S, Salotra P. Miltefosine as an effective choice in the treatment of post kala-azar dermal leishmaniasis. British Journal of Dermatology. 2011;165:411–4. https://doi.org/10.1111/j.1365-2133.2011.10402.x.

Rossi M, Fasel N. How to master the host immune system? Leishmania parasites have the solutions! International Immunology. 2018 Mar 10;30(3):103–11. https://doi.org/10.1093/intimm/dxx075.

Saurabh S, Mahabir M. Adverse ocular events on miltefosine treatment for post-kala-azar dermal leishmaniasis in India. Tropical Doctor. 2020;50:37–42. https://doi.org/10.1177/0049475519877317.

Soto J, Toledo J, Valda L, Balderrama M, Rea I, Parra R, et al. Treatment of bolivian mucosal leishmaniasis with miltefosine. Clinical Infectious Diseases. 2007; 44:350–6. https://doi.org/10.1086/510588.
Sundar S, More DK, Singh MK, Singh VP, Sharma S, Makharia A, et al. Failure of pentavalent antimony in visceral leishmaniasis in India: Report from the center of the Indian epidemic. Clinical Infectious Diseases. 2000;31:1104–7. https://doi.org/10.1086/318121.
Sundar S, Singh J, Singh VK, Agrawal N, Kumar R. Current and emerging therapies for the treatment of leishmaniasis. Expert Opinion on Orphan Drugs. 2024;12(1):19–32. https://doi.org/10.1080/21678707.2024.2335248
van Griensven J, Balasegaram M, Meheus F, Alvar J, Lynen L, Boelaert M. Combination therapy for visceral leishmaniasis. Lancet Infectious Diseases. 2010;10:184–94. https://doi.org/10.1016/S1473-3099(10) 70011-6.
Velez, R, Gállego, M. Commercially approved vaccines for canine leishmaniosis: A review of available data on their safety and efficacy. Tropical Medicine & International Health. 2020;25:540–57.
World Health Organization: Leishmaniasis. (2023) https://www.who.int/news-room/fact-sheets/detail/leishmaniasis
Younis, BM, Osman, M, Khalil, EAG, Santoro, F, Furini, S, Wiggins, R, Keding, A, Carraro, M, Musa, AEA, Abdarahaman, MAA, et al. Safety and immunogenicity of ChAd63-KH vaccine in post-kala-azar dermal leishmaniasis patients in Sudan. Molecular Therapy. 2021;29:2366–77.
Zhang SX, Yang GB, Sun JY, Li YJ, Yang J, Wang JC, Deng Y. Global, regional, and national burden of visceral leishmaniasis, 1990–2021: Findings from the global burden of disease study 2021. Parasites & Vectors. 2025 Apr 26;18(1):157.

Index

Note: Locators in *italics* represent figures and **bold** indicate tables in the text.

D

E

F

G

H

I

L

M

N

Q

R

S

T

U

V

W

Y